THE ROUTLEDGE REITS RESEARCH HANDBOOK

The Routledge REITs Research Handbook presents a cutting-edge examination of the research into this key global investment vehicle. Edited by internationally respected academic and REIT expert Professor David Parker, the book will set the research agenda for years to come. The handbook is divided into two parts, the first of which provides the global context and a thematic review covering: asset allocation, performance, sustainability, Islamic REITs, emerging sectors and behavioural finance. Part II presents a regional review of the issues with high level case studies from a diverse range of countries including the US, UK, Brazil, India, Australia, China, Singapore, Argentina and Poland, to name just a few.

This handbook redefines existing areas within the context of international REITs research, highlights emerging areas and future trends and provides postgraduates, professionals and researchers with ideas and encouragement for future research. It is essential reading for all those interested in real estate, international investment, global finance and asset management.

David Parker is an internationally recognized property industry expert on real estate investment trusts (REITs) and a highly regarded property academic, being a director and adviser to property investment groups including real estate investment trusts, unlisted funds and private property businesses (www.davidparker.com.au).

He is currently the inaugural Professor of Property at the University of South Australia, a Visiting Professor at the University of Reading, UK, and at Universiti Tun Hussein Onn, Malaysia, a Visiting Fellow at the University of Ulster, UK, an Acting Valuation Commissioner of the Land and Environment Court of New South Wales, and a Sessional Member of the SA Civil and Administrative Tribunal.

Author of the authoritative *Global Real Estate Investment Trusts: People, Process and Management* and *International Valuation Standards: A Guide to the Valuation of Real Property Assets*, he may be contacted at davidparker@davidparker.com.au.

THE ROUTLEDGE REITS RESEARCH HANDBOOK

Edited by David Parker

LONDON AND NEW YORK

First published 2018
by Routledge
2 Park Square, Milton Park, Abingdon, Oxon OX14 4RN

and by Routledge
605 Third Avenue, New York, NY 10017

First issued in paperback 2021

Routledge is an imprint of the Taylor & Francis Group, an informa business

© 2018 selection and editorial matter, David Parker; individual chapters, the contributors

The right of David Parker to be identified as the author of the editorial material, and of the authors for their individual chapters, has been asserted in accordance with sections 77 and 78 of the Copyright, Designs and Patents Act 1988.

All rights reserved. No part of this book may be reprinted or reproduced or utilised in any form or by any electronic, mechanical, or other means, now known or hereafter invented, including photocopying and recording, or in any information storage or retrieval system, without permission in writing from the publishers.

Trademark notice: Product or corporate names may be trademarks or registered trademarks, and are used only for identification and explanation without intent to infringe.

Publisher's Note
The publisher has gone to great lengths to ensure the quality of this reprint but points out that some imperfections in the original copies may be apparent.

British Library Cataloguing-in-Publication Data
A catalogue record for this book is available from the British Library

Library of Congress Cataloging-in-Publication Data
Names: Parker, David, 1961– editor.
Title: The Routledge REITs research handbook / edited by David Parker.
Other titles: REITs research handbook
Description: Abingdon, Oxon ; New York, NY : Routledge is an imprint of the Taylor & Francis Group, an Informa Business, 2019.
Identifiers: LCCN 2018021943| ISBN 9781138063112 (hbk : alk. paper) | ISBN 9781315161266 (ebk)
Subjects: LCSH: Real estate investment trusts.
Classification: LCC HG5095 .R68 2019 | DDC 332.63/247—dc23
LC record available at https://lccn.loc.gov/2018021943

ISBN 13: 978-1-03-209461-8 (pbk)
ISBN 13: 978-1-138-06311-2 (hbk)

Typeset in Bembo
by Apex CoVantage, LLC

To Elliot, Victoria and Ed – may the future recompense for that foregone.

CONTENTS

PREFACE

Real estate investment trusts (REITs) have been the most successful investment product of the last 60 years, starting from zero in 1960 and growing to a global market capitalization in excess of US$1 trillion by 2017, with REITs now existing in over 35 countries around the world.

The Routledge REITs Research Handbook brings together both academic and industry research into two thematic reviews – the first part of the book provides a thematic review of emerging issues in international REITs and the second part of the book provides a review of issues in the emerging, developing and developed REIT markets in each of the world's major regions.

A diverse range of emerging issues are identified and analysed by leading academics and REIT industry experts including REIT styles and asset allocation, emerging specialized sector REITs and Islamic REITS, the impact of sustainability and the role of behavioural risk in REIT management.

Within each of the world's principal regions including North America, Latin America, Europe, South East Asia, North Asia and Oceania, a developed, developing and emerging REIT market is analysed and common issues identified together with those issues idiosyncratic to specific countries which benefit or hinder the growth of REITs.

The book concludes by identifying directions for the future of international REITs and so posits a research agenda for the next five to ten years. *The Routledge REITs Research Handbook* is a comprehensive and contemporary guide to the latest in REIT research. REIT practitioners and analysts, academics and advanced students will find this book to be an essential guide to the latest global thinking in the dynamic world of REITs.

CONTRIBUTOR BIOGRAPHIES

Peter Barron graduated from Maastricht University in January 2017 with an MSc in International Business specializing in Sustainable Finance. Previously, he obtained a Bachelor's degree in Economics with German at The University of Nottingham. In May 2017 Peter joined TRIUVA Kapitalverwaltungsgesellschaft mbH in Frankfurt as an analyst. There he supports the Investor Relations team in developing real estate investment strategies and maintaining partnerships with institutional investors.

Brad Case, PhD, CFA, CAIA, is Senior Vice President, Research and Industry Information for NAREIT. He holds degrees from Williams College and the University of California at Berkeley and earned his PhD in Economics at Yale University.

Dr Case has researched residential and commercial real estate markets, domestically and globally, for more than 25 years. His research encompasses investment return characteristics including returns, volatilities and correlations with other assets; measuring appreciation in property values; inflation protection; use of DCC-GARCH and Markov regime switching models to measure and predict investment characteristics; the length of the real estate market cycle and the role of the investment horizon.

Dr Case has a particular interest in the relationship between listed and unlisted real estate markets. He holds patents as the co-inventor of the FTSE NAREIT PureProperty® index methodology and the backward-forward trading contract.

Hyunbum (Joe) Cho is a Lecturer in property at University of South Australia and an experienced investment professional who has worked across venture capital, property, construction, resources and international investment markets in Australia and Korea.

Hyunbum conducted his PhD research at Western Sydney University and completed his PhD thesis on the role of Asia-Pacific property in mixed-asset portfolios, being selected as the scholarship winner for the PRRES conference in 2015. Hyunbum's research has focused on the analysis of indirect property investments, such as REITs and unlisted property funds, including sub-sector analysis, in the Asia-Pacific region. While Hyunbum was undertaking his PhD, he also taught property investment, international property and introduction to property courses at the Western Sydney University.

Hyunbum's background in Korea included being an investment analyst for a venture capital company and a property analyst and marketing manager for a major property consulting firm where he conducted feasibility reports for more than $1 billion worth of residential and mixed-use development projects. Hyunbum also worked for one of Korea's largest construction/development companies, POSCO E&C (a subsidiary of the POSCO Group) where his roles included managing, contracting, negotiating and divesting of commercial properties.

His recent industry work has included facilitating investments into Australian property, resources and renewable energy projects primarily from Korean investors. He has also assisted Australian companies to enter the Korean market and Korean companies to enter the Australian market, having also published a book in Korea on the Australian property market.

Hyunbum is a Certified Investment Advisor for shares, bonds, futures and options in Korea and a Registered Land Agent/Auctioneer (SA) and Building Supervisor (SA) in Australia.

Joao da Rocha Lima and Claudio Tavares Alencar are both members of the University of São Paulo Polytechnic School Real Estate Research Group, having received their PhDs in 1985 and 1998, respectively.

Their research concentrates on the themes of planning, management, business, marketing, economics, sustainability and construction funding. Their research is focused on the development of systems to improve decision-making processes in the real estate sector, through the application of knowledge related to those planning techniques involved in all segments of the management process. Having regard to the main features of the sector, the specific features of its economic context, impacts on sustainability and the scale and duration of the real estate projects, their research investigates planning methods and routines with an emphasis on risk analysis for both projects and companies.

Piet Eichholtz is Professor of Real Estate and Finance at Maastricht University in the Netherlands. Most of his work concerns real estate markets, with a focus on real estate investments, long-term real estate market developments and sustainability. His research has been published in the top journals in economics and real estate finance. Besides his academic career, Eichholtz has also built up a solid reputation as an entrepreneur. After positions at pension fund ABP and at NIB Capital, Eichholtz started Global Property Research. In 2004, he was a co-founder of Finance Ideas, a financial consultancy company specializing in real estate finance and investments. In 2011, Piet Eichholtz was among the founders of GRESB, the Global Real Estate Sustainability Benchmark.

Chyi Lin Lee is an Associate Professor of Property at the Western Sydney University. His recent research on property investment has covered issues such as REIT futures, REIT volatility and inflation-hedging and has been published in leading property journals in the US and UK including the *Journal of Real Estate Finance and Economics*, *Housing Studies*, *Accounting and Finance*, *International Journal of Strategic Property Management* and the *Journal of Property Investment and Finance*.

Associate Professor Lee has undertaken and continues to undertake numerous research projects, externally funded by the Australian property industry including the Australian Housing and Urban Research Institute (AHURI), Transport for NSW, Property Investors' Alliance and overseas funding groups such as European Public Real Estate Association (EPRA) and European Association for Investors in Non-Listed Real Estate Vehicle (INREV). He has received numerous awards in recognition of his contributions to property research including the Australian Property Institute Achiever (Early Career Researcher) Award in 2015.

Stephen Lee is a Senior Lecturer in Real Estate Finance at the Cass Business School, Faculty of Finance at the City University London. Prior to joining the faculty Stephen was a Lecturer in Real Estate Investment at the University of Reading. He also previously served at the Metropolitan University London (formally the City of London Polytechnic) and was a visiting lecturer at the Syracuse University London.

Stephen's research focuses on the applications of Modern Portfolio Theory to commercial real estate portfolios. He has published over 100 articles in a number of refereed journals including the *Journal of Property Research*, the *Journal of Real Estate Portfolio Management*, the *Journal of Property Investment and Finance*, *Briefings in Real Estate Finance* and the *Journal of Strategic Property Management*. Stephen has also presented over 90 conference papers at conferences in Europe, the US and Asia. At Cass Business School, Stephen teaches courses on real estate portfolio investment, corporate real estate and international markets.

Stephen's work has been sponsored by the Corporation of London, RICS and INREV. Stephen has also received a number of prizes for his work from the American Real Estate Society (ARES) and the European Real Estate Society (ERES). He is currently on the board of the European Real Estate Society (ERES) and International Real Estate Society (IRES), being the current secretary of the European Real Estate Society.

Liow Kim Hiang is Professor of Real Estate at the Department of Real Estate in the School of Design and Environment, National University of Singapore. He received his PhD from the Manchester Business School, The Victoria University of Manchester in the United Kingdom in 1994. He teaches Real Estate Finance and Corporate Investment in Real Estate.

Professor Liow's research focuses on topics of economic globalization and international financial market integration, real estate finance and capital markets, global real estate companies and real estate investment trusts, financial economics of commercial real estate and strategic portfolio management in corporate real estate.

Professor Liow publishes widely in the key real estate and finance journals in the US, Europe and Asia and has won several international research awards. Professor Liow has been a Deputy Editor of Journal of the Property Research (JPR, UK) since 2006 and sits on several international and regional academic Editorial Advisory Boards.

Alex Moss has spent over 30 years specializing in the property sector, encompassing sell side research (BZW), corporate broking (CSFB), private equity (Apax Partners Capital) and asset management (M&G). Alex formed AME Capital in 2002, which developed a proprietary database and analytical tool for all listed real estate companies and real estate securities funds globally. This business was sold to Macquarie Securities in 2008, where Alex stayed for over three years as Head of Global Property Securities Analytics. In 2012 Alex founded Consilia Capital, a research and advisory company that specializes in implementing real estate strategies for corporate and institutional clients.

Alex is a Visiting Professor, Real Estate, at Cass Business School where he teaches the capital markets module for the MSc Real Estate Investment course and a Visiting Professor in the Real Estate and Planning Department at Henley Business School, University of Reading. A member of the Advisory Board of EPRA, Alex is also Chairman of the EPRA Research Committee and Chairman of the Investment Committee for the Investec Global Real Estate Securities Fund.

Joseph TL Ooi is a Professor of Real Estate at the National University of Singapore (NUS) where he is serving concurrently as Vice-Dean (Academic) of the School of Design and

Environment (SDE), Deputy Director (Development) of the Institute of Real Estate Studies and Director of the Graduate Certificate in Real Estate Finance (GCREF).

Professor Ooi has published more than 50 peer-reviewed papers on real estate finance, economics and development in top real estate journals, such as *Journal of Money, Credit & Banking, Journal of Regional Science, Regional Science & Urban Economics, Real Estate Economics, Journal of Real Estate Finance & Economics, Journal of Housing Economics, Journal of Real Estate Research* and *Urban Studies.*

A winner of multiple international research awards, Professor Ooi received the prestigious *Young Researcher Award* (2005) and the *Dean's Chair* (2015) from NUS. In 2017, Professor Ooi was inducted as a Fellow of the exclusive Weimer School of Advanced Studies in Real Estate and Land Economics for his outstanding and impactful scholarly accomplishments at the international level. He has also won many teaching excellence awards, including the *Outstanding Educator Award* (2012), which is the highest teaching award in NUS.

Professor Ooi is a board member of the Asian Real Estate Society as well as Associate Executive Director of the International Real Estate Society (2013–2018), he sits on the editorial board of ten real estate journals and was external examiner for the University of Reading's Masters programmes in real estate (2011–2014). In 2008, he was awarded the International Real Estate Society's *Achievement Award* for his outstanding contribution to research, education and practice at the international level.

Professor Ooi obtained his BSc (Estate Management) and MSc (Real Estate) degrees from NUS and his PhD in Real Estate Finance from the University of Manchester (*formerly UMIST*).

David Parker is an internationally recognized property industry expert on real estate investment trusts (REITs) and a highly regarded property academic, being a director and adviser to property investment groups including real estate investment trusts, unlisted funds and private property businesses (www.davidparker.com.au).

He is currently the inaugural Professor of Property at the University of South Australia a Visiting Professor at the University of Reading, UK and at Universiti Tun Hussein Onn, Malaysia, a Visiting Fellow at the University of Ulster, UK, an Acting Valuation Commissioner of the Land and Environment Court of New South Wales and a Sessional Member of the SA Civil and Administrative Tribunal.

Author of the authoritative *Global Real Estate Investment Trusts: People, Process and Management* and *International Valuation Standards: A Guide to the Valuation of Real Property Assets*, he may be contacted at davidparker@davidparker.com.au.

Paloma Taltavull de La Paz is a full Professor of Economics in the Faculty of Economics and Business Administration at the University of Alicante, Spain and a Visiting Scholar at the Goldman School of Public Policy, University of California at Berkeley (2011–2012) and at Georgia State University (2012, 2015).

Professor Paloma Taltavull de La Paz's main researchvv interest is in housing economics including macro and microeconomic perspectives of housing prices, demand and supply determinants. Real estate economics, education matters and energy efficiency in housing markets are further areas of research interest. She is the author of more than 50 academic articles and several books and has been a participant in various pan-European and international research projects, including more than 30 research projects in housing and urban economics, mostly with application to the Spanish economy.

Professor Paloma Taltavull de La Paz is a member of, and active participant in, several academic societies including the European Real Estate Society (ERES) since 1994, the American Real Estate Society (ARES) since 1992 and the International Real Estate Society (IRES) since

1999. She was ERES president in 2002 and IRES president during 2010. Academically recognized, she received the PhD Main Prize for her Thesis in 1997, the II Fundación Presidente de FICIA Prize in 1996, the 2006 IRES Service Award as well as four Outstanding Reviewer Awards in 2009, 2013, 2014 and 2015 from various journals and the Emerald Literary Club.

At the invitation of the Spanish Housing Ministry, Professor Paloma Taltavull de La Paz belonged to the I Observatorio de Vivienda en Alquiler during 2006–2009. She was principally responsible for the creation of the Biblioteca Virtual Miguel de Cervantes (Miguel de Cervantes Virtual Library) in 1997–2000, the first virtual library in Spanish. She was also the head of the real estate undergraduate course of the University of Alicante from 1997 to 2013 and organized several national and international academic congresses.

Muhammad Najib Razali is a full-time senior lecturer with the Department of Real Estate, Universiti Teknologi Malaysia, Malaysia. He has 15 years' experience in property economics and finance in academia. His research and writing interests include property investment analysis in developing countries, property portfolio management, international property markets, information technology application in real estate and knowledge management in property.

During the past year he has written and co-authored technical papers on various property topics and presented papers at both local and international conferences. He currently serves as a board member of the Pacific Rim Real Estate Society (PRRES). In UTM, he teaches real estate finance modules for both undergraduate and post-graduates courses.

Najib obtained a Doctor of Philosophy degree from Western Sydney University, Australia. He also holds a Master of Science degree in Information Technology and a Bachelor's degree in Property Management from Universiti Teknologi Malaysia.

Wejendra Reddy (Wejen) is a Lecturer in property investment at RMIT University, Melbourne, Australia. Wejen researches and teaches at RMIT in the areas of property investment concepts, securitized property, property accounting and capital markets.

Wejen has a PhD from RMIT University that investigated Australian managed funds' investment strategies and property allocation. He received his Bachelor's and Master's degrees in property from the University of the South Pacific (Fiji Islands) and post-graduate qualification in finance from the Financial Services Institute of Australasia (FINSIA, Australia).

Prior to joining RMIT, Wejen has worked as a capital markets analyst, REITs analyst and property academic in Australia and overseas. Wejen is a Fellow of FINSIA (F Fin) and Associate of the Australian Property Institute (AAPI) with Certified Property Practitioner (Education) certification.

Tien Foo Sing is Dean's Chair Associate Professor, Deputy Head (Admin & Finance) at the Department of Real Estate and Director at the Institute of Real Estate Studies (IRES). He serves as the Secretary of the Asian Real Estate Society (AsRES) and has been a member of the AsRES Board since 2000 and of the Global Chinese Real Estate Congress (GCREC) Board since 2008. He serves as an editorial board member for the *Journal of Real Estate Research*, *International Real Estate Review*, *Pacific Rim Real Estate Journal* and *Journal of Housing Studies* (Taiwan). He has also served as special issues guest editor and referee for several international refereed journals.

He currently serves as a member of the panel of assessors under the Appeal Board (Land Acquisition), Ministry of Law, Singapore (2010–2012). He is the co-founder and principal investigator for the Singapore's NUS-REDAS real estate sentiment index. He has been invited by the Control Yuan, the Republic of China (Taiwan), as an expert witness in the review of land prices in Greater Taipei metropolitan area (August 2010).

In NUS, he teaches real estate finance modules for both undergraduate and post-graduate courses. His research interests include real options, real estate finance and securitization, REITs and housing price dynamics.

Tien Foo obtained a Doctor of Philosophy degree from University of Cambridge, UK under the Cambridge Commonwealth Trust scholarship and the Overseas Research Students award. He also holds a Master of Philosophy degree in Land Economy from University of Cambridge, UK and a Bachelor of Science degree with first class honors in Estate Management from National University of Singapore.

Michael Troilo is the Wellspring Associate Professor of International Business in the Collins College of Business at the University of Tulsa. His research interests focus on the effects of institutions like rule of law on entrepreneurial activity across countries, particularly in transition economies such as China. He has worked as a consultant to the United Nations in the Asia-Pacific region, and he has received a Fulbright grant to study economic development in Portugal. Mike teaches an economics class to MBAs, international finance and international management to undergraduates and short-term study-abroad classes across levels.

Mike has worked in a variety of accounting and finance positions throughout his career. He was a senior auditor for Arthur Andersen LLP and a cost accountant and pricing analyst for IBM Microelectronics Division. He also served as a senior financial analyst for a British company investing in high-tech new ventures and as a controller for an internet start-up. He currently advises students who are trying to launch their own firms.

Mike holds a PhD in business economics and international business from the University of Michigan, an MBA and an MA in East Asian Studies from the University of Virginia and a BBA in accounting from the College of William and Mary. He is a licensed CPA in Virginia.

Larry Wofford is the Davis D. Bovaird Endowed Chair of Entrepreneurial Studies at the University of Tulsa. He is a founder and past-president of both the American Real Estate Society and the International Real Estate Society. Larry has published extensively in real estate and real estate-related fields. He has served on numerous editorial boards and was editor of the *Journal of Real Estate Research*. His research interests include the interaction between real estate academicians and practicing professionals and the behavioural issues affecting problem-solving and decision-making in real estate.

Larry earned a PhD in finance and real estate from the University of Texas at Austin and a Post-Graduate Diploma from the University of Oxford. He holds the AICP designation from the American Institute of Certified Planners, is a Fellow of the Royal Institution of Chartered Surveyors and is a Distinguished Fellow of the NAIOP Foundation.

As an entrepreneur, Larry Wofford developed and operated a number of successful enterprises. He has served on a number of boards of for-profit and not-for-profit organizations. Currently, he is a member of the board of directors of Cancer Treatment Centers of America Global.

Woei-Chyuan Wong is an Associate Professor in the School of Economics, Finance and Banking, Universiti Utara Malaysia. He received his PhD in Real Estate from the National University of Singapore.

Associate Professor Wong's research interests include REIT corporate finance and housing markets. He has published in primary real estate journals including *Real Estate Economics*, *Journal of Real Estate Finance and Economics*, *Real Estate Literature*, *Journal of Property Investment & Finance* and *Pacific Rim Property Journal*.

Erkan Yönder is an Assistant Professor of Finance and Real Estate at Ozyegin University, Istanbul. He received his PhD degree in Finance and Real Estate at Maastricht University in 2013 and holds a second PhD degree in Economics from Middle East Technical University. While Erkan's research interests are mainly in real estate finance and corporate finance, he has also been actively conducting sustainability research in those fields. Dr Yönder's sustainability research has been recognized and funded by leading academic and private institutions such as the Real Estate Research Institute (RERI) and European Public Real Estate Association (EPRA). His research on sustainable investment by real estate companies won the Principles for Responsible Investment (PRI) award in 2012 and the Nick Tyrell Research Prize in 2017. Erkan has published papers in the leading journals in the field.

Huang Yuting is currently pursuing her PhD research in real estate finance at the Department of Real Estate, National University of Singapore.

Ali Zaidi heads the Index & Research department at EPRA and is responsible for the management of the FTSE EPRA/NAREIT Index series. The Index team at EPRA engages with active and passive investors to develop standard and custom versions of the FTSE EPRA/NAREIT series.

Ali graduated from the University of Amsterdam prior to joining EPRA and earned the Executive MBA degree from Rotterdam School of management in 2016. Ali started his career at EPRA in Amsterdam working on the creation of Emerging Market REIT index in 2008 and on the Corporate Governance report for the European listed property sector.

As head of research at EPRA, Ali is engaged in undertaking and facilitating research initiatives demonstrating the merits of listed real estate and REITs for portfolio allocation. The research initiatives include academic studies and industry whitepapers mandated by the Research Committee. Ali is a board member of the European Real Estate Society and is also responsible for EPRA's investor outreach in the US in partnership with NAREIT.

PART I

International REITs

Thematic review of emerging issues

1

CRITICAL CONTEXTUAL ISSUES IN INTERNATIONAL REITS

David Parker, Stephen Lee, Alex Moss, Ali Zaidi and Brad Case

1.1 Introduction

This book aims to identify key areas for research in the REIT discipline for the next five to ten years by surveying the current state of the REIT discipline around the world and identifying emerging and cutting-edge research areas through a thematic review of current contextual issues and a regional analysis based on case studies.

This book comprises two parts, the first part being a thematic review of emerging and cutting-edge global research into current contextual issues in REITs internationally and the second part being a regional analysis of REITs around the world, each written by authoritative academic authors from the world's leading Universities and REIT industry experts.

Following an extensive review of academic journals relevant to the field of REIT research, the following themes were identified as resonating with researchers around the world:

- the impact on the REIT market and on REIT industry practice of the rapid rate of evolution of the REIT sector following the wave of legislative changes in the US in the 1980s, leading to the modern REIT era of the 1990s, followed by the global financial crisis of the 2000s leading to what has become known as the post-modern REIT era and which is considered in detail in Chapter 2;
- the growing role of specialized property market sectors as a source of assets for REITs. Where, previously, REITs focused on commercial, retail and industrial property as core assets for both diversified REITs and sector-specific REITs, this is now widening to include specialized property market sectors such as self-storage or timberland and thus potentially creating a new risk-return profile in the REIT sector, which is considered in Chapter 3;
- the role of sustainability in both the property market sector and the REIT sector has grown considerably over the last decade, with major performance implications for those physical property assets that comprise the primary investment for REITs and for REITs themselves, as considered in Chapter 4;

- the future role of Islam features prominently in both global and national politics as well as in the REIT sector, given the massive depth of investable funds available in the Islamic world and the specific investment requirements of the Islamic world, as considered in detail in Chapter 5;
- as REITs become massive global businesses, a key risk that emerges is the human risk arising from the actions or inactions of human beings as REIT managers and operatives, making the world of behavioural finance and cognitive risk increasingly relevant to REIT management, as considered in Chapter 6; and
- the perennial questions: *are REITs real estate?* and *what place do REITs have in a mixed-asset portfolio?* continue to challenge both academics and REIT experts around the world with recent research reviewed in Chapter 7.

With Part I of this book providing a thematic review of emerging and cutting-edge global research into current contextual issues in REITs internationally, Part II comprises an analysis of REITs around the world grouped by region with the consideration of a developed, developing and emerging REIT sector in each:

Chapter	*Region*	*Developed REIT Sector*	*Developing REIT Sector*	*Emerging REIT Sector*
8	North America	US	Canada	Mexico
9	Latin America	Brazil	Argentina	Uruguay
10	Europe	UK	Spain	Poland
11	South East Asia	Singapore	Malaysia	Thailand
12	North Asia	Japan	Hong Kong	China
13	Oceania	Australia	South Africa	India

concluding with Chapter 14 which considers *Directions for the Future of International REITs.*

Before considering those global cutting-edge research themes identified in the following chapters, this chapter seeks to identify critical contextual issues in international REITs including the defining characteristics of REITs and their structure, the role of REITs relative to property companies and unlisted property funds, the evolution of REITs through a US lens, the global REIT markets and aspects of REIT return, risk and correlation.

1.1.1 Defining characteristics of a REIT

There is no single international REIT model, as there are significant differences in real estate markets across different countries with each country adopting a variation of a range of defining REIT characteristics.

A general definition of a REIT is provided by the OECD:

> A widely held company, trust or contractual or fiduciary arrangement that derives its income primarily from long-term investment in immovable property, distributes most of that income annually and does not pay income tax on the income related to immovable property that is so distributed.

Regional industry bodies have defined REITs as follows:

> **European Public Real Estate Association (EPRA, Europe):**
>
> REITs are defined as publicly listed property investment companies that own, operate, develop and manage real estate assets for obtaining returns from rental income and capital appreciation. REITs obtain a special "tax-transparent" status in return for meeting certain obligations, most importantly high distribution requirements.
>
> (www.epra.com/regulation-and-reporting/taxation/)

> **National Association of Real Estate Investment Trusts (NAREIT, USA):**
>
> A REIT, or Real Estate Investment Trust, is a company that owns or finances income-producing real estate. Modelled after mutual funds, REITs provide investors of all types regular income streams, diversification and long-term capital appreciation. REITs typically pay out all their taxable income as dividends to shareholders. In turn, shareholders pay the income taxes on those dividends.
>
> (www.reit.com/investing/reit-basics/what-reit)

> **Asia Pacific Real Estate Association (APREA, Asia):**
>
> A REIT is a collective investment vehicle that invests in a diversified pool of professionally managed, investment grade real estate. In its simplest form, a REIT provides ownership of a portfolio of properties in units that are held by investors as a way of securitising property. Most of the income from the properties, typically 90–95%, will be paid directly to investors as a dividend on a regular basis. Conditional on the high dividend pay-out, most countries waive corporate income tax on the trusts.
>
> (www.aprea.asia/file/The%20Impact%20of%20REITs%20on%20Asian%20Economies.pdf)

By comparing the definitions, a REIT has some core characteristics (tax exemption, focus on real estate and pay-out requirement) for which there are only minor differences between countries. One of the main differences between countries is related to specific real estate activities that may be undertaken by the REIT. Unlike developed markets, which are characterized by a higher proportion of property owners focused on managing and leasing properties, emerging markets are often focused on property development, which may be considered to be a key point of inconsistency. Property owners are more likely to see a stable income arising from leases and then being able to distribute a significant proportion of their profits on a regular basis, through dividends, as being a relatively lower-risk activity. On the other hand, developers are typically dependent of the project's stage and economic cycle, where early stages are characterized by a less-stable income stream and relatively high leverage, being a relatively higher risk activity.

1.1.2 Structure of a REIT

From the definitions given previously, five key structural aspects may be identified:

- corporate structure – widely held company, trust, or other legal form;
- investment focus – investing in immovable property;

- time horizon – for the long-term;
- profits distribution – distributes most of the profits annually; and
- tax treatment – does not pay income tax on the income related to immovable property that is so distributed,

which may often be accompanied by some level of limitation on debt or gearing.

From these five points, it is evident that there is a wide scope for interpretation in different countries as required by each market's characteristics and stage of development.

1.1.2.1 Corporate structure

Most of the REIT regimes are oriented towards the *collective investment vehicle* model although the specific structure depends on the legal framework in each country. Many REIT regimes are designed in a similar way to *mutual funds*, although the traditional corporate structure remains relevant in other cases with some adjustments to ensure a long-term approach. Generally, to be listed on a public stock exchange is not a requirement in the Americas and Asia-Pacific regions while it is in the EMEA region.

In terms of preference for internal or external management, the debate is still ongoing. Although the US historically adopted an internally managed REIT model, mainly trying to avoid concerns about manager's conflicts of interest, the implementation of *Taxable REIT Subsidiaries* (TRS) has introduced some flexibility for REITs to develop real estate for third parties. In contrast, Australia has gradually moved from an externally managed structure towards an internally managed structure.

The rest of the Asia-Pacific region is mainly focused on externally managed REITs, principally because of significant differences across the countries with developed economies but small populations and restricted availability of development land stock. For example, Singapore and Hong Kong are interested in diversifying their real estate activities through the use of third-party management schemes. Europe is closer to the internally managed structure with specialized sectors. For example, in Germany, REITs may only provide secondary activities via a 100%-owned REIT service company whereas, in the UK, the parent company must own at least 75% of a subsidiary company for the subsidiary to be a member of the UK REIT group.

1.1.2.2 Investment focus

The traditional investment focus of REITs is, as the name suggests, real estate. While this started as a focus on commercial (office), retail and industrial property, or as a combination in a diversified portfolio, as time passed, REITs have begun to broaden such sectors to include hotels, residential and bulky goods retail centres.

Further, as the global availability of stock for acquisition by REITs decreases, the focus is now moving to specialized sectors such as self-storage, data centres and prisons as well as a range of non-real estate investments from debt products to infrastructure.

1.1.2.3 Time horizon

In some countries, the trust underlying a REIT has a finite life of, for example, 60 or 100 years whereas in other countries it may be perpetual. This is consistent with the concept of REITs as a long-term, secure, low-risk investment which buys and holds properties for long-term income growth and capital appreciation, rather than some form of trading vehicle that seeks to acquire and dispose of properties on a regular basis as its principal source of revenue.

1.1.2.4 Profits distribution

Most of the developed countries have adopted a REIT regime that has a profit distribution requirement within the range of 80% to 100%, usually as a pre-requisite for receiving favourable tax treatment, though Italy appears to be an exception at 70%. Some countries apply the threshold to the company's net income while others apply it to the company's rental income. Interestingly, there is no clear consensus for emerging economies where some have adopted a higher requirement (Brazil, Mexico, Malaysia, Thailand) and some have adopted a lower requirement (South Africa, Greece) or even a discretionary requirement (Turkey).

1.1.2.5 Tax treatment

Corporate tax exemption or reduction is one of the main incentives used by regulators to promote the development of a country's real estate market through the use of REIT structures, usually linked to a requirement to distribute a specified proportion of income to investors, a minimum percentage for real estate activities within the REIT and/or a specified proportion of assets to be invested in real estate.

However, tax exemptions vary around the world with countries like Australia, Canada, Japan and the US contemplating a full effective exemption over net income distribution in contrast with countries like Brazil, France, Singapore and UK where only rental income is exempt. Capital gains and profits from the disposal of real estate are partially or totally exempt in most countries, however REIT regimes do sometimes have special treatment for domestic investors and may have Foreign Ownership Restrictions (FOR) in place.

Withholding tax requirements also differ across the regions. Further, in the Asia-Pacific region most countries enable a tax recapture from offshore investors. In many cases, the national tax authority is the final regulatory entity who determines eligibility for the REIT designation.

1.1.2.6 Debt and leverage

In terms of leverage, there are no significant differences between emerging and developed economies. Some use debt limits between 40% and 75% of the properties' value, while some apply specific leverage ratios and others have no specific limit. Differences seem to be more associated with geography, with most of the countries in the Americas region having no restrictions while many European countries use their own-specified leverage ratios.

1.1.3 REITs, property companies and unlisted property funds

As at September 2016, EPRA (2016) notes that there were 35 countries with REIT regimes in place, as shown in Table 1.1.

However, it should be noted that, whilst the legislation and regulation for a REIT regime may be in place, not all of these countries have a trading REIT market. The largest listed global

Table 1.1 Countries with REIT Legislation: September 2016

Europe	*Americas*	*Asia Pac*	*Africa*
Belgium	Brazil	Australia	South Africa
Bulgaria	Canada	Dubai	
Finland	Costa Rica	Hong Kong	
Germany	Mexico	India	
Greece	Puerto Rico	Japan	
Hungary	United States	Malaysia	
Ireland		New Zealand	
Israel		Pakistan	
Italy		Philippines	
Lithuania		Singapore	
Luxembourg		South Korea	
Netherlands		Taiwan	
Spain		Thailand	
Turkey			
United Kingdom			

Source: Lee and Moss based on EPRA (2016)

Table 1.2 Market Capitalization of the Largest REIT Markets

	Market Cap *$USm*	*Percentage of* *REIT Market*
US REITs	746,748	68.3%
Japan REITs	81,694	7.5%
Australia REITs	79,781	7.3%
UK REITs	54,149	5.0%
Canadian REITs	34,995	3.2%
Netherland REITs	28,327	2.6%
French REITs	20,456	1.9%
Singapore REITs	18,179	1.7%
Hong Kong REITs	17,449	1.6%
Belgium REITs	6,667	0.6%
German REITs	2,392	0.2%
New Zealand REITs	1,242	0.1%
Italy REITs	889	0.1%
Total	1,092,968	100.0%

Source: Lee and Moss

REIT markets, as measured by free float market capitalization in US$ millions at the end of January 2017, is shown in Table 1.2.

The dominance of the US REIT market in the world is particularly noticeable, accounting for two-thirds of the total global market capitalization. This does reduce, however, when including the non-REITs (i.e. developers and property companies) in Hong Kong and Japan in a calculation of the global listed real estate market. Including such non-REITs means that the US percentage falls to 55% of the total, while the percentage in Japan and Hong Kong rises to 11% and 7.7%, respectively.

It is important to note that, before the advent of REITs, many countries had very successful property companies and unlisted property trusts operating. Property companies were common in the UK, Hong Kong and Australia and had, for decades, operated profitably in both the listed and unlisted environments, developing, owning and managing income-producing property. Their key difference to REITs is, of course, that they do not enjoy the beneficial tax status of REITs but are subject to the relevant jurisdictional taxes on corporate profits. Further, property companies often had a profile of being either a developer or a trader of property with a short to medium-term outlook, rather than being a holder of property with a long-term outlook. Despite the beneficial tax treatment of REITs, property companies remain popular in such countries as Hong Kong, and a related vehicle, the *fideicomiso*, remains preferable to REITs through much of Latin America.

For holding property with a long-term outlook, unlisted property trusts were very common in various countries prior to the advent of REITs. As a trust structure, unlisted property trusts have some beneficial tax features but their principal difference to REITs and their principal shortcoming is that they are unlisted and therefore illiquid, with their units unable to be traded on a stock exchange. However, unlisted property trusts remain popular with many large international wholesale investors and pension funds who often perceive them to offer a purer form of property pricing and returns than REITs.

1.2 Evolution of REITs – the US experience

REITs were first established by the US Congress under the 1960 *Real Estate Investment Trust Act* to make investment in large-scale income-producing real estate accessible to smaller investors. In other words, the REIT structure is designed to provide a similar vehicle for investment in real estate to that which mutual funds provide for investment in stocks.

1.2.1 Types of REITs in the US

There are three types of REITs classified by the National Association of Real Estate Investment Trust (NAREIT): Mortgage REITs, Hybrid REITs and Equity REITs.

Mortgage REITs (MREITs) invest in housing and commercial mortgage debt markets. Some companies only buy mortgages that are backed by a federal agency like Fannie Mae, Freddie Mac or Ginnie Mae. They choose these mortgages because they are backed by a federal guarantee and thus have a lower risk of default, which means that they are also less profitable. Other MREITs specialize in mortgages that are not guaranteed by a federal agency, as they tend to pay higher dividends, largely because there is a higher risk of default. At the end of 2015 there were 41 listed MREITs worth $52 billion.

Hybrid REITs (HREITs) are real estate investment trusts that offer, effectively, a combination of Equity REITs which own properties and MREITs which invest in mortgage loans or mortgage backed securities. By diversifying across both types of investments, HREITs aim to obtain the benefits of both with less risk than if invested in either one or the other. Historically, HREITs represented the smallest class of REITs and in 2010 NAREIT reclassified the remaining four remaining HREITs as MREITs.

Equity REITs (EREITs) typically specialize in investing in, actively managing and developing portfolios of commercial property, but with some advantages compared to private real estate investment. First, EREITs are operating entities investing in real estate assets, designed to reduce or eliminate corporate income tax. In return, EREITs are required to distribute at least 90% of their income, which is then taxed at the investor's tax rate. Second, EREITs offer both ownership shares to the public and, as a result of trading on a stock exchange, the benefit of daily trading, i.e. liquidity. At the end of 2015, there were 182 listed EREITs worth $886 billion, accounting for 82% of the market by number and 95% by value. EREITs are often the preferred route for small, non-specialized investors who want some commercial real estate exposure in their portfolio. Larger, more sophisticated investors will typically have the ability to access the private real estate sector but may also gain some benefit from using REITs.

1.2.2 REITs and evolving tax law in the US

Since inception, the US REIT market has developed anything but evenly with a number of structural changes occurring in the industry linked to changes in the tax laws and regulations governing REITs. For instance, prior to 1986, REITs were mostly MREITs heavily involved in land, construction and construction loans. Indeed, in 1972, short-term construction and development loans constituted over 50% of all MREIT assets. So, as interest rates rose in 1972 following the "oil price shock", MREITs experienced cash flow problems leading to many MREITs becoming insolvent.

In addition, prior to 1986, REITs had to be passive investors in real estate as they were prohibited from both operating and managing their own real estate. However, following the 1986 *Tax Reform Act*, this restriction was removed, providing REITs with the ability to manage most types of income-producing commercial property with greater management flexibility under a less restrictive tax environment and with changes in depreciation allowances.

The early 1990s led to even greater changes in the REIT industry as a result of the introduction of the Umbrella Partnership REIT (UPREIT) structure and the 1993 *Revenue Reconciliation Act*. Instead of selling the property, the owner contributed it to an UPREIT in exchange for securities called "operating partnership units" or "OPUs"These partnership units would be equivalent to the value of the contributed property. Unlike a disposal of the property asset, this transaction would not create a taxable event.

The UPREIT concept was first implemented in 1992 with the Taubman REIT IPO. Taubman had a large debt position that was maturing but rather than sell off part of the portfolio to repay the debt, which would have triggered significant tax liabilities, by doing an IPO as an umbrella partnership Taubman was able to raise the crucial capital needed to repay the loan without triggering capital gains tax liabilities. In the next few years, the UPREIT structure proved so popular that the majority of REIT IPOs in the 1990s were UPREITs while a number of existing REITs converted to the UPREIT form.

Prior to 1993, to be qualified as a REIT, more than 50% of its shares could not be owned by five or fewer individuals. However, a pension fund was treated as a single entity, even if it had many members. Effectively, therefore, pension funds were restricted from holding REITs in any substantial quantity. The 1993 *Revenue Reconciliation Act* changed the "five-or-fewer" rule by treating all the individuals in a pension fund as investors. Thus, the interest by pension funds and other institutional investors in the REIT market increased substantially resulting in REITs growing as a mainstream investment option and leading to the dawn of the "Modern REIT Era".

The next major change affecting the development of REITs was the 1999 *REIT Modernization Act*, which reduced the income distribution requirement for US REITs from 95% to 90% of taxable income and allowed REITs to own up to a 100% controlling stake in taxable REIT subsidiaries (TRS). A TRS provides services to REIT tenants but does not qualify for the same tax exemption status as the REIT. The *REIT Modernization Act* was the most significant amendment in REIT legislation as it meant that REITs were now more like an operating firm than a mutual fund, which is evidenced by the inclusion of a number of REITs in various S&P indices from October 2001.

1.2.3 REITs classification

The latest development in the REIT industry occurred in August 2016 when S&P Dow Jones Indices and MSCI moved Equity REITs and other listed real estate companies from the Financials Sector of their Global Industry Classification Standard (GICS®) to a new Real Estate Sector. This should increase the visibility of real estate as a distinct asset class, prompting investors, managers and advisors to more actively consider real estate when developing their investment strategy. The reclassification of REITs has also led to the introduction of new Exchange-Traded Funds (ETFs), which should also increase investor interest in the sector.

Consequently, the market for US REITs has continuously evolved and has shown tremendous growth since the start of the new millennium, providing the template for REIT structures across the world, albeit with slight modifications.

1.3 The global REIT market

After its introduction in the US in 1960, the REIT regime remained almost exclusively limited to this market for three decades, with only a few countries like the Netherlands (1969) and Australia (1971) taking the initial steps to establish a similar regime. The global expansion of REITs effectively began during the early 1990s, almost simultaneously in Europe (Belgium, 1995) and the Americas (Brazil and Canada, 1993). In Asia, Singapore introduced the REIT framework in 1999 (effective in 2002) followed by Japan (2001), Malaysia (2005) and Hong Kong (2005).

The three major European economies adopted a REIT regime at the turn of the century (France, 2003, and Germany and UK, 2007). In the last ten years, more countries have successfully adopted or amended this structure, particularly emerging economies like South Africa (2013) and Mexico (2004 and 2010), with many more working towards launching a version of a REIT framework that will attract international investors while addressing the property market dynamics of the local market.

1.3.1 REITs risk and return

The implementation of REIT regimes has supported the expansion of property markets around the world. Gradually, REITs have become a key component of the listed real estate markets, reaching a total market capitalization of over US$1 trillion in March 2017 and representing 34% of the global listed real estate industry. Considering only developed economies, the proportion is even higher at 44.8%, with around 76% of the total number of REITs around the world showing a high level of diversity in terms of size, sector and geographical allocation.

At the same time, emerging economies show a contrasting picture with REITs representing only 4.4% of the listed real estate market, mainly as a consequence of some structural differences in terms of the real estate activities compared with developed economies but also reflecting the evolutionary stage of their markets and predominance of property development.

Significantly, the average REIT market capitalization in emerging markets is around US$192 million while it stands over US$1,760 million in developed markets (Table 1.3).

Table 1.3 Size of the Commercial Real Estate Market

Size of the total commercial real estate market												
Dec-16	*Dec-16*	*Dec-16*	*Mar-17*	*Mar-17*	*Mar-17*	*Mar-17*	*Mar-17*	*Mar-17*	*Mar-17*	*Mar-17*	*Mar-17*	*Mar-17*
GDP per Capita	*GDP**	*Commercial Real Estate*	*Total Listed Real Estate*	*Number of Companies*	*Of which REITs*	*Full M.Cap REITs*	*REITs / Listed RE*	*Index** Market Cap*	*No. of Index Cons.*	*Stock Market Size*	*RE/Stock Market*	*Listed RE / Total RE*
($)	*($ Bln.)*	*($ Bln.)*	*($ Bln.)*	*#*	*#*	*($ Bln.)*	*%*	*($ Bln.)*	*#*	*($ Bln.)*	*%*	*%*
Total Emerging Markets	**23,810.70**	**6,915.77**	**809.95**	**899**	**185**	**35.45**	**4.38%**	**140.39**	**148**	**14,701.94**	**5.51%**	**11.71%**
Total Developed Markets	**44,627.26**	**20,638.35**	**2,338.58**	**1,343**	**595**	**1,047.17**	**44.78%**	**1,289.33**	**338**	**54,156.26**	**4.32%**	**11.33%**
Total Global Markets	**68,437.96**	**27,554.12**	**3,148.53**	**2,242**	**780**	**1,082.63**	**34.39%**	**1,429.72**	**486**	**68,858.20**	**4.57%**	**11.43%**

* *GDP values and estimated CRE size based on Dec-15 data*
** *FTSE EPRA/NAREIT Global Index*

Source: Zaidi based on FTSE SPRA/NAREIT Global Index

In terms of total returns and market risk, REITs have traditionally outperformed other listed property companies, in part because of higher dividend yields but also because of lower volatility. As evidenced by the FTSE EPRA/NAREIT REITs and Non-REITs indices, during the ten-year period between January 2007 and December 2016, REITs showed an annualized total return of 5.5% compare to 1.1% for Non-REITs, where both type of real estate vehicles presented a similar volatility (Figure 1.1).

In this case, the risk-adjusted return reached 0.33 and 0.07, respectively, which was, of course, highly influenced by the effects of the global financial crisis of 2007–2009 as well as the early stage of many REIT regimes around the world, particularly in emerging markets. The post-crisis period has evidenced a dynamic expansion and improvement for REIT regimes and markets in many countries with the three-year period between January 2014 and December 2016 recording a total return of 19.8% and volatility of 13.7% for REITs, which represents a risk-adjusted return of 1.44 compared to 0.55 for the Non-REITs (total return 6.70% and volatility 12.20%). (Note: all performance figures are annualized.)

This pattern can also be observed across the regions. In the Asia-Pacific region, REITs outperformed Non-REITs by 163bps in the last ten years and by 1,371bps in the last three years. On a risk-adjusted basis, the difference is even more significant with ten years at 0.25 vs 0.14 and three years at 1.51 vs 0.31 respectively (Table 1.4).

The EMEA region shows a similar outcome for ten years' total return where REITs outperformed Non-REITs by 31bps and also in terms of risk-adjusted return (0.07 vs 0.05). The exception comes in the last three year comparison in which REITs underperformed by 554bps the Non-REITs in terms of total return and 0.51 units in terms of risk-adjusted return, mainly as a result of a continuous depreciation of the British pound and a more stable performance of real estate companies listed in countries without a REIT regime.

Finally, although the same comparison basis in the Americas region shows exactly the same pattern, being a significant outperformance of REITs against Non-REITs, such analysis is not appropriate for this region as a consequence of the fundamental differences between developed and emerging countries, particularly in terms of market size and liquidity, that affect the weight of REITs from specific countries, like the US, in most of the benchmarks.

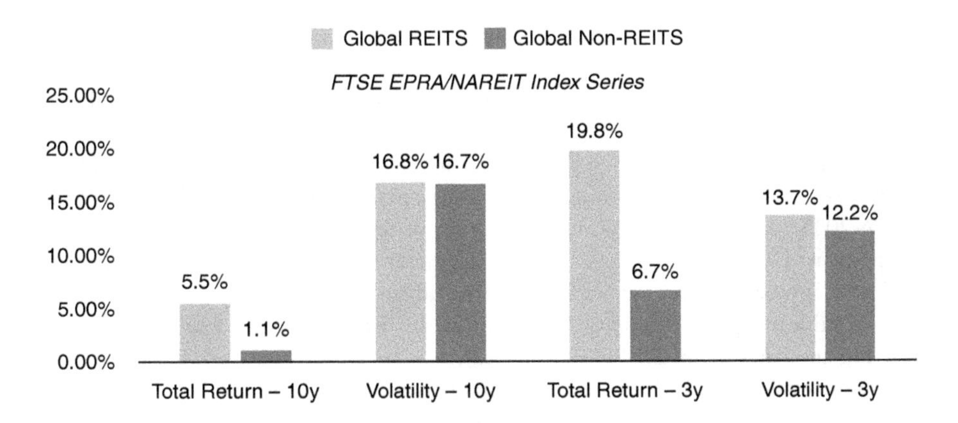

Figure 1.1 Comparative Performance 2007–2016

Source: Zaidi based on FTSE EPRA/NAREIT Index Series

Table 1.4 Comparative Performance 2007–2016

COMPARATIVE PERFORMANCE: 2007–2016						
	Total Return – 10y	*Volatility – 10y*	*Risk-Adjusted Return: 10y*	*Total Return – 3y*	*Volatility – 3y*	*Risk-Adjusted Return:3y*
Asia-Pacific REITs	4.40%	17.48%	0.25	18.10%	12.00%	1.51
Asia-Pacific Non- REITs	2.77%	19.34%	0.14	4.39%	14.28%	0.31
EMEA REITS	1.21%	17.50%	0.07	10.45%	14.94%	0.70
EMEA Non-REITs	0.90%	17.42%	0.05	15.99%	13.26%	1.21
Americas REITs	7.12%	19.22%	0.37	22.33%	15.65%	1.43
Americas Non-REITs	-6.83%	25.89%	-0.26	1.09%	22.35%	0.05

Source: Zaidi based on FTSE EPRA/NAREIT Index Series; Based on the *FTSE EPRA/NAREIT Index Series, EUR*

1.3.2 REITs overview – North America

The North America region represents the largest listed real estate market globally with US$1,188 billion in market capitalization, of which 60.4% are REITs. With almost six decades of evolution, the US REIT regime is the most mature market in the world with 246 highly specialized REITs. In Canada, REITs represent 48% of the listed real estate market, which is US$66 billion in market capitalization (Table 1.5).

Because of the great success of the REIT industry in the US, most of the biggest and most liquid listed real estate entities have adopted this structure as their main operational basis, guiding the country-level benchmarks to show a very small participation of Non-REITs companies. By the end of 2016, Non-REITs represented just 0.61% of the FTSE EPRA/NAREIT US Index. Being the market with the longest history, the US listed real estate market is usually presented as one reference point for a comparative analysis with many other countries.

Following the initial stage of introduction of the REIT regime in the 1960s and the later deregulation and adjustment processes to both financial and property markets until the mid-1990s, the US market has been able to generate an annualized total return rate of 10.87% during the last 20 years, showing three main historical stages: market expansion and REIT consolidation (1997–2005); market correction in the context of the global financial crisis (2006–2009); and the post-crisis "*New REIT Era*". This last stage seems to be one of the most dynamic in terms of new approaches and participants to take advantage of the REIT structure benefits, bringing a significant specialization of sectors inside the listed real estate markets including student housing and data-centre REITs (Figure 1.2).

The recent evolution of the listed real estate market in the US has allowed REITs to gradually improve their performance in terms of return and market risk. In the US, the risk-adjusted return rose from 0.3 in the last ten years to 1.46 in the last three years, where

Table 1.5 Size of the Total Commercial Real Estate Market – Canada and US

Size of the total commercial real estate market-Developed Markets

	Dec-16	*Dec-16*	*Dec-16*	*Mar-17*	*Mar-17*	*Mar-17*	*Mar-17*	*Mar-17*	*Mar-17*	*Mar-17*	*Mar-17*	*Mar-17*	*Mar-17*
	GDP per Capita	*GDP**	*Commercial Real Estate*	*Total Listed Real Estate*	*Number of Companies*	*Of which REITs*	*Full M.Cap REITs*	*REITs / Listed RE*	*Index** Market Cap*	*No. of Index Cons.*	*Stock Market Size*	*RE/Stock Market*	*Listed RE/ Total RE*
	($)	*($ Bln.)*	*($ Bln.)*	*($ Bln.)*	*#*	*#*	*($ Bln.)*	*%*	*($ Bln.)*	*#*	*($ Bln.)*	*%*	*%*
Canada	**42,319.07**	**1,532.34**	**689.55**	**66.66**	**84**	**51**	**32.00**	**48.01%**	**37.09**	**19**	**2.073.40**	**3.21%**	**9.67%**
United States	**57,293.79**	**18,561.93**	**8,352.87**	**1,121.32**	**331**	**246**	**685.36**	**61.12%**	**699.18**	**134**	**26,535.72**	**4.23%**	**13.42%**
Total North America		20,094.28	9,042.42	1,187.98	415	297	717.36	60.39%	736.27	153	28,609.12	4.15%	13.14%

Source: Zaidi based on FTSE EPRA/NAREIT Index Series

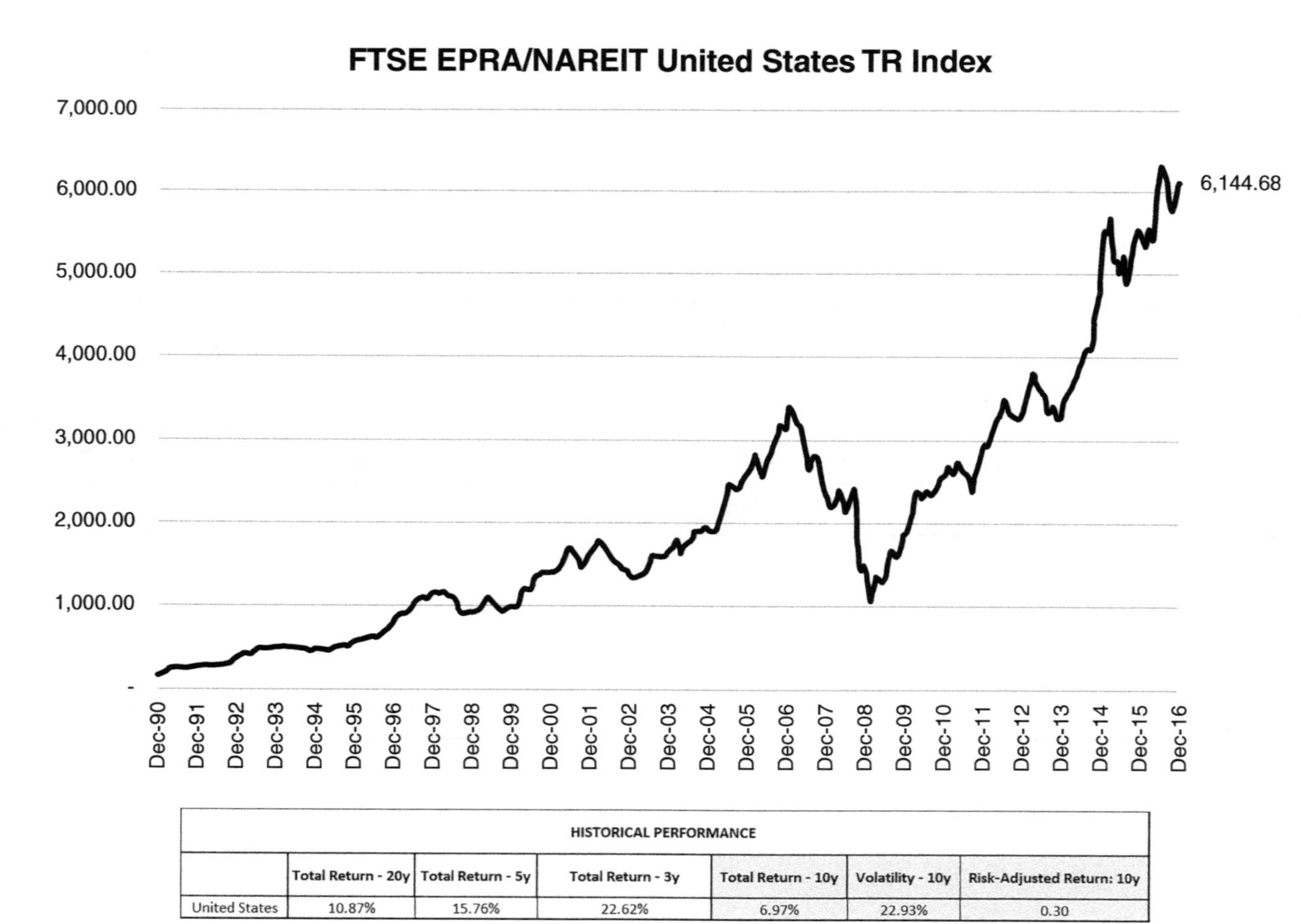

HISTORICAL PERFORMANCE						
	Total Return - 20y	Total Return - 5y	Total Return - 3y	Total Return - 10y	Volatility - 10y	Risk-Adjusted Return: 10y
United States	10.87%	15.76%	22.62%	6.97%	22.93%	0.30

Figure 1.2 Historical Performance

Source: Zaidi based on FTSE EPRA/NAREIT Index Series

Table 1.6 Comparative Performance – Canada and US

COMPARATIVE PERFORMANCE						
COUNTRY	*REITs*			*Non-REITs*		
	Total Return – 3y	*Volatility – 3y*	*Risk-Adjusted Return:3y*	*Total Return – 3y*	*Volatility – 3y*	*Risk-Adjusted Return:3y*
Canada	**8.69%**	**13.22%**	**0.66**	**14.36%**	**14.37%**	**1.00**
United States	**23.67%**	**16.23%**	**1.46**	**15.68%**	**21.01%**	**0.75**

Source: Zaidi based on FTSE EPRA/NAREIT Index Series

REITs also outperformed Non-REITs by 799bps. However, as the market fluctuations affect both types of real estate companies in different ways, Non-REITs can also show a better performance, as was the case in Canada during the last three years where REITs underperformed on both total return and risk-adjusted return (567bps and 0.34 units respectively) (Table 1.6).

1.3.3 REITs overview – Europe

European countries mostly introduced REITs in the last 20 years, with a few early adopters. The continental market capitalization of the listed real estate market reached US$383 billion and REITs represent 33.9% (Table 1.7). However, there are significant differences inside the European region, with countries like Belgium, France, Italy, Netherlands and the UK having mature REIT regimes with more than ten years of history and representing above 50% of the listed market.

At the same time, countries like Finland, Ireland and Spain are in the developing stage after the recent introduction of a REIT regime. Being relatively recent, Germany remains a special case with real estate companies focusing on the residential sector, which has a low presence of REITs. The growth of the listed residential sector in Germany has outpaced other sectors in Europe due to strong fundamentals and high demand from international investors seeking exposure to that specific sector. Finally, there is a third group of countries such as Austria, Sweden and Switzerland that have not adopted a REIT regime.

Unlike the US, where listed real estate markets are more mature and showed strong performance in the post-crisis stage, Europe is still working on the consolidation of REIT regimes. Real estate markets show differences across various countries, including those with an absence of a REIT framework, with many companies still focused on a country-level strategy rather than a continental one, which implies a temporal limitation but also a significant potential for growth. Although the entire continent is now recovering from one of the deepest economic crises in decades, many real estate companies have successfully adopted the REIT structure and are improving their performance, with risk-adjusted return now reaching 0.61 during the last three years compared to 0.11 in the last ten years (Table 1.8).

In fact, the introduction of REIT regimes during the last decade has helped the continent to close the gap against the North American markets. This is evident in changes in Sharpe ratios for the FTSE EPRA/NAREIT Developed Europe Index and the FTSE

Table 1.7 Size of the Total Commercial Real Estate Market – Europe

Size of the total commercial real estate market-Developed Markets

	Dec-16	Dec-16	Dec-16	Mar-17	Mar-17	Mar-17	Mar-17	Mar-17	Mar-17	Mar-17	Mar-17	Mar-17	Mar-17
	GDP per Capita	*GDP**	*Commercial Real Estate*	*Total Listed Real Estate*	*Number of Companies*	*Of which REITs*	*Full M.Cap REITs*	*REITs / Listed RE*	*Index** Market Cap*	*No. ofIndex Cons.*	*Stock Market Size*	*RE/Stock Market*	*Listed RE / Total RE*
	($)	*($ Bln.)*	*($ Bln.)*	*($ Bln.)*	#	#	*($ Bln.)*	%	*($ Bln.)*	#	*($ Bln.)*	%	%
Austria	**44,561.31**	**387.30**	**174.28**	**11.13**	**11**	-	-	**0.00%**	**4.13**	**3**	**108.24**	**10.28%**	**6.38%**
Belgium	**41,491.12**	**470.18**	**211.58**	**14.63**	**26**	**17**	**8.39**	**57.39%**	**6.95**	**8**	**416.78**	**3.51%**	**6.91%**
Denmark	**53,242.91**	**302.57**	**136.16**	**1.47**	**10**	**1**	-	**0.00%**	-	-	**366.03**	**0.40%**	**1.08%**
Finland	**43,492.07**	**239.19**	**107.63**	**4.24**	**6**	**1**	**0.08**	**1.77%**	**2.26**	**3**	**221.55**	**1.91%**	**3.94%**
France	**38,536.71**	**2,488.28**	**1,119.73**	**56.18**	**57**	**33**	**30.15**	**53.67%**	**19.09**	**7**	**1,959.64**	**4.04%**	**7.07%**
Germany	**42,326.03**	**3,494.90**	**1,572.70**	**69.06**	**64**	**4**	**2.34**	**3.39%**	**41.27**	**13**	**1,938.15**	**3.56%**	**4.39%**
Ireland	**65,870.83**	**307.92**	**138.56**	**2.43**	**3**	**3**	**2.15**	**88.53%**	**2.21**	**3**	**107.56**	**2.26%**	**1.76%**
Italy	**30,294.08**	**1,852.50**	**833.62**	**2.77**	**10**	**4**	**1.87**	**67.35%**	**0.93**	**2**	**535.85**	**0.52%**	**0.33%**
Luxembourg	**105,829.05**	**60.98**	**27.44**	-	-	-	-	-	-	-	**19.67**	**0.00%**	**0.00%**
Netherlands	**45,210.24**	**769.93**	**346.47**	**28.74**	**12**	**4**	**24.40**	**84.92%**	**26.03**	**5**	**508.34**	**1.13%**	**1.66%**
Norway	**71,497.29**	**376.27**	**169.32**	**5.26**	**7**	-	-	**0.00%**	**1.14**	**2**	**252.46**	**2.08%**	**3.10%**
Portugal	**19,758.74**	**205.86**	**82.60**	**0.02**	**2**	**1**	-	**0.00%**	-	-	**60.36**	**0.04%**	**0.03%**
Spain	**27,012.16**	**1,252.16**	**557.60**	**14.87**	**25**	**8**	**7.52**	**50.54%**	**7.69**	**5**	**651.07**	**2.28%**	**2.67%**
Sweden	**51,603.94**	**517.44**	**232.85**	**43.29**	**43**	-	-	**0.00%**	**15.40**	**12**	**701.20**	**6.17%**	**18.59%**
Switzerland	**79,577.56**	**662.48**	**298.12**	**46.72**	**42**	-	-	**0.00%**	**11.70**	**4**	**1,535.84**	**3.04%**	**15.67%**
United Kingdom	**40,411.71**	**2,649.89**	**1,490.56**	**82.31**	**84**	**43**	**52.84**	**64.20%**	**62.61**	**37**	**3,189.56**	**2.58%**	**5.52%**
Total Europe		16,037.85	7,499.23	383.10	402	**119**	129.74	33.87%	201.39	104	12,572.31	3.05%	5.11%

Source: Zaidi based on FTSE EPRA/NAREIT Index Series

Table 1.8 Comparative Performance – Europe

COMPARATIVE REITS PERFORMANCE – DECEMBER 2016						
	Total Return – 10y	Volatility – 10y	Risk-Adjusted Return: 10y	Total Return – 3y	Volatility – 3y	Risk-Adjusted Return:3y
Dev. Europe	1.58%	14.97%	0.11	9.61%	15.68%	0.61

Source: Zaidi based on FTSE EPRA/NAREIT Index Series

Table 1.9 Comparative Performance – Europe – REITs and Non-REITs

COMPARATIVE PERFORMANCE						
COUNTRY	*REITs*			*Non-REITs*		
	Total Return – 3y	*Volatility – 3y*	*Risk-Adjusted Return:3y*	*Total Return – 3y*	*Volatility – 3y*	*Risk-Adjusted Return:3y*
Austria				19.42%	13.79%	1.41
Belgium	14.88%	11.75%	1.27			
Finland				7.57%	17.44%	0.43
France	10.37%	17.44%	0.59			
Germany	13.46%	14.11%	0.95	24.43%	17.22%	1.42
Ireland *	-5.40%	14.39%	-0.38			
Italy	8.46%	25.15%	0.34			
Netherlands	10.64%	17.57%	0.61			
Norway				14.28%	18.00%	0.79
Spain **	-9.79%	17.49%	-0.56	2.73%	30.25%	0.09
Sweden				19.40%	16.72%	1.16
Switzerland				16.54%	16.41%	1.01
United Kingdom	7.49%	18.89%	0.40	2.79%	17.80%	0.16

*Corresponds to Mar/2015 – Dec/2016
**Corresponds to Dec/2015 – Dec/2016

Source: Zaidi based on FTSE EPRA/NAREIT Index Series

EPRA/NAREIT North America Index where the Sharpe ratio for the last 25 years was 0.26 in Developed Europe and 0.52 in North America, while the same comparison for the last five years shows 1.14 and 1.04, respectively, with a similar pattern also observable in the Asia-Pacific region.

Finally, comparisons between European countries do not show strong conclusions because of the differences in REIT regime evolution stages, macroeconomic fundamentals and market size. However, countries with more stable economic growth, like Austria, Belgium, Germany or Sweden, have shown a better risk-adjusted performance in the recent years (Table 1.9).

1.3.4 REITs overview – Asia-Pacific

Countries in the Asia-Pacific region are mainly in a mature stage, but with variations in terms of REIT participation mainly associated with differences in market size and activities. The market capitalization of the real estate industry reached US$741 billion, with 27% REIT participation (177 companies) (Table 1.10).

The region has also gradually closed the gap against the US market. The 25-year Sharpe ratios for the region were 0.29 compared with 0.52 from North America while the five years ratios were 1.05 and 1.04, respectively. Most of the countries have benefitted from the implementation of REIT regimes during the last two decades, making the listed real estate markets more dynamic since REITs have brought stability, efficiency and some standardization in a geographical region with significant differences. The last ten years evidenced a significant improvement in REIT performance with higher returns and lower volatility leading to the three-year, risk-adjusted return reaching 1.51 in 2016 compared with the ten years figure of 0.32 (Table 1.11).

Finally, this evolution can also be observed at a country level. For those cases where Non-REITs still have a significant role, like Hong Kong, Japan and Singapore, REITs showed a considerable outperformance both in terms of absolute return (1,774bps, 2,161bps and 927bps, respectively) and risk-adjusted return (1.1 units, 1.47 units and 0.74 units, respectively).

The diversification strategy adopted by many REITs in the region is interesting, in terms of geographical exposure and activities. Many of them have adopted a horizontal expansion strategy, mostly by adding properties from other real estate sectors such as lodging/resorts or offices to their traditional business model. In addition, the expansion of the Chinese property market has also guided the geographical diversification through the region (Table 1.12).

1.3.5 REITs overview – emerging markets

Listed real estate markets are less homogenous across emerging countries. Most of the emerging countries in Europe are in an early stage of REIT introduction (Greece, Russia, Hungary) or are taking the initial steps towards the implementation of a REIT regime (Poland). Some have already implemented a REIT structure but are still considered "*frontier markets*" by global investors (Bulgaria, Lithuania) with promising potential to be reclassified as emerging in the coming years.

On the other hand, some countries in the Middle East and African sub-region (South Africa, Turkey) have managed to adopt a modified version of a REIT structure, which makes it possible to include a wider classification of real estate activities specific to the local market. Turkey, for example, requires only 51% of assets to be deployed in real estate activities and has a minimal pay-out ratio requirement. The listed real estate market in the EMEA region has a total market capitalization of US$100 billion with REITs participation of 19.9% (Table 1.13).

In the Americas region, only Brazil and Mexico have established REIT regimes, with specific adjustments according to their emerging economy's characteristics. In both cases, REIT participation is over 45% with 38 companies in total. Some other countries, like Chile and Colombia, are still in an early implementation stage and Costa Rica has already implemented a similar structure although is still considered a "*frontier market*" (Table 1.14).

On the other hand, the Asia-Pacific region is collectively at an early stage. Malaysia and Indonesia have made significant progress towards REIT regime consolidation while some other

Table 1.10 Size of the Total Commercial Real Estate Market – Asia-Pacific

Size of the total commercial real estate market-Developed Markets

	Dec-16	*Dec-16*	*Dec-16*	*Mar-17*	*Mar-17*	*Mar-17*	*Mar-17*	*Mar-17*	*Mar-17*	*Mar-17*	*Mar-17*	*Mar-17*	*Mar-17*
	GDP per Capita	*GDP**	*Commercial Real Estate*	*Total Listed Real Estate*	*Number of Companies*	*Of which REITs*	*Full M.Cap REITs*	*REITs / Listed RE*	*Index** Market Cap*	*No. of Index Cons.*	*Stock Market Size*	*RE/Stock Market*	*Listed RE/ Total RE*
	($)	*($ Bln.)*	*($ Bln.)*	*($ Bln.)*	#	#	*($ Bln.)*	%	*($ Bln.)*	#	*($ Bln.)*	%	%
Australia	51,592.91	1,256.64	565.49	109.13	77	53	76.05	69.69%	77.53	13	1,247.88	8.74%	19.30%
Hong Kong	42,963.40	316.07	284.46	272.38	144	12	18.50	6.79%	102.80	14	4,341.63	6.27%	95.75%
Japan	37,304.14	4,730.30	2,128.64	228.34	152	59	80.26	35.15%	137.06	40	5,307.60	4.30%	10.73%
New Zealand	38,065.93	179.36	80.71	5.57	9	6	1.13	20.33%	1.21	1	78.41	7.10%	6.90%
Singapore	53,053.26	296.64	266.98	123.96	74	41	24.12	19.46%	30.91	12	515.70	24.04%	46.43%
South Korea	27,632.84	1,404.38	630.14	2.04	8	6	1.50	73.52%	-	-	1,317.01	0.15%	0.32%
Total Asia-Pacific		**8,183.39**	**3,956.41**	**741.42**	**464**	**177**	**201.57**	**27.19%**	**349.51**	**80**	**12,808.24**	**5.79%**	**18.74%**

Source: Zaidi based on FTSE EPRA/NAREIT Index Series

Table 1.11 Comparative Performance – Developed Asia

COMPARATIVE REITS PERFORMANCE – DECEMBER 2016						
	Total Return – 10y	*Volatility – 10y*	*Risk-Adjusted Return: 10y*	*Total Return – 3y*	*Volatility – 3y*	*Risk-Adjusted Return:3y*
Developed Asia	4.36%	13.53%	0.32	18.34%	12.11%	1.51

Source: Zaidi based on FTSE EPRA/NAREIT Index Series

Table 1.12 Comparative Performance – Asia-Pacific

COMPARATIVE PERFORMANCE						
COUNTRY	*REITs*			*Non-REITs*		
	Total Return – 3y	*Volatility – 3y*	*Risk-Adjusted Return:3y*	*Total Return – 3y*	*Volatility – 3y*	*Risk-Adjusted Return:3y*
Australia	20.51%	49.90%	0.41			
Hong Kong	25.88%	17.01%	1.52	8.14%	19.18%	0.42
Japan	15.94%	13.95%	1.14	-5.67%	17.40%	-0.33
New Zealand	18.72%	14.44%	1.30			
Singapore	11.50%	13.13%	0.88	2.23%	17.00%	0.13

Source: Zaidi based on FTSE EPRA/NAREIT Index Series

countries, like India and Pakistan, are working towards implementation. The integration of Chinese companies via listings through the Hong Kong and Singapore markets has already allowed several companies to become part of the listed real estate sector, though few of them through a REIT structure. The regional market capitalization of the listed real estate market is US$657 billion, although REITs only represent 2% (Table 1.15).

REITs in emerging markets are gradually becoming more significant in the global real estate benchmarks. By the end of 2016, REITs represented 24.4% of the FTSE EPRA/NAREIT Global Emerging Index, which represents a significant improvement compared with the 20.8% observed one year before. Countries like Mexico, Turkey and South Africa are becoming more visible for international investors, while their property markets still show significant potential for new market participants.

Finally, in terms of performance, REITs still have significant potential for future improvement. The gap against Non-REIT companies in emerging markets and REITs in developed markets is still negative, but this has been reduced in the recent years. REITs in emerging markets underperformed their geographical Non-REIT competitors by 345bps during the last five years, although the same difference reached just 115bps in the last three years and even turned in an outperformance of 1,259bps in 2016. On a risk-adjusted basis, REITs underperformed Non-REITs only by 0.06 units over the 2014–2016 period. In the same way, REITs from emerging markets underperformed those from developed markets by 1,142bps during the last three years on a total return basis and by 0.97 units in a risk-adjusted return basis (Table 1.16).

Table 1.13 Size of the Total Commercial Real Estate Market – Emerging Markets – Europe

Size of the total commercial real estate market – Emerging Markets

	Dec-16	*Dec-16*	*Dec-16*	*Mar-17*	*Mar-17*	*Mar-17*	*Mar-17*	*Mar-17*	*Mar-17*	*Mar-17*	*Mar-17*	*Mar-17*	*Mar-17*
	GDP per Capita	*GDP**	*Commercial Real Estate*	*Total Listed Real Estate*	*Number of Companies*	*Of which REITs*	*Full M.Cap REITs*	*REITs / Listed RE*	*Index** Market Cap*	*No. of Index Cons.*	*Stock Market Size*	*RE/Stock Market*	*Listed RE / Total RE*
	($)	*($ Bln.)*	*($ Bln.)*	*($ Bln.)*	*#*	*#*	*($ Bln.)*	*%*	*($ Bln.)*	*#*	*($ Bln.)*	*%*	*%*
Czech Republ	18,325.89	193.54	75.73	-	-	-	-	-	-	-	25.84	0.00%	0.00%
Greece	18,077.64	195.88	76.30	2.61	11	3	0.75	28.8%	0.31	1	35.37	7.37%	3.41%
Hungary	11,902.79	117.07	39.67	0.16	3	-	-	-	-	-	23.23	0.71%	0.41%
Poland	12,309.30	467.35	160.15	4.76	41	-	-	-	-	-	160.18	2.97%	2.97%
Russian Fede	8,838.23	1,267.75	389.01	4.86	9	1	-	-	-	-	580.65	0.84%	1.25%
Total Europe		**2,241.58**	**740.86**	**12.39**	**64**	**4**	**0.75**	**6.06%**	**0.31**	**1**	**825.27**	**1.50%**	**1.67%**

(*Continued*)

Table 1.13 (Continued)

Size of the total commercial real estate market – Emerging Markets

	Dec-16	*Dec-16*	*Dec-16*	*Mar-17*	*Mar-17*	*Mar-17*	*Mar-17*	*Mar-17*	*Mar-17*	*Mar-17*	*Mar-17*	*Mar-17*	*Mar-17*
	GDP per Capita	*GDP*	*Commercial Real Estate*	*Total Listed Real Estate*	*Number of Companies*	*Of which REITs*	*Full M.Cap REITs*	*REITs / Listed RE*	*Index Market Cap*	*No. of Index Cons.*	*Stock Market Size*	*RE/Stock Market*	*Listed RE / Total RE*
	($)	*($ Bln.)*	*($ Bln.)*	*($ Bln.)*	#	#	*($ Bln.)*	%	*($ Bln.)*	#	*($ Bln.)*	%	%
Egypt*	3,709.65	330.16	76.17	4.78	28	-	-	-	0.24		2 39.51	12.10%	6.27%
Qatar	60,732.69	156.60	70.47	17.72	4	-	-	-	-	-	161.15	10.99%	25.14%
South Africa	5,018.22	280.37	71.24	32.76	41	30	17.73	54.11%	17.51	13	448.91	7.30%	45.99%
Turkey	9,316.76	735.72	229.76	6.96	32	24	3.72	53.37%	1.72	4	178.92	3.89%	3.03%
United Arab E	38,050.17	375.02	168.76	37.07	13	3	-	-	3.37	3	243.56	15.22%	21.97%
Total MEA		**1,877.86**	**616.40**	**99.29**	**118**	**57**	**21.44**	**21.59%**	**22.84**	**20**	**1272.05**	**9.26%**	**16.11%**
Total EMEA		**4,119.44**	**1,357.25**	**111.68**	**182**	**61**	**22.19**	**19.87%**	**23.15**	**23**	**1,897.32**	**5.89%**	**8.23%**

Source: Zaidi based on FTSE EPRA/NAREIT Index Series

Table 1.14 Size of the Total Commercial Real Estate Market – Emerging Markets – Latin America

Size of the total commercial real estate market – Emerging Markets													
	Dec-16	*Dec-16*	*Dec-16*	*Mar-17*	*Mar-17*	*Mar-17*	*Mar-17*	*Mar-17*	*Mar-17*	*Mar-17*	*Mar-17*	*Mar-17*	*Mar-17*
	GDP per Capita	*GDP*	*Commercial Real Estate*	*Total Listed Real Estate*	*Number of Companies*	*Of which REITs*	*Full M.Cap REITs*	*REITs / Listed RE*	*Index Market Cap*	*No. of Index Cons.*	*Stock Market Size*	*RE/Stock Market*	*Listed RE / Total RE*
	($)	*($ Bln.)*	*($ Bln.)*	*($ Bln.)*	#	#	*($ Bln.)*	%	*($ Bln.)*	#	*($ Bln.)*	%	%
Brazil	8,586.55	1,769.60	537.80	22.24	50	24	10.86	48.81%	9.03	16	826.97	2.69%	4.14%
Chile	12,909.81	234.90	81.78	2.66	11	-	-	-	1.38	1	223.92	1.19%	3.25%
Colombia	5,623.27	274.14	72.35	-	-	-	-	-	-	-	103.11	0.00%	0.00%
Mexico	8,698.59	1,063.61	324.64	15.69	18	14	8.61	54.90%	7.01	7	330.41	4.75%	4.83%
Peru	5,726.93	180.29	47.87	0.42	6	-	-	-	-	-	81.40	0.52%	0.88%
Total Americas		**3,522.54**	**1,064.45**	**41.01**	**85**	**38**	**19.47**	**47.5%**	**17.42**	**24**	**1,565.80**	**2.62%**	**3.85%**

Source: Zaidi based on FTSE EPRA/NAREIT Index Series

Table 1.15 Size of the Total Commercial Real Estate Market – Emerging Markets – Asia-Pacific

Size of the total commercial real estate market – Emerging Markets													
	Dec-16	*Dec-16*	*Dec-16*	*Mar-17*	*Mar-17*	*Mar-17*	*Mar-17*	*Mar-17*	*Mar-17*	*Mar-17*	*Mar-17*	*Mar-17*	*Mar-17*
	GDP per Capita	*GDP*	*Commercial Real Estate*	*Total Listed Real Estate*	*Number of Companies*	*Of which REITs*	*Full M.Cap REITs*	*REITs / Listed RE*	*Index Market Cap*	*No. of Index Cons.*	*Stock Market Size*	*RE/Stock Market*	*Listed RE / Total RE*
	($)	*($ Bln.)*	*($ Bln.)*	*($ Bln.)*	*#*	*#*	*($ Bln.)*	*%*	*($ Bln.)*	*#*	*($ Bln.)*	*%*	*%*
China	8,140.98	11,181.56	3,352.34	491.10	181	.	8.09	1.65%	66.95	52	6,842.64	7.18%	14.65%
India	1,718.69	2,250.99	400.17	14.82	111	-	-	-	2.38	5	1,760.64	0.84%	3.70%
Indonesia	3,635.81	940.95	214.74	21.80	48	1	1.32	6.07%	6.33	11	436.88	4.99%	10.15%
Malaysia	9,545.52	302.75	95.31	28.95	96	17	3.85	13.30%	5.99	13	373.19	7.76%	30.38%
Pakistan*	1,427.56	271.05	45.48	0.28	4	1	-	-	-	-	92.97	0.30%	0.62%
Philippines	2,991.38	311.69	66.65	43.85	44	-	-	-	9.90	6	245.15	17.89%	65.79%
Taiwan	22,044.05	519.15	216.04	15.55	33	5	-	-	0.17	1	1,060.30	1.47%	7.20%
Thailand	5,662.31	390.59	103.32	40.90	115	62	-	-	8.09	13	427.05	9.58%	39.59%
Total Asia-Pacific		**16,168.72**	**4,494.06**	**657.26**	**632**	**86**	**13.26**	**2.02%**	**99.81**	**101**	**11,238.82**	**5.85%**	**14.63%**

Source: Zaidi based on FTSE EPRA/NAREIT Index Series

Table 1.16 Comparative Performance – Emerging Markets

EMERGING MARKETS – COMPARATIVE PERFORMANCE						
Region	*REITs*			*Non-REITs*		
	Total Return – 3y	*Volatility – 3y*	*Risk-Adjusted Return:3y*	*Total Return – 3y*	*Volatility – 3y*	*Risk-Adjusted Return:3y*
EMEA	16.25%	23.32%	0.70	6.41%	22.36%	0.29
Americas				-7.13%	30.30%	-0.24
Asia-Pacififc				13.48%	19.04%	0.71
Global Emerging	**8.77%**	**18.03%**	**0.49**	**9.92%**	**17.34%**	**0.57**
Global Developed	**20.19%**	**13.87%**	**1.46**	**5.38%**	**11.44%**	**0.47**

Source: Zaidi based on FTSE EPRA/NAREIT Index Series

1.4 REITs – return, risk and correlations

Among the most important and interesting set of research questions concerning the REIT approach to real estate investing are those focused on whether the REIT structure affects the investment attributes of the real estate asset class. For example, whether REIT and non-REIT investments in real estate generally differ in terms of the total returns that accrue to investors and in terms of their diversification benefits (volatilities and correlations, as well as betas and alphas), or in other significant ways. Many of these questions were first addressed using empirical data from real estate markets in the US, but the research has since been expanded to include markets in Europe and Asia.

1.4.1 Long-term average returns

The first two empirical studies comparing the returns on REIT and non-REIT investments were by Riddiough, Moriarty and Yeatman (2005) and Pagliari, Scherer and Monopoli (2005). Both used returns data from exchange-traded US equity REITs and from the National Council of Real Estate Investment Fiduciaries (NCREIF). Both studies corrected for the use of leverage by REITs and restricted their analysis to REITs with property portfolios concentrated in one of the four property types (apartment, industrial, office and retail) for which NCREIF reported institutional returns. Given the similarities in their data and methodology, it is unsurprising that the two studies reached nearly identical conclusions, that property-level (unlevered) returns from exchange-traded REITs exceeded property-level returns from non-REIT institutional real estate investments by an average of about three percentage points per year.

Pagliari, Scherer and Monopoli (2005) conducted a test of statistical significance and concluded that they were unable to reject the null hypothesis that the returns of equivalent REIT and non-REIT investments were the same. Their difference-in-averages test, however, incorporated an implicit assumption that the correlation between the two forms of real estate investment was slightly negative. However, such an assumption has never been supported theoretically and is contradicted by numerous empirical studies including those by Morawski, Rehkugler and Füss (2008), Oikarinen, Hoesli and Serrano (2011), Pedersen et al. (2012) and Boudry et al. (2012).

Tsai (2007) and Ling and Naranjo (2015) conducted studies similar in methodology and data sources to those of Riddiough, Moriarty and Yeatman (2005) and Pagliari, Scherer and Monopoli (2005) and reached very similar conclusions. Ling, Naranjo and Scheick (2016) also controlled for differences in the geographic distribution of property investments, but that methodological change did not explain away the performance difference documented in the earlier studies. Fisher and Hartzell (2016), Andonov, Kok and Eichholtz (2013) and Ang et al. (2014) applied very different methodologies with different data sources but again reached similar conclusions.

Baum, Fear and Colley (2011) was the first empirical study to compare REIT and non-REIT investments outside the US. Using data bases of cash flows from private equity real estate funds with global mandates, Baum, Fear and Colley (2011) as well as Alcock et al. (2013) reached the same conclusion as the US-focused empirical studies, being that real estate investments through exchange-traded equity REITs outperformed otherwise similar but non-REIT real estate investments. Kiehelä and Falkenbach (2015) applied a similar methodology to private equity funds with a focus on European real estate and reached the same general conclusion.

In short, some dozen empirical studies employing different methodologies, different data sources and different historical time periods but with the same objective of comparing the returns to similar real estate investments made through REITs and non-REITs, both within and outside the US, have all reached the same conclusion, being that REIT returns have outperformed the returns on otherwise similar but non-REIT real estate investments.

1.4.1.1 Contributors to REIT outperformance – manager costs

Academic research has not yet, however, convincingly explained the observed outperformance by REITs. The extant studies suggest that the difference in performance cannot be attributed to differences in the use of leverage, the distribution of holdings by property type or the geographic distribution of holdings.

Several studies have noted that the difference between net-to-investor total returns and what are described as "gross" total returns is much greater for non-REIT investments than for REIT investments. For example, in the US, the difference seems to average approximately 52bps pa to cover the costs of an actively managed mutual fund holding US REITs, whereas the difference for institutional core real estate investments (direct, through separate accounts, or through a private equity real estate fund following a core strategy) averages about twice as much. However, this discrepancy seems to be explained primarily by the fact that "gross" returns for REIT investments are measured net of general and administrative (GandA) expenses, whereas gross returns for non-REIT investments are measured truly gross of GandA.

Unpublished empirical research suggests that "all-in" investment costs are actually quite similar for REIT and non-REIT investments in similar assets, suggesting that REIT outperformance may have to be attributed to something other than a difference in investment costs. Empirical data from private equity real estate funds following value-add and opportunistic strategies shows that their investment costs are much larger than for either REIT investments or non-REIT core investments, but the difference in investment costs does not seem to be large enough to fully explain the difference in performance.

1.4.1.2 Contributors to REIT outperformance – liquidity

At least part of the outperformance by REITs can be explained by the illiquidity and non-granularity ("lumpiness") of private equity real estate investments. Capital committed to exchange-traded REITs can be deployed almost immediately and fully. For example, an average trading volume of exchange-traded REITs in the US of more than $7 billion per day, coupled with individual share prices no greater than a few hundred US dollars, means that any amount of capital committed to REIT investments will start earning real estate returns typically starting on the same day.

In contrast, even a relatively simple investment in private equity real estate will typically not be concluded for about six months after capital has been committed, with private equity real estate funds generally not finishing deploying capital until several years after it has been committed. Since limited partners (LPs) can suffer severe monetary and non-monetary penalties for failing to meet a capital call from the general partner (GP), capital committed to private equity real estate funds most commonly earns only cash returns while awaiting deployment.

Meads, Morandi and Carnelli (2016) note a range of penalties "likely to" be imposed on an LP failing to meet a capital call:

> In some cases, the LP will be subject to a penal rate of interest until such time as it is able to meet its cash obligation. Other, more severe penalties may include the LP being forced to sell its position in the fund to other investors, potentially at a steep discount to fair value, or the LP being forced to give up its entire stake in the fund and for its position to be carved up amongst the other LPs. This is a particularly penal, but not uncommon, remedy that can be very expensive for a LP if the default on the drawdown request occurs towards the end of a fund's investment period. By this time a LP may already have paid in significant amounts of cash to meet earlier drawdowns; as a result, all of this built-up value would be lost .... If a LP wishes to avoid default but does not have adequate cash on hand, in a funding emergency it may choose to generate the required amount of cash through a sale of other assets within its overall portfolio, if available. The quicker this cash is required, the more a LP may risk having to conduct a fire sale, (meaning) that these assets may be sold only at high discounts to their intrinsic worth.

Moreover, given the large per unit prices of institutional-grade properties, it is rare for capital committed to private equity real estate investment (whether through funds or directly) to be deployed fully with some capital remaining as cash until it can be "soaked up" by another property investment. This is especially true for open-end funds, which also maintain a cash buffer to meet an expected pace of share redemptions. For example, the Open-End Diversified Core Equity (ODCE) Fund Index published by NCREIF shows that an average of only 95% of capital collected by open-end core funds is actually invested in real estate assets, with the other 5% remaining as uninvested cash. Unpublished research by Case suggests that this "cash drag" may reduce the returns of investments through core private equity real estate funds by an average of 0.58% pa over the available period Q1.2000–Q1.2017.

1.4.1.3 Contributors to REIT outperformance – leverage discretion

Manager discretion over the use of financial leverage offers another possible explanation for the performance difference between REIT and non-REIT investments in real estate. NCREIF data, for example, show a negative correlation between the quarterly average leverage measured for private equity real estate funds (whether employing core, value-add or opportunistic strategies) and the funds' quarterly average returns, suggesting that managers have tended to use greater leverage when returns were declining (thereby magnifying negative returns) and to use less leverage when returns were increasing (thereby minimizing positive returns). On a somewhat related issue, Shilling and Wurtzebach (2012) found that differences in the returns on individual properties in the NCREIF database could be explained by differences in leverage but not by other recorded differences, suggesting that manager discretion over the use of leverage has tended to reduce risk-adjusted returns by increasing performance-based management fees without any countervailing benefit to the investor.

1.4.1.4 Contributors to REIT outperformance – buy low, sell high

Perhaps the most convincing explanation for the observed underperformance of private equity real estate investments relative to REIT-based real estate investments is provided by evidence comparing acquisition and disposition activity on the private and public sides of the real estate market. Data on transaction activity in the US from sources such as Real Capital Analytics, CoStar Group and 10-K forms filed by exchange-traded equity REITs suggests that REITs have tended be more active net acquirers of property during off-peak periods in the real estate market cycle and more active net sellers (or at least less active net acquirers) during market peaks.

In contrast, non-REIT institutional investors (primarily private equity real estate investment funds and institutions investing either directly or through separate accounts) have tended to be more active net acquirers during market peaks and more active net sellers during off-peak periods. For example, the largest real estate deal in history was the sale of Equity Office Properties Trust, an exchange-traded equity REIT, to a private equity real estate fund managed by Blackstone on 6 February 2007, the exact day that the real estate market reached its peak before its 2007–2008 decline and its 2008–2009 crash.

At least with respect to private equity real estate funds, this "buy-high, sell-low" tendency may be prompted by their funding model. Capital invested through private equity investment funds is expected to be deployed rapidly to purchase assets and investment managers risk losing a substantial management fee (typically 2% of either committed capital or invested capital) as well as an additional performance-based fee (typically 20% of returns exceeding an 8% hurdle) if they fail to deploy the committed capital. Since investors have a tendency to "follow returns", this funding model encourages fund managers to deploy capital following significant asset value increases – that is, near market peaks. In contrast, capital invested through exchange-traded REITs typically purchases shares already held by other investors, so has no appreciable effect on the REIT's decision whether to purchase assets. Conversely, the same return-following tendency among investors leads them to seek to redeem shares in a private equity investment funds following significant asset value declines – that is, near market troughs – thus reducing the capital that fund managers would otherwise be able to use to make asset purchases at favourable prices.

The explanations offered for the observed outperformance of exchange-traded equity REITs relative to institutional investments in private equity real estate are speculative for now, but increasing availability of data, especially regarding institutional private equity investments, suggest that future empirical evidence to support or refute the various possibilities may help answer the question going forward.

1.4.2 Illiquidity, volatility and other risks

A second major question is whether real estate investments made through ownership of stock in exchange-traded REITs carries the same risks as otherwise equivalent but Non-REIT real estate investments. Perhaps the two most prominent risks associated with real estate investing are illiquidity and volatility.

Real property, especially of institutional investment quality, is among the most illiquid assets generally held for investment purposes. The risks associated with such a high level of illiquidity, many of which can just as validly be thought of as illiquidity costs, include the following:

- *uncertainty* regarding the duration of the acquisition period, the ultimate price at which the asset can be purchased, the duration of the marketing period at disposition and the ultimate price at which the asset can be sold even after pre-sale marketing has begun;

- *opportunity costs* of other investments that cannot be made because capital is tied up, other assets that can be purchased but not at the most favourable time, committed capital that must be held as cash while a transaction is finalized and capital that may have to be held as cash after disposition while a new investment is identified;
- *transaction impacts* including the direct effect of offering an expensive property for sale – that is, materially increasing the supply of real estate in the active transaction market – on the market value of other properties, the information cost of offering an asset for sale and therefore revealing that the owner has information suggesting that and similar assets may be overvalued and the "latent" cost of transactions that do not happen because of concern over transaction impacts; and
- *direct costs* associated with marketing an asset, negotiating terms, and completing a transaction.

1.4.2.1 Illiquidity

Institutional investors generally understand that they should not invest in illiquid real estate without being compensated for these additional risks and costs relative to otherwise equivalent but liquid real estate investments, but little empirical research has yet been done to quantify the required illiquidity premium. Cheng, Lin and Liu (2013) concluded that for typical investors, liquidity risk is to such magnitude that the conventional return volatility should be adjusted upward by as much as 97% in order to fully reflect the liquidity risk. Ang, Papanikolaou and Westerfield (2014) modelled the costs of illiquidity as a function of the expected holding period of the asset and concluded that the liquidity premium for rebalancing once per ten years, on average, is an extremely large 0.06. Other research that considers the costs and risks of illiquidity generally, without specific reference to real estate, includes Kinlaw, Kritzman and Turkington (2013), Hayes, Primbs and Chiquoine (2015) and Lindsey and Weisman (2016).

Continuing to commit capital to private equity real estate investments despite the fact that it, primarily, exposes them to very significant illiquidity risks while, secondarily, providing lower returns than otherwise similar but liquid real estate investments could be justifiable if illiquid real estate were substantially less volatile than liquid real estate – that is, if risks other than illiquidity were low enough that illiquid real estate could still be expected to outperform on a risk-adjusted basis. There is ample anecdotal evidence that institutional investors do, in fact, believe the volatility of private real estate to be extremely low.

For example, the benchmarks that institutional investors seem to follow most closely – such as the NCREIF Property Index (NPI) and the Open-End Diversified Core Equity (ODCE) Index, both published by NCREIF, as well as the Moody's/Real Capital Analytics (RCA) Commercial Property Price Index (CPPI) and the Cambridge Associates Real Estate Fund Index – all report quarterly (or, in the case of the Moody's/RCA CPPI, monthly) returns that suggest extremely low real estate volatility roughly equivalent to that of US Treasury bonds with seven-year maturities.

Moreover, marketing materials published by private equity real estate investment managers frequently highlight what they call the low volatility of private real estate. An example that is typical in content though perhaps not in tone is by Principal Real Estate Investors, which published a series of three articles entitled "*Market Volatility and Private Equity Real Estate: The Visual Case for Private Equity Real Estate*" (11 November 2015), "*Have Concerns About Today's Market Volatility? Consider Diversifying with Commercial Real Estate*" (11 February 2016) and "*Commercial Real Estate May Help Provide a Smoother Ride on the Road to Your Investment Goals*" (20 May 2016).

1.4.2.2 Volatility – REITs vs non-REITs

It has been recognized, at least since the 1980s, however, that the volatility implied by the NPI, ODCE and similar indices quite significantly understates the true volatility of private real estate investments: see Hoag (1980), Ibbotson and Siegel (1984), Diermeier, Freundlich and Schlarbaum (1986), Firstenberg, Ross and Zisler (1988), Lee (1989), McMillan (1989), Geltner (1989, 1990), Giliberto (1990), Gyourko and Keim (1990) and Miles, Cole and Guilkey (1990). The most important reason for this is the use of appraisals to measure property values and therefore to measure periodic returns. Appraisals may "smooth" reported returns in at least three ways:

- first, appraisers are thought to be affected by the cognitive bias known as *anchoring*, meaning that in estimating the market value of a property at any given time they tend to attach too much weight to known values that were attached to the same property in previous times, including previous appraised values;
- second, because appraisals are expensive and therefore are conducted quite infrequently, the quarterly value of an appraisal-based index includes quite a few "stale" appraisals for properties whose values were most recently estimated up to three years previously; and
- third, because almost no appraisals are conducted on any given day, appraisals conducted at any time during a given quarter are aggregated to estimate what amounts to an "average" value for the portfolio of constituent properties over all days during that quarter.

While appraisal-based indices seem to suffer most severely from "appraisal smoothing", even those indices based on property transactions suffer from what may be termed "illiquidity smoothing". As with appraisals, almost no property transactions are conducted on any given day, so transactions concluded at any time during a given quarter (or month) are aggregated to estimate what amounts to an "average" transaction price for transacting constituent properties over all days during that quarter (or month).

The available empirical evidence suggests that the volatility of institutional investment quality private real estate is either virtually identical to or even somewhat greater than the volatility of otherwise similar but liquid investments in REITs. Specifically, the FTSE NAREIT Pure-Property® Index Series provides daily estimates of the returns of *unlevered* investments in liquid real estate-based on REIT stock transactions, while the NCREIF Transaction Based Index (NTBI) provides quarterly estimates of the returns of unlevered investments in illiquid real estate-based on property transactions. Over the commonly available period Q4.2002–Q1.2017, the annualized volatility of quarterly capital appreciation rates implied by the PureProperty index was 11.0% while the annualized volatility implied by the NTBI was very slightly lower at 10.3%. Interestingly, both forms of real estate investment had extremely low income volatility, so the volatility of total returns was only marginally higher at 11.1% and 10.4% respectively.

As noted, an index such as the NTBI based on property transactions suffers from return smoothing because all transactions occurring at any time during a given quarter (or month) must be aggregated to estimate the "average" change in transaction prices for that quarter (month). It is possible to derive a rough analogue for this "illiquidity smoothing" by averaging across all daily values of the PureProperty index during a given quarter, rather than using the value of the index as of the last day of each quarter. The annualized volatility implied by such an "analogously smoothed" quarterly version of the PureProperty index over the available historical period Q1.2003–Q1.2017 was just 9.8% for price appreciation and 9.9% for total returns, suggesting that liquid real estate investments may actually be very slightly less volatile than otherwise similar but illiquid real estate investments.

1.4.2.3 Volatility – cash flow and discount rate

In addition to the empirical evidence, it is worth considering the theoretical basis for believing that there might be any difference in volatilities between otherwise similar real estate investments made through exchange-traded REITs or privately. The *value* of any asset, including a property, can be modelled as the present-discounted value of the future stream of net cash flows produced by that asset. For a management-intensive asset such as real estate, the future stream of cash flows depends on the management capabilities of the owner, meaning that the value will be specific to each owner. Moreover, the discount rate is also specific to each prospective owner as it depends on that prospective owner's risk aversion, alternative uses of capital and so forth. Finally, since neither the future stream of cash flows nor the discount rate can be known with certainty, the value of a given property to any prospective owner depends on that prospective owner's current expectations regarding each in the future. The market value, therefore, is bracketed by the top two from among all owner-specific values.

For the purpose of comparing volatilities, the important point of the foregoing discussion is that not a single one of the determinants of value – the future stream of owner-dependent cash flows, the future owner-dependent discount rate and the owner-dependent current expectations regarding each – depends on whether the ownership of that property is direct or securitized, i.e. whether the property is owned privately or whether ownership is through shares in a publicly traded REIT. In other words, owner-specific values will change over time as the progression of new information changes each prospective owner's expectations regarding future cash flows and future discount rates, but the progression of new information is exactly the same for liquid and illiquid real estate. This means that there is little theoretical basis for expecting any difference in the volatility of owner-specific values.

As noted, the market value of a property will be bracketed by the top two owner-specific values for that property. When each prospective owner is confident in the information available to himself or herself and when the information available to each prospective owner can easily be revealed through transactions, that free flow of information will tend to reduce the volatility of market values. In contrast, when investors have less information and less confidence in the available information, the volatility of market values tends to increase, as has often been noted during market crises.

In short, volatility is, in some sense, a measure of uncertainty regarding asset values. It will typically be subdued in situations where better information reduces uncertainty and elevated in situations where a lack of information increases uncertainty. Because value-relevant information is revealed much more efficiently through a liquid market than through an illiquid market, there is a theoretical basis for believing that the volatility of true market values will actually be lower for assets trading in a liquid market than for the same assets trading in an illiquid market, even if the volatility of owner-specific values is the same.

1.4.2.4 Summary – illiquidity, volatility and other risks

To summarize:

- there is no theoretical reason to believe that the volatility of illiquid real estate investments would be lower than the volatility of otherwise similar but REIT-based real estate investments;
- there is no valid empirical evidence suggesting that the volatility of illiquid real estate investments is in fact lower than the volatility of otherwise similar but REIT-based real estate investments;
- there is some theoretical reason to believe that the volatility of illiquid real estate investments should actually be higher than the volatility of otherwise similar but REIT-based real estate investments; and

- there is some empirical evidence suggesting that the volatility of illiquid real estate investments is in fact higher than the volatility of otherwise similar but REIT-based real estate investments.

1.4.3 REITS and non-REITs – correlation and diversification benefits

The third major question is whether real estate investments made through ownership of stock in exchange-traded REITs provide the same diversification benefits as otherwise equivalent but non-REIT real estate investments. Continuing to commit capital to private equity real estate investments despite the fact that it has provided lower returns than otherwise similar but liquid real estate investments, even on a risk-adjusted basis, could be justifiable if illiquid real estate provided substantially better diversification benefits than liquid real estate. As with the question of relative volatilities, there is ample anecdotal evidence that institutional investors do, in fact, believe the diversification benefits of private real estate to be substantially better than the diversification benefits of REITs.

All US private equity real estate benchmarks report returns that have very low correlations with the broad US stock market, in fact all imply very low correlations with the US REIT market as well. Moreover, marketing materials published by private equity real estate investment managers frequently highlight what they call the stronger diversification benefits of private real estate relative to REITs. Typical examples include "*Commercial Real Estate*" by JD Bowden and MR Smith of Merrill Lynch Wealth Management, which reported a correlation of just 0.23 between REITs, and private real estate and J.P. Morgan Asset Management's "*2016 Long-Term Capital Market Assumptions*", which reported the correlation at just 0.40.

1.4.3.1 Long-run correlations

There is both theoretical and empirical evidence, however, that the actual long-term correlation between liquid and illiquid real estate is very close to 100%, and therefore that the correlation between private real estate and the broad stock market is essentially the same as the correlation between REITs and the broad stock market. In fact, as with volatilities, it has been recognized at least since the 1980s that the low reported correlations between illiquid real estate and other assets are primarily artefacts of shortcomings in the reporting of private equity real estate returns: see Diermeier, Freundlich and Schlarbaum (1986), Firstenberg, Ross and Zisler (1988), Geltner (1989, 1990), Giliberto (1990) and Gyourko and Keim (1990).

As noted, the progression of new information that affects current expectations regarding the future streams of both cash flows and discount rates does not depend on whether a given property is owned privately or by an exchange-traded REIT. In the liquid REIT market, any new information that changes property values will generally have an almost immediate effect on the stock prices of REITs that own such properties. For example, new statistical data indicating that office employment growth (and therefore the demand for office space) will be stronger than previously expected can generally be expected to manifest promptly in higher stock prices for office REITs, other factors being equal.

In the illiquid real estate market, however, the same new information – and the same effect on property values – will not become manifest until transactions are completed at the new prices. Because it takes about six months, on average, to transact a property, the returns measured by indices based on property transactions tend to lag behind actual changes in market value by about two quarters. Because appraisals must be based on completed transactions – meaning new information may not even begin to be incorporated into appraised values until after the six-month average transaction lag has passed – and because most properties are not appraised each quarter, the returns measured by appraisal-based indices tend to lag behind actual changes in market value by about four to five quarters. For example, a real estate market peak was measured in Q4.1989 by REIT-based indices, but only four quarters later, in Q4.1990, by appraisal-based private real estate indices. There was a second market peak measured in Q1.2007 by REIT-based indices but three to five quarters later by various appraisal-based private real estate indices.

CEM Benchmarking (2016) evaluated return reporting lag at the level of individual private equity real estate investment funds and reported that "the distribution of lags... (showed) two peaks, one at 6–8 months and another at 14–16 months" with an overall average of 242 trading days, concluding that "the majority of pension fund real estate valuations are either a bit more than half a year or a bit more than a whole year stale". After correcting for reporting lag, CEM Benchmarking concluded that "the correlation between listed equity REITs and unlisted real estate increased 10-fold from 9 percent to 91 percent" and pointed out that "the high correlation is not surprising given both asset classes invest in fundamentally the same assets".

1.4.3.2 Term structure of correlations

Another simple way to assess the long-term correlation between public and private real estate is to compute the term structure of correlations – that is, how the computed correlation coefficient changes depending on the period over which returns are measured. For example, the NPI and other appraisal-based indices of private real estate returns are reported on a quarterly basis, so correlation coefficients are commonly computed using returns measured over quarterly periods but it is equally possible and equally valid to compute correlation coefficients using returns measured over longer periods such as semi-annual (two-quarter) periods, three-quarter periods, annual (four-quarter) periods and so forth.

Figure 1.3 shows the term structure of correlations between the NPI and the FTSE NAREIT Equity REIT index, based on returns as reported, using the historical period Q1.1978–Q1.2017. The term structure is very strongly upward-sloping, indicating that, while return measurement problems may produce a sharp divergence between the returns reported for private and public real estate during any given quarter, over longer-investment horizons the two return series respond to the same underlying drivers. Figure 1.3 shows the same analysis after lagging REIT returns by four quarters, as a very simple adjustment to replicate the typical reporting lag in private real estate indices. The term structure of lag-corrected returns is equally strongly upward-sloping, but the computed correlation coefficients are higher across the board.

Cointegration analysis provides a third means to estimate whether two data series respond to a common set of return drivers even if the response of one series is delayed relative to the other. Multiple cointegration analyses including those by Morawski, Rehkugler and Füss (2008), Kutlu (2010), Oikarinen, Hoesli and Serrano (2011) and Hoesli and Oikarinen (2013) have found that REIT returns and private real estate returns are cointegrated, suggesting (as noted by Oikarinen, Hoesli and Serrano (2011)) that "the correlation between NAREIT and NCREIF returns approaches one as the investment horizon lengthens".

The fact that the implied true correlation between public and private real estate is very close to 100% may indicate that the two categories of real estate investment returns have essentially identical correlations with any other assets. As noted, real estate investment through exchange-traded REITs has produced higher long-term average net total returns than otherwise similar

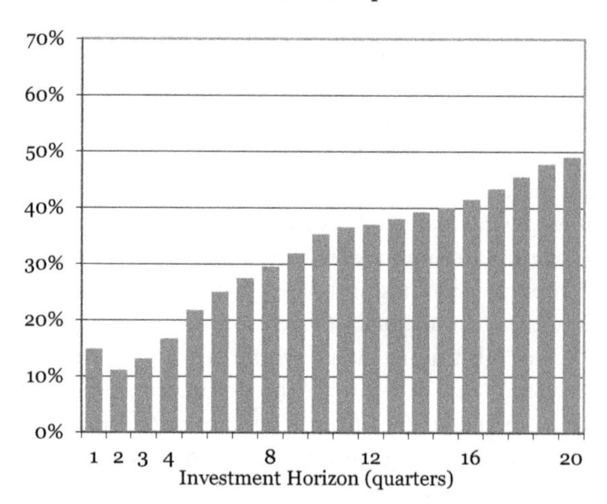

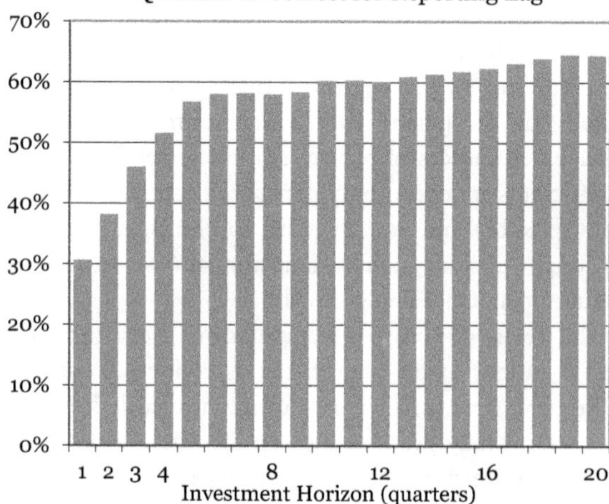

Figure 1.3 Term Structure of Correlations between Public and Private Real Estate

Source: Case based on NPI and FTSE NAREIT Equity REIT Index

but illiquid real estate investments, including on a risk-adjusted basis. The contention may be, therefore, that investors may gain greater diversification benefits from exchange-traded REITs than from private real estate.

1.5 Summary

This chapter sought to identify critical contextual issues in international REITs, including the defining characteristics of REITs, the global evolution of REITs and the risk-return characteristics of real estate investment through REITs relative to non-REIT real estate investment.

While the details of REIT regimes may vary around the world, there may be contended to be a high level of commonality in the defining characteristics of structure, investment focus, time horizon, profits distribution, tax treatment and the use of debt or leverage. It is interesting that, while REITs provide significant tax benefits, traditional property companies remain more popular in countries such as Hong Kong, *fideicomiso* remain more popular in Latin America and unlisted property trusts remain more popular with pension funds.

As an investment structure, REITs are incredibly dynamic. That structure established in the US in 1960 has evolved through the decades, partly in response to property and financial market changes and partly as a catalyst for property and financial market changes. As with any fast-moving consumer good, be it beer or confectionery, while the underlying fundamentals of REITs remain constant, the peripheral attributes change over time to meet consumer demand and to drive consumer demand. It is, however, a testament to the durability of the underlying fundamentals of REITs that, in less than 60 years, the sector has gone from non-existence in 1959 to a market capitalization of US$1.4 trillion in March 2017 (Table 1.3), justifying its own GICS sector.

While the risk and return profiles of the REIT markets in the 35 countries that have REIT regimes vary significantly, there appears to be consensus that REIT returns have outperformed the returns on otherwise similar but Non-REIT investments and that the volatility of investment quality private real estate and REITs is very similar in the long-run. Significantly, therefore, over the last 50 years, the portfolio management argument has clearly moved from "*why would you include REITs in a mixed-asset portfolio?*" to "*why wouldn't you include REITs in a mixed-asset portfolio?*".

Part I continues with six chapters each reviewing a current theme of REIT evolution through the lens of contemporary research. Chapter 2, the *Post-Modern REIT Era*, examines the evolution of the global REIT market, what is deemed to be best current market practice and how the market is expected to change going forward, with Chapter 3, *Emerging Sector REITs*, investigating such sectors as self-storage and data-centres. Chapter 4, *Sustainable REITs*, analyses REIT environmental performance and the cost of equity with Chapter 5, *Islamic REITs*, examining the evolution of global Islamic REITs through a study of the Malaysian market. Chapter 6, *Behavioural Risk in REITs*, addresses the management of behavioural risk in global REITs and Part I concludes with Chapter 7 which reviews recent research into *REIT Asset Allocation*.

Part II then includes six chapters each analysing REITs in a region of the world, using case studies from a developed, developing and emerging REIT sector in the region, concluding with Chapter 14 which considers *Directions for the Future of International REITs*.

References

Alcock, J., Baum, A., Colley, N. and Steiner, E. 2013, 'The role of financial leverage in the performance of private equity real estate funds', *Journal of Portfolio Management*, Vol. 39, No. 5, pp. 99–110.

Andonov, A., Kok, N. and Eichholtz, P. 2013, 'A global perspective on pension fund investments in real estate', *Journal of Portfolio Management*, Vol. 39, No. 5, pp. 32–42.

Ang, A., Chen, B., Goetzmann, W. and Phalippou, L. 2014, *Estimating private equity returns from limited partner cash flows*, Working Paper 2014–8, Saïd Business School, Oxford.

Ang, A., Papanikolaou, D. and Westerfield, M. 2014, 'Portfolio choice with illiquid assets', *Management Science*, Vol. 60, No. 11, pp. 2737–2761.

Baum, A., Fear, J. and Colley, N. 2011, *Have property funds performed?*, ULI Europe.

Boudry, W., Coulson, E., Kallberg, J. and Liu, C. 2012, 'On the hybrid nature of REITs', *Journal of Real Estate Finance and Economics*, Vol. 44, Nos. 1–2, pp. 230–249.

CEM Benchmarking 2016, *Asset allocation and fund performance June 2016*, available at www.cembenchmarking.com/Files/Documents/Asset_Allocation_and_Fund_Performance_June_2016.pdf.

Cheng, P., Lin, A. and Liu, Y. 2013, 'Liquidity risk of private assets: Evidence from real estate markets', *Financial Review*, Vol. 48, No. 4, pp. 671–696.

Diermeier, J., Freundlich, J. and Schlarbaum, G. 1986, 'Appendix: the role of real estate in a multi-asset portfolio', *ICFA Continuing Education Series*, Vol. 1986, No. 1.

EPRA 2016, *Global REIT report September 2016*, EPRA, Brussels.

Firstenberg, P., Ross, S. and Zisler, R. 1988, 'Real estate: the whole story', *Journal of Portfolio Management*, Vol. 14, No. 3, pp. 22–34.

Fisher, L. and Hartzell, D. 2016, 'Class differences in real estate private equity fund performance', *Journal of Real Estate Finance and Economics*, Vol. 52, No. 4, pp. 327–346.

Geltner, D. 1989, 'Estimating real estate's systematic risk from aggregate level appraisal-based returns', *Real Estate Economics*, Vol. 17, No. 4, pp. 463–481.

Geltner, D. 1990, 'Return risk and cash flow risk with long-term riskless leases in commercial real estate', *Real Estate Economics*, Vol. 18, No. 4, pp. 377–402.

Giliberto, M. 1990, 'Equity real estate investment trusts and real estate returns', *Journal of Real Estate Research*, Vol. 5, No. 2, pp. 259–263.

Gyourko, J. and Keim, D. 1990, *The risk and return characteristics of stock market-based real estate indexes and their relation to appraisal-based returns*, Working Paper, available at www.researchgate.net/profile/Joseph_Gyourko/publication/5071205_The_Risk_and_Return_Characteristics_of_Stock_Market_Based_Real_Estate_Indexes_and_Their_Relation_to_Appraisal-Based_Returns/links/00b4952c1912040a72000000.pdf.

Hayes, M., Primbs, J. and Chiquoine, B. 2015, 'A penalty cost approach to strategic asset allocation with illiquid asset classes', *Journal of Portfolio Management*, Vol. 41, No. 2, pp. 33–41.

Hoag, J. 1980, 'Towards indices of real estate value and return', *Journal of Finance*, Vol. 35, No. 2, pp. 569–580.

Hoesli, M. and Oikarinen, E. 2013, *Are public and private asset returns and risks the same? Evidence from real estate data*, Working Paper available at http://papers.ssrn.com/sol3/papers.cfm?abstract_id=2348591, November 2013.

Ibbotson, R. and Siegel, L. 1984. 'Real estate returns: A comparison with other investments', *Real Estate Economics*, Vol. 12, No. 3, pp. 219–242.

Kiehelä, S. and Falkenbach, H. 2015, 'Performance of non-core private equity real estate funds: A European view', *Journal of Portfolio Management*, Vol. 41, No. 6, pp. 62–72.

Kinlaw, W., Kritzman, M. and Turkington, D. 2013, 'Liquidity and portfolio choice: A unified approach', *Journal of Portfolio Management*, Vol. 39, No. 2, pp. 19–27.

Kutlu, V. 2010, *The long-term relation between indirect and direct real estate*, unpublished thesis available at http://arno.uvt.nl/show.cgi?fid=111132.

Lee, S. 1989, 'Property returns in a portfolio context', *Journal of Valuation*, Vol. 7, No. 3, pp. 248–258.

Lindsey, R. and Weisman, A. 2016, 'Forced liquidations, fire sales, and the cost of illiquidity', *Journal of Private Equity*, Vol. 20, No. 1, pp. 45–57.

Ling, D. and Naranjo, A. 2015, 'Returns and information transmission dynamics in public and private real estate markets', *Real Estate Economics*, Vol. 43, No. 1, pp. 163–208.

Ling, D., Naranjo, A. and Scheick, B. 2016, 'Geographic allocations, property selection, and performance attribution in public and private real estate markets', *Real Estate Economics*, November 2016.

McMillan, D. 1989, 'Measurement of risk in real estate', *ICFA Continuing Education Series*, available at www.cfapubs.org/doi/pdf/10.2469/cp.v1989.n1.5.

Meads, C., Morandi, N. and Carnelli, A. 2016, 'Cash management strategies for private equity investors', *Alternative Investment Analyst Review*, Vol. 4, No. 4, pp. 31–43.

Miles, M., Cole, R. and Guilkey, D. 1990, 'A different look at commercial real estate returns', *Real Estate Economics*, Vol. 18, No. 4, pp. 403–430.

Morawski, J., Rehkugler, H. and Füss, R. 2008, 'The nature of listed real estate companies: Property or equity market?', *Financial Markets and Portfolio Management*, Vol. 22, No. 2, pp. 01–126.

Oikarinen, E., Hoesli, M. and Serrano, C. 2011, 'The long-run dynamics between direct and securitized real estate', *Journal of Real Estate Research*, Vol. 33, No. 1, pp. 73–103.

Pagliari, J., Scherer, K. and Monopoli, R. 2005, 'Public versus private real estate equities: A more refined, long-term comparison', *Real Estate Economics*, Vol. 13, No. 1, pp. 147–187.

Pedersen, N., He, F., Tiwari, A. and Hoffmann, A. 2012, *Modeling the risk characteristics of real estate investments*, PIMCO, Newport Beach.

Riddiough, T., Moriarty, M. and Yeatman, P. 2005, 'Privately versus publicly held asset investment performance', *Real Estate Economics*, Vol. 13, No. 1, pp. 121–146.

Shilling, J. and Wurtzebach, C. 2012, 'Is value-added and opportunistic real estate investment beneficial? If so, why?', *Journal of Real Estate Research*, Vol. 34, No. 4, pp. 429–461.

Tsai, J. 2007, *A successive effort on performance comparison between public and private real estate equity investment*, Thesis submitted for Masters in Real Estate Development, Massachusetts Institute of Technology.

2
POST-MODERN REIT ERA
Strategies for REIT style and significance

Alex Moss

2.1 Introduction

This book aims to identify key areas for research in the REIT discipline for the next five to ten years by surveying the current state of the REIT discipline around the world and identifying emerging and cutting edge research areas through a thematic review of current contextual issues and a regional analysis based on case studies.

This book comprises two parts, the first part being a thematic review of emerging and cutting edge global research into current contextual issues in REITs internationally and the second part being a regional analysis of REITs around the world, each written by authoritative academic authors from the world's leading Universities and REIT industry experts.

Part I includes six chapters each reviewing a current theme of REIT evolution through the lens of contemporary research. Chapter 1 focused on critical contextual issues in international REITs while this chapter, the *Post-Modern REIT Era*, examines the evolution of the global REIT market, what is deemed to be best current market practice and how the market is expected to change going forward.

There are a number of reasons to suggest that listed real estate is entering a new era, which may be termed the Post-Modern REIT Era, where some of the historical assumptions have started to come under pressure and may no longer be applicable. It may be contended that such changes have implications for both corporate strategies and investment strategies, affecting all levels of the REIT model including the cyclicality and structural obsolescence of certain underlying assets, the type of debt and equity funding available and the characteristics of the preferred corporate vehicle.

In particular, the following questions are pertinent to this Chapter:

- what are the key drivers behind the growth of global REIT markets and what are the current market conditions and practices that are determining the growth of REIT markets?;
- how relevant is the REIT sector relative to the rest of the equity market?; and
- what are the current asset allocation implications for REITs?

Chapter 3, *Emerging Sector REITs*, then investigates such sectors as timber and data centres. Chapter 4, *Sustainable REITs*, analyses REIT environmental performance and the cost of equity with Chapter 5, *Islamic REITs*, examining the evolution of global Islamic REITs through a study of the Malaysian market. Chapter 6, *Behavioural Risk in REITs*, addresses the management of behavioural risk in global REITs and Part I concludes with Chapter 7, which reviews recent research into *REIT Asset Allocation*.

Part II includes six chapters each analysing REITs in a region of the world, using case studies from a developed, developing and emerging REIT sector in the region:

Chapter	*Region*	*Developed REIT Sector*	*Developing REIT Sector*	*Emerging REIT Sector*
8	North America	USA	Canada	Mexico
9	Latin America	Brazil	Argentina	Uruguay
10	Europe	UK	Spain	Poland
11	South East Asia	Singapore	Malaysia	Thailand
12	North Asia	Japan	Hong Kong	China
13	Oceania	Australia	South Africa	India

concluding with Chapter 14 which considers *Directions for the Future of International REITs*.

2.2 Post-modern era: growth in global REIT markets and current market practices

Why should this current period be referred to as a Post-Modern REIT Era? To understand the current context it is worth re-visiting the evolution of the modern REIT market, which started in the US but has now spread globally. As is well known, although US REITs had been in existence since the 1960s, the so-called Modern REIT Era, when REITs became an investible universe of significance, only started in 1992/3. Indeed, in 1989 the US REIT market was smaller than both the UK and the European market. During the course of the 1990s the relative weighting of the US in the global index went from 6% to 46%. Why the delay, and why the growth? The circumstances behind this evolution were manifold.

2.2.1 Post-modern era: key drivers of growth

In summary, although market conditions were benign, they were not sufficient to explain the growth during the Modern REIT Era which arose from a combination of structural changes and beneficial market imbalances including the following:

- the creation of modern investible corporate real estate stock for REITs from the prior use of Real Estate Limited Partnership ("RELPs");
- regulatory change which led to the emergence of a liquid debt capital market which REITs could access;
- changes in tax laws making REITs more effective than competitor vehicles;
- legislative changes which ensured that institutions were free to invest and vendors had a tax-efficient way of injecting real estate assets into REITs. As a result both demand for REITs and supply of assets increased exponentially;
- a structural imbalance which meant that institutions needed to double their holdings to match their general equity weightings; and
- sufficient momentum behind the sector to allow recent REITs to have follow on offerings and move from using equity to refinancing to pursue the acquisition of distressed assets.

In the current era, it may be contended that there now exist a number of inter-related factors which will change both the reality and perception of the listed real estate sector and lead to a change in the way that listed real estate companies are operated and their securities used for asset management purposes. This evolution may, therefore, be regarded as the next phase of development, being the Post-Modern REIT Era. Whilst this may not lead to significant changes in the size of the sector, as happened in the Modern REIT Era, there may well be lasting changes to the underlying composition of the sector over the coming years as these factors exert an influence.

Relevant issues include, but are not limited to, the following (not in order of importance):

- a rising risk-free rate;
- maturity of sector awareness;
- introduction of MiFID II;
- increased use of passive products, such as ETFs;
- continued demand for "non-correlated" asset classes;
- requirements for increasing maintenance capex to maintain values;
- a structural change in the perception of shopping centre and office assets;
- a blurring of the distinction between companies and funds; and
- increased focus on solutions and specific investment characteristics rather than generic sector allocations

Therefore, what are the implications for the REIT sector and their corporate and fund management teams?

2.2.1.1 Rising risk-free rate

Something of an old chestnut has threatened the sector since the Taper Tantrum on 22 May 2013, but is certainly worthy of re-consideration on a regular basis. After a 35-year downward trend in US and UK bond yields, it is conceivable that an inflection point has been reached and that, at best, yields have reached a level from which the next sustained linear movement is likely to be upwards.

What seems likely is that the operational margin enhancement and dividend growth which has been created by management teams over the last few years through refinancing at lower rates on debt finance cannot be sustained. Therefore, at best, future earnings and thus dividend growth will have to come from rental growth, or an improvement in operational efficiency, rather than a reduction in interest costs. At worst, rising refinancing costs will dampen income growth from the underlying assets, leading to pressure on dividends.

2.2.1.2 Maturity of sector awareness

One of the key drivers of the Modern REIT Era in the early 1990s was the fact that institutions were structurally underweight in this "new" sector. The separation of the sector from Financials by MSCI and S&P to become the eleventh GICS sector in September 2016 marked the maturity of the sector relative to the equity market. There was some expectation at the time that underweight institutions would increase their weightings, leading to extra demand for the sector. In the event, this proved not to be the case and the US sector has underperformed the rest of the equity market since that date. Partly this can be explained by positioning ahead of the change and partly by the stage of the cycle, with REITs having had a strong period of performance post global financial crisis.

Although other benchmark providers such as Russell have now adopted the same policy of separation, it is fair to assume that there are unlikely to be further structural tailwinds and benefits to the sector from this area. However, the lasting impact appears to be that more generalists are meeting with companies on roadshows and have adapted their equity valuation models to incorporate REITs as part of the overall screening/weighting process. Therefore the influence of generalists on price setting going forward may well remain at an elevated level. Interestingly, as a rule of thumb, the percentage ownership of (US) REITs by specialists or dedicated, fund managers such as Cohen & Steers is thought to be around 25%, which is the same level as that owned by passive fund managers such as Vanguard.

2.2.1.3 Introduction of MiFID II

The introduction of MiFID II in Europe in January 2018 may be expected to have two specific impacts upon the sector, being research coverage and insurance company allocations.

Concerning research coverage, from January 2018, fund managers in Europe will have to pay and account for research separately. This unbundling is expected to be adopted as market practice globally and may eventually lead to a reduction in the number of external analysts providing forecasts for listed companies. Potentially this may lead to less price discovery and greater volatility as data points emerge and a reduction in quality of "consensus" forecasts for valuation purposes. However, companies may counteract this by either commissioning more paid-for research or (subject to regulatory compliance) giving more granular guidance.

Concerning insurance company allocations, one of the biggest obstacles to European insurers investing in listed real estate companies is the heavy capital weightings imposed by Solvency II. The European REIT advocacy group EPRA is strongly petitioning the European Commission to cut this burden from around the industry's neck under the Capital Markets Union Action Plan and the Solvency II review. Solvency II deters insurance investors from a key source of quality assets and management, liquidity and transparency for their real estate portfolios. EPRA contends that, if insurers are able to appropriately weight listed real estate in their investment asset allocations, the market capitalization of the sector in Europe could possibly double.

The EU's Solvency II rules became fully applicable at the start of 2016 and are designed to ensure that insurance companies can meet their obligations to policy holders in the event of major losses. The rules fix the amount of capital they must hold against their investments for regulatory purposes and these vary according to the risk of potential losses for each asset class. The capital risk weighting required for listed real estate is 39%, along with equities, contrasting with only 25% for direct property investments due to the lower perceived risk, or volatility, of physical bricks and mortar assets.

An extensive body of academic research has concluded, however, that the performance of listed real estate converges with direct property over the longer term, meaning there is no market rationale for treating the two as separate investment asset classes from a relative risk perspective. In the latest and most comprehensive study conducted by MSCI in 2017, the researchers found listed real estate sheds the influence of the general equities market after just 18 months and that there are strong correlations in performance across individual property assets, non-listed funds and securities, particularly over three- and five-year periods.

Europe's €10 trillion insurance investment industry is the largest single pool of institutional capital in the EU, but it has very limited exposure to listed real estate. The situation in the US differs markedly where pension funds are the sector's dominant source of investment. Analysts and investors agree that Solvency II plays an important role in the insurance sector's severe underweight allocation to property stocks compared with other forms of real estate investment.

2.2.1.4 Increased use of passive FM strategies

The growth in the use of ETFs has been well documented and the real estate sector (albeit predominantly in the US) has seen a significant growth in the provision and usage of ETFs. In addition to the standard, long only sector ETFs there are also reverse sector ETFs available, which aim to provide exactly the opposite to the market index (e.g. -3% vs +3% and vice versa) as well as leveraged ETFs aiming to providing two or three times the underlying index return.

It may be contended that the next phase of growth will include Smart Beta 2.0 style and automated trading strategies which aim to provide an element of active management and superior performance, but at closer to the required lower "passive" cost.

2.2.1.5 Continued demand for "non-correlated" asset classes

Historically there has been a case for real estate, including listed real estate, on the basis of the lack of or reduced correlation with equities and bonds. As the sector has become more mainstream and therefore more correlated, this has led to an increase in demand for REITs with alternative assets, such as health care, student accommodation and so forth.

It may be contended that this is part of a longer-term trend with those sectors which can demonstrate lower levels of correlation and therefore increased diversification benefits continuing to attract support. In particular, it is noticeable that demand for traditional cyclical sectors such as shopping centres (see 2.2.1.7,) and CBD offices has reduced significantly. This is important, as numerous REIT indices have had very high weightings to these asset classes, sectors and sub-sectors.

2.2.1.6 Requirements for disclosure on maintenance capex

In the UK and Europe, there has been a general assumption that no, or little, depreciation is required for freehold assets. However, market conditions have changed and it is now clear that a level of depreciation/obsolescence or maintenance capital expenditure is required to maintain values. The question is how much?

Advanced research is currently being undertaken by EPRA and various sell side analysts to clarify this issue, but the implications are clear. Whatever level of maintenance capital expenditure that may be required will come out of free cash flow and therefore potential dividends, also requiring set off against any forecast valuation improvements. It may be anticipated that companies will be expected to provide this metric going forward and it will become a key measure of portfolio quality and income duration.

2.2.1.7 A structural change in the perception of shopping centre assets

The asset class which has faced the most headwinds recently is that which, historically, has been the most stable: shopping centres. It is very important to note that the listed sector is significantly overweight shopping centres relative to their weighting in a real estate market portfolio. Therefore, negative sentiment towards shopping centres will impact the sector disproportionately, regardless of the quality of the underlying assets.

The headwinds are well understood and include the growth of e-commerce, the saturation of retail space per capita (particularly in the US) and the speed with which major tenants are cutting back stores (voluntarily or involuntarily). This suggests a structural shift (albeit not as widespread or within as short a timescale as some headlines would suggest) not a cyclical change which could lead to a recalibration of pricing and the relative valuation of shopping centre assets with obvious implications for the future composition of a benchmark portfolio.

2.2.1.8 Blurring of the distinction between companies and funds

Traditionally it has been easy to distinguish between corporate structures, REITs and funds. More recently, however, there has been a blurring of the lines. In particular, a number of listed REITs now exhibit specific characteristics which are traditionally associated with unlisted funds. Current examples include the use of external management structures (such as Tritax in the UK), changing remuneration structures (such as Merlin in Spain) and finite life structures (such as Hispania in Spain). All of these companies have performed well, so a fund style structure within a REIT is not necessarily a negative subject, obviously, to management quality and asset appeal.

2.2.1.9 Increased focus on solutions and investment characteristics rather than sector allocations

One of the biggest trends that has emerged is an increasing focus on the specific investment characteristics of individual companies and a decreasing tendency towards general sector allocations. Recent examples of this include Secure Income REIT in the UK and indeed the plethora of IPOs seeking to follow that example. Given previous comments on correlation, structural changes in asset preferences and the blurring of the distinction between companies and funds, there could be increasing numbers of REITs seeking to provide an investment solution (possibly via a non-traditional or emerging sector exposure) rather than a diversified sector exposure.

2.2.2 Post-modern era: initial conclusions

Initial conclusions arising from an analysis of growth in global REIT markets and current market practices in the Post-Modern Era include the following:

- increased reliance on rental growth for dividend growth rather than reduced financing costs;
- the influence of generalists on price setting going forward may well remain at an elevated level;
- a reduction in the number of external forecasts, but possibly increased levels of paid-for research and/or (subject to regulatory compliance) corporate guidance;
- a second generation of passive funds, using Smart Beta 2.0 style and automated trading strategies, to provide an element of active management and superior performance but at closer to the required lower "passive" cost;
- continued demand for "alternative" non-correlated sectors that can demonstrate lower levels of correlation and therefore increased diversification benefits;
- maintenance capex to be a key metric going forward with increased levels of disclosure;
- a decline in the overweight holding of shopping centres within benchmark portfolios;
- more fund style structures and practices within a listed format; and
- increasing numbers of REITs seeking to provide an investment solution (possibly via a non-traditional or emerging sector exposure) rather than diversified sector allocation.

2.3 How relevant is the REIT sector relative to the rest of the equity market? – implications of the new GICS classifications

On 1 September 2016 MSCI extracted Real Estate to be a separate sector in their Global Industry Classification Standard (*GICS* – GICS was developed by MSCI and Standard & Poor's to offer an efficient investment tool to capture the breadth, depth and evolution of industry sectors), which is used by asset allocators for benchmarking weightings of individual sectors. S&P followed suit on 16 September 2016. There were ten sectors in the *MSCI All Country World Index* (ACWI), which are shown in Table 2.1, together with their weight in the Index pre and post the change.

As can be seen from Table 2.1, Financials have the largest weighting accounting for a fifth of this global equity index. It should be noted that sector weightings between different equity indices can vary, according to the Index Provider criteria. Prior to the change Real Estate sat within Financials, as shown in Table 2.2:

The Index Classification hierarchy includes, within the Real Estate category, a further division into REITs and real estate management and development companies.

Table 2.1 Sectors in the MSCI All Country World Index (ACWI)

	Pre Sep 1	*Post Sep 1*
Consumer Discretionary	12.29%	12.29%
Consumer Staples	10.45%	10.45%
Energy	6.75%	6.75%
Financials	20.25%	16.85%
Health care	11.86%	11.86%
Industrials	10.49%	10.49%
Information Technology	15.67%	15.67%
Materials	5.11%	5.11%
Real Estate	-	3.40%
Telecommunication Services	3.81%	3.81%
Utilities	3.32%	3.32%
Total	**100.00%**	**100.00%**

Source: MSCI

Table 2.2 Financials Sector in the MSCI All Country World Index (ACWI)

	Pre Sep 1	*Post Sep 1*
Banks	9.29%	9.29%
Diversified Financials	3.74%	3.79%
Insurance	3.78%	3.78%
Real Estate	3.44%	-
Total	**20.25%**	**16.86%**

Source: MSCI

REITs are then further split into the following:

Diversified REITs
Health care REITs
Hotel and Resort REITs
Industrial REITs
Mortgage REITs
Office REITs
Residential REITs
Retail REITs
Specialized REITs

Real estate management and development companies are then further split into the following:

Diversified Real Estate Activities
Real Estate Development
Real Estate Operating Companies
Real Estate Services

Typically, generalist fund managers would allocate an active weighting to the top-level (10 now 11) industry codes, relative to their benchmark weighting. The sub-sectors provide generalists with an appreciation of the variety of specialization and differentiation within the sector, although weightings can be more subject to asset allocation preferences rather than Index weightings.

As at 1 September 2016, Real Estate became the eleventh separate sector and was no longer grouped as a sub-sector in the Financials category. Since the formation of GICS in 1999, this is the first and only sector to be separated out in this manner. Whilst there may have been a short-term boost ahead of and immediately after this event, it may be contended that the real benefits will continue to flow to the sector over time, including:

- that the separation clearly provides third-party validation for the growth, institutional demand, investment characteristics and "independence" of listed real estate, ensuring that it will receive an increased level of attention from generalists on an ongoing basis;
- that a short-term boost will come from investors positioning to a more market weight ahead of the new classification;
- that over the ensuing months and quarters, enhanced levels of activity and liquidity in the sector may be expected as asset allocators determine the portfolio attribution impact of the sector weighing; and
- that not all Index Providers are making the change, with FTSE Russell having not yet decided to change. If and when they and others do, it may be contended that this would provide a further longer term stimulus to profile and, more importantly, funds committed to the sector.

It may be further contended that the following three key areas are particularly relevant, with their potential impact analysed later:

- sector independence, being the rationale behind the new classification;
- key potential implications of the new classification with a number of claims made, the validity of which are considered below; and

- why listed real estate is being (and may in future be) used in practice by generalist fund managers. Given the new separation, will generalist fund managers choose to utilize a separate listed real estate allocation?

2.3.1 Sector independence

It is important to understand the reasons why the real estate sector is being separated out from the financials grouping. The first factor relates to size. The move marks a major step in recognizing that the sector is now sufficiently large to warrant separate allocations and dedicated resources.

The second factor relates to the fact that asset allocators are able to formulate and execute real estate strategies incorporating listed real estate outside of a standard (c. 3–4%) equity market allocation. In particular the larger pension funds and sovereign wealth funds have been using a combination of direct real estate, joint ventures, unlisted funds and listed real estate to achieve their objectives. Typically, real estate receives an explicit allocation vs equities and bonds, but listed real estate does not always do so and can be used to form part of the real estate allocation. Allocations can vary enormously, from 0–20%, but are more commonly in the 5–10% band. These apply not only to benchmarked funds but also to absolute return funds and those using an opportunity cost model. As a result, a number of smaller funds are re-assessing the role that listed real estate can play in their portfolio allocations.

The third factor is the unique structure of REITs, particularly in a market environment of low inflation and bond yields. REITs now account for around 70% of the EPRA Global Developed Index. The structure means they are comparable to underlying real estate in the cash flows they produce and have unique characteristics because of the obligatory (typically around 90%) pay-out ratio. The sector has, therefore, found increasing favour with asset allocators as they seek to combine income and growth as the market adjusts to expected rate rises and more normalized bond yield levels.

2.3.2 Growth of the sector

In terms of absolute size, the sector can now be considered significant. At the trough of the latest market cycle (end of February 2009) the free float market capitalization of the EPRA Global Index was US$297bn and the sector represented just 1.1% of the equity market. Fast forward to December 2015 and, through a combination of equity fund raising and strong investment performance, the free float market capitalization of the EPRA Global Developed Market Index was US$1,284bn, (a fourfold increase), currently representing c.3.5% of the global equity market. Therefore, in global terms, it is easy to understand why the sector has become worthy of a separate classification.

With regard to the specific issue of the size of the listed real estate sector relative to the size of the financials sector, it is important to understand the regional variations to determine the likely initial impact of the change in classification. Table 2.3 provides the breakdown by region of the overall equity market weighting, the weighting of financials in the index, the weighting of real estate in the index and as a percentage of financials as well as a breakdown of the number of companies in the index, broken down by sector.

As can be seen, North America is by far the largest region of the total equity index and accounts for around 57% of the total weighting, followed by Europe at 22.5%. Within the regions there is a huge variation in the financials weighting, ranging from 35% in Asia Pac ex Japan to 18% in Japan and the US with Europe at 21%.

Within financials, the weighting of real estate may be considered relative to the overall equity market. Japan has the highest weighting for real estate within financials at 28% and Europe the

Table 2.3 Regional Weightings (May 2016)

		Europe	*North America*	*Pacific Ex Ja*	*Japan*	*Africa/ Mideas*	*Latin America*
ACWI Weight	Weight in ACWI	22.5	56.7	10.8	7.6	1.1	1.3
Financials	Weight of financials in index	21.29%	18%	35%	18%	34%	30%
Real Estate	Weight of real estate in index	1.51%	3.27%	7.04%	4.99%	8.16%	1.81%
Real Estate %	Real estate % of financials in index	7.1%	18.5%	20.1%	28.2%	24.1%	6.0%
	Number of companies in index	527	714	702	318	93	121
Financials	Number of financials in index	123	143	178	58	45	28
Real Estate	Number of real estate in index	14	43	58	18	14	3

Source: MSCI

Table 2.4 European Weightings

	Comment	*United Kingdom*	*France*	*Germany*	*Netherlands*	*Belgium*	*Sweden*
ACWI Weight	Weight in ACWI	6.6	3.3	3.0	1.0	0.5	1.0
Financials	Weight of financials in index	22%	18%	18%	19%	17%	32%
Real Estate	Weight of real estate in index	1.70%	3.94%	2.50%	0.00%	0.00%	0.00%
Real Estate %	Real estate % of financials in index	7.9%	21.8%	14.1%	0.0%	0.0%	0.0%
	Number of companies in index	113	73	54	23	11	30
Financials	Number of financials in index	27	14	8	3	3	7
Real Estate	Number of real estate in index	5	5	2	0	0	0

Source: MSCI

lowest at 7.1%. Finally, in terms of real estate as a total of the whole equity market, this is highest in Asia Pac ex Japan at 7.04% and lowest in Europe at 1.5%.

Therefore, what are the implications for the separate classification of the index? Initially, it would appear that the greatest initial impact will be on the biggest overall market (the US) as well as the regions with greatest percentage of real estate in the financials index (Asia-Pacific and Japan). Thus far, in terms of press coverage and literature, it is the US which has received the most attention regarding the likely impact. By contrast, the attention Europe has received is currently relatively small.

In terms of the breakdown within Europe (Table 2.4) and using the same analysis, it can be seen that it is the largest markets of the UK, France and Germany that are most likely to be affected.

2.3.3 Lack of correlation with the financials sector

One of the reasons given for the separation has been that the listed real estate sector does not have a significant correlation with the rest of the financials sector. Obviously this can vary over

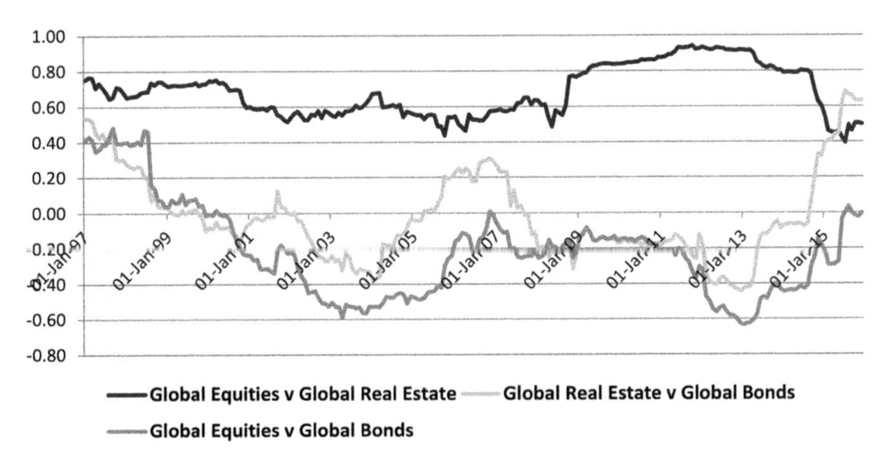

Figure 2.1 REIT Correlations with Equities and Bonds

Source: **EPRA**

time, but Cohen & Steers (2016) estimated that over the period 1990–2015 US REITs had a correlation of 55% with the S&P 500 Index, whereas the US financials sector had a correlation of 84%, suggesting that there are very different drivers of performance over the longer term. The implication is, therefore, that some of the volatility attributed to real estate stocks by being part of the financials sector may erode once they are classified as a separate sector.

EPRA monitors the rolling correlation of the global listed real estate sector with the global equity market and the global bond market (Figure 2.1). What may surprise some generalist investors, who regard listed real estate as "pure" equities, is that over the last two years the sector is more correlated with the bond market than the equity market. This has important, positive diversification implications for the sectors' role within multi-asset portfolios, particularly as global markets appear to be in long-term "lower for longer" phase, where investors have become more income sensitive.

When considered in terms of the sector beta, Figure 2.2 shows that this has been reducing significantly since the GFC.

Another popular misconception by some generalist investors is that the listed sector is highly leveraged and dependent upon bank debt. In fact, post global financial crisis, the sector has been reducing leverage (despite ever lower costs of debt capital) as well as diversifying away from bank sources to the debt capital markets. For the European sector, EPRA produces a monthly LTV Monitor which shows the trend in LTVs, broken down by country, as well as all the equity and debt capital issues that have occurred. In particular, this report shows the leading European listed real estate companies have been able to transition away from dependence upon debt provided by banks through the use of a wide range of instruments available in the debt capital markets. This has allowed them to enhance earnings (and therefore dividends) via the lower interest costs available on debt refinancing, diversify risk away from individual lenders, reduce the level of secured debt in their portfolio, extend maturities and achieve a balanced loan portfolio with a reduced level of risk.

2.3.4 Alternative ways of accessing the asset class

Following on from the declining correlation with the equity market, the question that generalists, who will be seeking to take an explicit weighting in the sector for the first time, will be

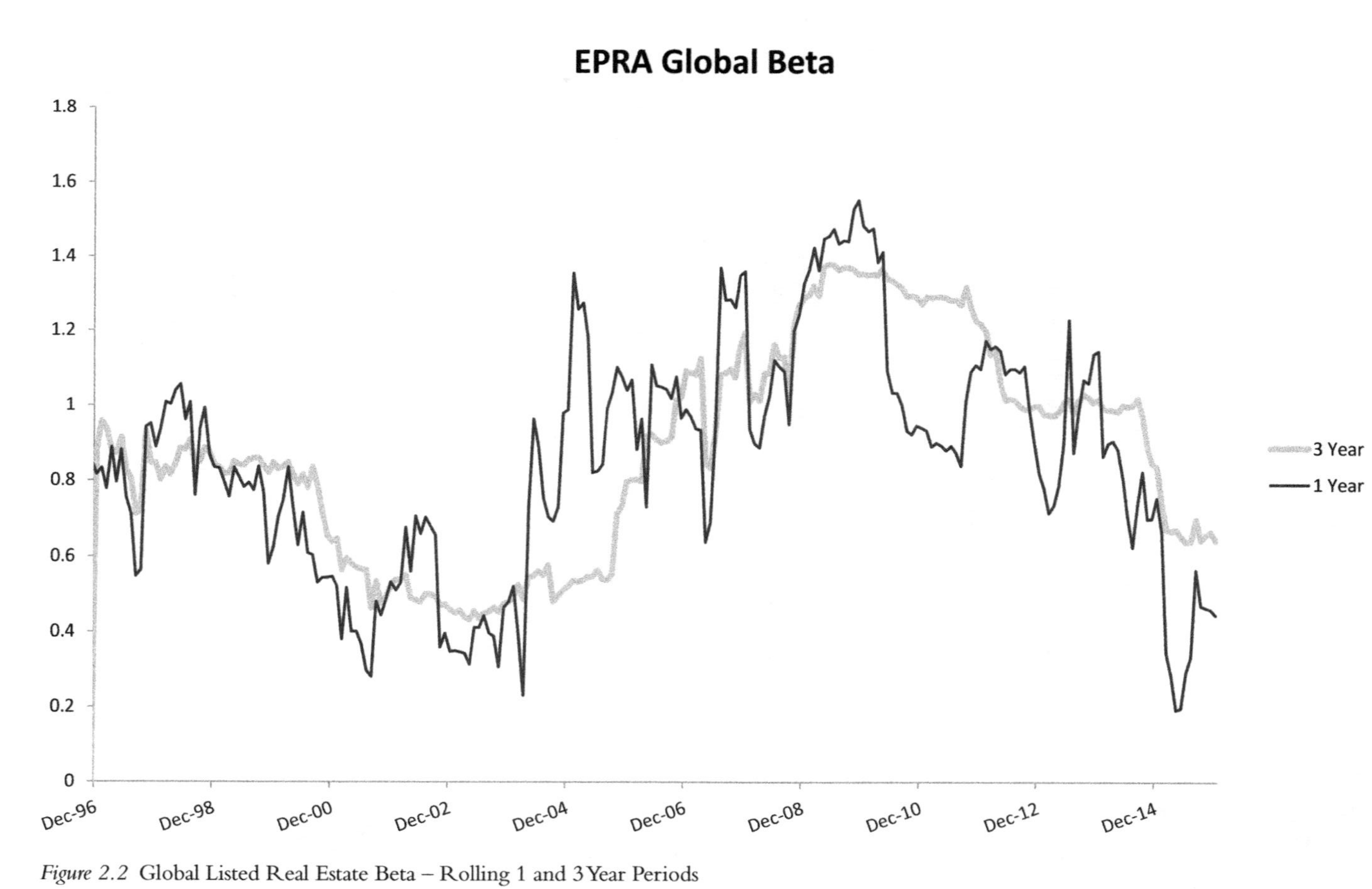

Figure 2.2 Global Listed Real Estate Beta – Rolling 1 and 3 Year Periods

Source: Consilia Capital

asking is this: to what extent would an exposure to the listed sector provide effective exposure to the underlying asset class?

There are two ways of responding to this question as will be considered in Chapter 7. First, by comparing the performance of the listed sector to a direct property benchmark and second by determining the time it takes for listed real estate returns to correlate more closely to direct property returns rather than equity market returns. As will be confirmed in Chapter 7, the separation of the sector can be expected to provide generalists with a clear way of gaining exposure to the return characteristics of the underlying assets.

2.4 Asset allocation implications

Concerning the implications for asset allocation, the key questions that arise include whether:

2.4.1 there will be an increase in demand over time from "generalist" fund managers who may be underweight the sector?;
2.4.2 there will be a decline in volatility?;
2.4.3 fund managers will have to make an explicit weighting decision on the sector for the first time?; and
2.4.4 fund managers will choose to utilize a separate listed real estate allocation?

2.4.1 Increase in demand?

There has been a lot of debate during 2017 as to both the extent to which generalist fund managers may be underweight real estate equities and the extent to which they may seek to reduce this underweight position.

So why might generalists be underweight? The reason lies in the fact that a fund may have a market weighting in financials but be long banks and insurance companies and underweight real estate. Research by JP Morgan found that the US long-only 1940 Act equity funds have an average real estate weight of 2.3%, compared to 4.4% for their benchmark, representing a 2.1% underweight. Translating this to absolute amounts, it is estimated that this would require US$100bn of inflows (c. 12% of the value of the US sector) to move to a neutral position. Clearly there is unlikely to be a move of that proportion in the short term, but the analysis does illustrate a potential positive impact arising from the sector having a separate classification.

2.4.2 Decline in volatility?

With regard to volatility there is an argument that, with increased visibility and by implication liquidity, the real estate sector should exhibit lower volatility as it "de-merges" from the historically more volatile financials sector and becomes more related to the lower volatility of the underlying real estate. Intuitively this makes sense and Table 2.5 illustrates the superior returns and lower risk of real estate compared to financials.

Although it is often claimed that direct real estate has a significantly lower level of volatility, it is interesting to note that, over a ten-year period to June 2016, the annualized volatility of the IPD Total Return All Property Index was 15.0% and so not significantly lower than that for real estate securities, whilst the annualized total returns were 4.3%, which is significantly lower than that for real estate securities (IPD, 2016).

Table 2.5 Risk and Return: 2000–2015

Asset Class	*Annualized Return*	*Standard Deviation*
Real Estate Securities	9.70%	18.90%
Bonds	5.00%	5.70%
Equities	3.30%	15.80%
Financials Sector	2.00%	20.80%

Source: Investec Asset Management (2016)

In the Expert Report prepared for the Norwegian Government Pension Fund Global (GPFG, 2015), the authors looked at the long-term performance of listed real estate and found that, over the period 1999–2015, global core real estate had an average annualized return of 9.8% and an annualized volatility of 17.6%. It can, therefore, be seen that listed real estate typically exhibits superior raw returns with only marginally higher levels of volatility, but far greater levels of actual liquidity.

2.4.3 Explicit weighting decision required?

One of the issues surrounding the separation of the sector is clearly that generalists will be required to make an explicit weighting to real estate securities. Though this is potentially positive, the interesting question is how that weighting will be determined. Whilst the fundamental outlook for the underlying real estate will obviously be considered, it is important to note that there are typically three broad valuation methodologies utilized for listed real estate, namely earnings (or Adjusted Funds From Operations; *AFFO*)-based, dividend-based and NAV-based metrics.

Whilst the earnings and dividend data can be accessed from the same sources as the other ten GICS sectors, typically listed real estate specialists focus on AFFO and NAV data which is not commonly available or used for other sectors. This may, therefore, create greater liquidity in the sector as multiple valuation methods are applied. To help generalists in this area, EPRA publishes two reports for its members which relate to the NAVs of European listed real estate companies, comprising monthly published NAV data and discount to NAV data.

2.4.4 Will listed real estate be used in practice by generalist fund managers?

Given the new separation, will generalist fund managers choose to utilize a separate listed real estate allocation? As the role that listed real estate can play in portfolio management continues to evolve, this section highlights some of the key topics that may be contended to be of interest to generalist fund managers new to the sector in determining a separate listed real estate allocation.

To ensure up to date and independent data, this section draws on the conclusions of GPFG (2015), a recent expert report commissioned by the Norwegian Ministry of Finance (*GPFG Report*), concerning the following:

2.4.4.1 does listed real estate provide true real estate exposure?;
2.4.4.2 performance – has listed real estate performance justified a separate allocation?;
2.4.4.3 weighting for real estate – what is an appropriate figure?; and
2.4.4.4 how beneficial is adding a global listed exposure to a multi-asset portfolio?

2.4.4.1 Listed real estate: true real estate exposure?

Academic evidence has established that listed (public) and unlisted (private) real estate markets have the same return characteristics over the long run. GPFG (2015) suggests that there is no evidence for superior performance or reduced risk from unlisted property or evidence of diversification benefits from adding unlisted property to listed property and that the volatility of unlisted property is similar to listed property after adjusting for smoothing and extending the time horizon.

2.4.4.2 Performance of global listed real estate: sufficient to justify a separate allocation?

Listed real estate has, on average, outperformed private real estate by 3%pa over the period 1994–2015 (i.e. from the beginning of the Modern REIT Era). Over all three periods studied in GPFG (2015), public real estate outperformed private real estate, which outperformed stocks, which outperformed bonds. The evidence is clear that listed real estate has generated sufficient levels of return across the short, medium and long term to warrant inclusion of at least market weighting.

2.4.4.3 Weighting: appropriate figure?

The Norwegian Ministry of Finance recently determined to increase maximum allocation to real estate from 5% to 7%, with a weighting range of 3–10% consistent with broader institutional real estate allocations. GPFG (2015) estimates that real estate represents about 6% of the world market portfolio with listed real estate comprising 15% of the real estate universe. REITs make up only a small fraction of (12–13%) of total real estate investment by pension funds.

2.4.4.4 How beneficial is adding a global listed exposure to a multi-asset portfolio?

A number of academic studies have highlighted the benefits to risk-adjusted returns of adding a global listed real estate allocation to a multi-asset portfolio. Alongside studies which focus on a permanent long-only exposure, there has been increasing interest in adding automated trading strategies which can increase/decrease allocations to the sector.

A recent study by Moss et al. (2015) investigated the impact of adding global REITs to a multi-asset portfolio, using momentum and trend-following strategies, finding that the main improvements arose when a broad index is replaced with one of four trend-following (TF) strategies. The portfolios deliver similar returns but volatility is reduced by up to a quarter to the 8–9% range, the Sharpe ratios increase by 0.1 to 0.5 with the main benefit being the reduction in the maximum drawdown to under 30% compared to 43% when the broad index was used. The authors concluded that a combined momentum and trend-following global REIT strategy can be beneficial for both a dedicated REIT portfolio and adding REITs to a multi-asset portfolio.

2.4 Summary

Chapter 1 sought to identify critical contextual issues in international REITs, including the defining characteristics of REITs, the global evolution of REITs and the risk-return characteristics of real estate investment through REITs relative to non-REIT real estate investment with this chapter, the *Post-Modern REIT Era*, examining the evolution of the global REIT market, what is deemed to be best current market practice and how the market is expected to change going forward through the following:

- the key drivers behind the growth of global REIT markets, current market conditions and practices that are determining the growth of REIT markets;
- the relevance of the REIT sector relative to the rest of the equity market; and
- the current asset allocation implications for REITs.

It is contended that the REIT sector is entering a new Post-Modern Era, with a number of specific factors influencing its style, growth and development including these:

- a rising risk-free rate;
- maturity of sector awareness;
- introduction of MiFID II;
- increased use of passive products, such as ETFs;
- continued demand for "non-correlated" asset classes;
- requirements for increasing maintenance capex to maintain values;
- a structural change in the perception of shopping centre and office assets;
- a blurring of the distinction between companies and funds; and
- increased focus on solutions and specific investment characteristics rather than generic sector allocations.

The growth in the global listed real estate market is due largely to the expansion of tax-efficient REIT regimes, which now make up 70% of the Global Developed Market Index. REITs distribute most of their rental income cash flows as dividends, so gaining favour with investors seeking income and capital value growth in the prevailing low interest rate environment.

It may be contended that the separation of listed real estate as the eleventh equity sector is significant, as a stand-alone real estate sector is recognition that listed property companies are an important investment sector in their own right, standing shoulder to shoulder with telecommunication services, health care and other mainstream industries. This underpins what academic and industry research has highlighted for many years, being that the inclusion of real estate stocks in a portfolio means that, over the medium term, investors can access the returns of the direct property market with the added advantages of much greater liquidity and lower costs.

One of the reasons for removing real estate from the GICS financials sector (the largest of the ten industry sectors with a 20.12% weighting under the former treatment) was the low correlation in the performance of property stocks relative to other equities classified as financials, such as banks. This is supported by the relative volatility of listed real estate and financial stock indices. The ten-year volatility of the FTSE EPRA/NAREIT Developed Europe Index is 19.4%, whereas for the MSCI Europe/Financials it is 29.52% for the same period. Changing the listed property sector's status should, therefore, reduce volatility to levels that are closer to that of the underlying direct property.

The volatility of listed real estate securities has been the main reason referred to by European institutional investors for why they do not allocate to listed property. The decoupling from financials is expected to improve the risk profile of REITs and start attracting new investment allocations. The potential for these additional investment capital flows is enormous, even though attracting the money would be a gradual process.

With the GICS move replicated by other major equity indices providers and given the momentum of growth in the global listed real estate sector, there is clear evidence of the increasing maturity of listed real estate as an asset class within the equities sector, becoming a credible and sizeable complement to fixed income and general equity investments.

Part I continues with chapters each reviewing a current theme of REIT evolution through the lens of contemporary research. Chapter 3, *Emerging Sector REITs*, investigates such sectors as timber and data centres. Chapter 4, *Sustainable REITs*, analyses REIT environmental performance and the cost of equity with Chapter 5, *Islamic REITs*, examining the evolution of global Islamic REITs through a study of the Malaysian market. Chapter 6, *Behavioural Risk in REITs*, addresses the management of behavioural risk in global REITs and Part I concludes with Chapter 7 which reviews recent research into *REIT Asset Allocation*.

Part II then includes six chapters each analysing REITs in a region of the world, using case studies from a developed, developing and emerging REIT sector in the region, concluding with Chapter 14 which considers *Directions for the Future of International REITs*.

References

Cohen & Steers 2016, *Real estate in a class of its own*, Cohen & Steers, London.

GPFG 2015, *A review of real estate and infrastructure investments by the Norwegian Pension Fund Global December 2015*, Norwegian Ministry of Finance, Oslo.

Investec Asset Management 2016, *Real estate securities: Securing its own space*, Investec Asset Management, London.

IPD 2016, *IPD monthly property index June 2016*, IPD, London.

Moss, A., Clare, A., Thomas, S. and Seaton, J. 2015, 'Trend following and momentum strategies for global REITs', *Journal of Real Estate Portfolio Management*, Vol. 21, No. 1, pp. 21–31.

3

EMERGING SECTOR REITS

Wejendra Reddy and Hyunbum Cho

3.1 Introduction

This book aims to identify key areas for research in the REIT discipline for the next five to ten years by surveying the current state of the REIT discipline around the world and identifying emerging and cutting edge research areas through a thematic review of current contextual issues and a regional analysis based on case studies.

This book comprises two parts, the first part being a thematic review of emerging and cutting edge global research into current contextual issues in REITs internationally and the second part being a regional analysis of REITs around the world, each written by authoritative academic authors from the world's leading Universities and REIT industry experts.

Part I includes six chapters each reviewing a current theme of REIT evolution through the lens of contemporary research. Chapter 1 focused on critical contextual issues in international REITs while Chapter 2, the *Post-Modern REIT Era,* examined the evolution of the global REIT market, what is deemed to be best current market practice and how the market is expected to change going forward, with this chapter, *Emerging Sector REITs*, investigating such sectors as residential, health care, self-storage, timber, infrastructure and data centres generally and then specifically through a case study of US REITs.

Chapter 4, *Sustainable REITs*, then analyses REIT environmental performance and the cost of equity with Chapter 5, *Islamic REITs*, examining the evolution of global Islamic REITs through a study of the Malaysian market. Chapter 6, *Behavioural Risk in REITs*, addresses the management of behavioural risk in global REITs and Part I concludes with Chapter 7 which reviews recent research into *REIT Asset Allocation.*

Part II includes six chapters each analysing REITs in a region of the world, using case studies from a developed, developing and emerging REIT sector in the region:

Chapter	*Region*	*Developed REIT Sector*	*Developing REIT Sector*	*Emerging REIT Sector*
8	North America	USA	Canada	Mexico
9	Latin America	Brazil	Argentina	Uruguay
10	Europe	UK	Spain	Poland

11	South East Asia	Singapore	Malaysia	Thailand
12	North Asia	Japan	Hong Kong	China
13	Oceania	Australia	South Africa	India

concluding with Chapter 14 which considers *Directions for the Future of International REITs*.

3.1.1 Specialized sector REITs

Since their inception, REITs have either been equity REITs or mortgage REITs. Equity REITs invest in tangible real estate assets and manage the assets to generate income, whereas mortgage REITs invest in property secured mortgages, mezzanine loans, subordinated loans, construction loans and commercial mortgage backed securities (CMBS). As of November 2016, around 94% of the US REITs were equity REITs (NAREIT, 2016a).

Traditionally, equity REITs were either diversified or sector specific. Diversified equity REITs evolved to hold massive diversified portfolios comprising a combination of commercial, retail and industrial property. Sector-specific equity REITs evolved to hold massive sector specific portfolios of either commercial, retail or industrial property. Over time, commercial sector–specific REITs broadened to hold both CBD and suburban office property, retail sector–specific REITs broadened to hold shopping centres and out-of-town retail parks and industrial sector–specific REITs broadened to hold industrial property and logistics centres.

However, the evolution of the REIT sector has now seen broadening beyond the traditional commercial, retail and industrial property sectors to embrace a range of other specialist sectors such as residential, health care, self-storage, timber, infrastructure and data centres which are considered generally in this chapter and then specifically in the context of the US market, being one of deepest and most mature REIT markets in the world.

The growth of the global property market, competition from other investors, availability of attractive assets and the strong need for diversified investment strategies are the main incentives for property fund managers and institutional investors to consider more specifically targeted asset classes, with the last few years witnessing significant movement in sector-specific allocations in global real estate investment.

A notable feature of recent years was that the office sector, which was traditionally the principal investment sector in the REIT universe, has been gradually losing its market share and recorded only a 35% market share in H1 2015, being a significant decrease from 46% market share in 2007. However, during the same period, hotel and other specialist sector REITs have grown significantly and increased their market share. Also, the period has seen the industrial and logistics sector becoming significantly more popular with investors (CBRE, 2015).

Table 3.1 and Figure 3.1 show the decomposition of FTSE EPRA/NAREIT index, dividing the market by developed and emerging markets. As of November 2016, although the diversified sector is the major portion of the market, many sector-specific real estate investors are very active in some countries, especially in the developed markets. The developed markets have well-distributed sector portfolios throughout the traditional diversified (27.4%), retail (24.3%), office (11.8%) and industrial (6%) sectors, together with residential (13.1%), health care (7.3%), self-storage (3.7%) and resorts (3.7%), but the emerging markets have a skewed distribution to diversified (53.2%), retail (11.3%), industrial (5.1%) and residential (28.6%). This suggests that, once a market becomes mature, then the demand for sector-specific and specialist sector investment will increase. The CBRE survey also showed similar results, with the degree of sector change in recent years found to vary from one region to another (CBRE, 2015).

Table 3.1 FTSE EPRA/NAREIT Sector Breakdown: 2016

	Developed Market			*Emerging Market*		
Sector	*No of Cons.*	*Market Cap (USD M)*	*Weight %*	*No of Cons.*	*Market Cap (USD M)*	*Weight %*
Diversified	96	359,116	27.4%	87	70,583	53.2%
HEALTH CARE	21	95,412	7.3%	1	453	0.3%
Industrial	23	78,739	6.0%	6	6,740	5.1%
Industrial/Office	14	34,632	2.6%	2	450	0.3%
Lodging/Resorts	19	48,952	3.7%	1	305	0.2%
Office	53	155,145	11.8%	3	1,052	0.8%
Residential	37	172,367	13.1%	32	37,987	28.6%
Retail	64	319,313	24.3%	15	15,045	11.3%
Self-storage	6	48,821	3.7%			
Totals	**333.00**	**1,312,496**	**100%**	**147**	**132,615**	**100%**

Source: Author's compilation from FTSE Russell (2016a, 2016b)

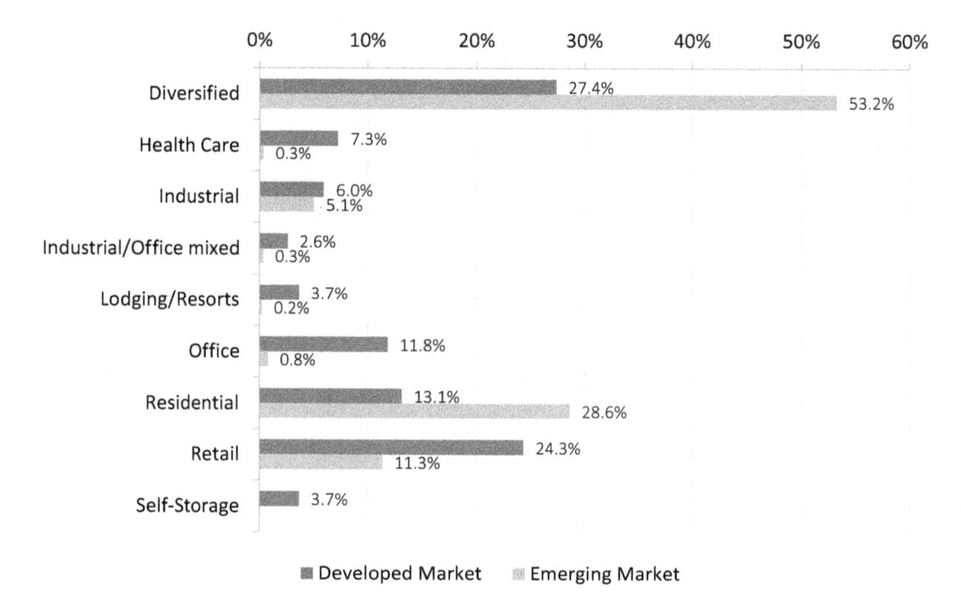

Figure 3.1 FTSE EPRA/NAREIT Index Sector Breakdown: 2016

Source: Author's compilation from FTSE Russell (2016a, 2016b)

3.2 Specialist sector REITs

With REIT investment in the traditional commercial, retail and industrial property sectors well established and researched, this section seeks to investigate REIT investment in other specialist sectors of the property market, notably the following:

3.2.1 Residential REITs
3.2.2 Health care REITs
3.2.3 Self-storage REITs

3.2.4 Hotels and lodging
3.2.5 Timber REITs
3.2.6 Infrastructure REITs
3.2.7 Data centre REITs
3.2.8 Other specialty REITs

generally, with the following section then seeking to investigate REIT investment in such sectors specifically in the context of the US REIT market.

3.2.1 Residential REITs

Residential REITs generally own and manage apartments in diverse locations, often within or in close proximity to CBDs. Some REIT managers also specialize in development, allowing them to build residential properties in desirable cities or suburbs with strong rental and occupancy rates. Depending on property location, quality (age, condition, fittings, energy ratings), financial market conditions and supply/demand factors, US capitalization rates (for example) range from 5% to 7.5% (Block, 2011).

Residential REITs perform optimally during expanding economic conditions when there is a greater creation of jobs resulting in better income and thus formation of new households. Therefore, conditions in the housing market, including cost and affordability (household income vs mortgage repayments) are critical to residential REIT performance. In addition, conditions in the single-family "individual" housing market are also important. Despite the currently prevailing global low interest rate environment, apartments continue to capture a large percentage of prospective residents. The appetite for single-family dwelling is generally declining in the face of rising house prices.

The rate of supply also impacts the performance of residential REITs. For example, higher levels of construction of similar style new apartments in an area with abating demand could force owners of existing residential complexes to reduce rents or offer incentives. Such market conditions would generally result in lower occupancy rates and reduced income. Inflation also impacts residential REIT performance, with rising inflation causing increases in operating expenses, including items such as insurance, regular repairs and maintenance, which cannot be passed along to the tenants. On the other hand, rising inflationary pressures may justify owners of newly constructed apartments charging higher rents to offset increased construction costs. If new apartment complexes are at full occupancy, owners of existing apartments may then also negotiate to raise their own rents at rent review or lease renewal.

The major risks for residential REITs are similar to those common to all other REITs and to property markets generally. Regardless of the state of national economy, regional or local economies can suffer periods of decline or even recession, pushing occupancy levels lower and causing rents to flatten or even decline significantly. This economic risk is more profound for residential REITs which often lack geographic diversity. Poor management, including failure to respond to market changes, can also adversely impact the performance of residential REITs. As the property market is cyclical, during periods of economic boom, overbuilding may occur. This is particularly common when both land and finance are cheap, with excess supply pushing rents down and adversely impacting residential REIT performance.

3.2.2 Health care REITs

Health care REITs do not engage in the operation of health care businesses but rather specialize in providing leased space for businesses that provide such services as medical facilities, hospitals, nursing homes, seniors housing and medical science laboratories. Similar to commercial properties, the revenues generated by health care REITs are derived from lease payments from independent tenants. As such, leases tend to be long-term, with rent reviews over fixed terms or based on inflationary movements, such as CPI. The lease arrangement is normally on a 'triple-net' basis, where the master lessee bears all property operating expenses and taxes. The lease is structured to provide stable income to the health care REIT and protection from the usual upturns and downturns in the tenant's business, whilst capturing rental growth through the lease structure.

To maximize revenue and reduce tenancy risk, REIT managers are likely to seek to lease to several health care providers within one building, for example, medical facilities, radiologists and pharmacists within one medical centre. Although health care REIT investors experience modest income growth, the investment upside is that these funds generally offer a low-risk profile when compared to other REITs (Block, 2011). Health care REITs are likely to trade at lower multiples of free cash flow, but they generally offer higher dividend yields than other REIT sectors.

The key risk faced by health care REITs concerns changes in government health care policies such as medical rebate or reimbursement for certain procedures performed by their tenant businesses. Reduction in medical rebates is likely to result in a lower number of patients due to higher out of pocket expenses for individuals. Increased medical rebates, on the other hand, may translate into modest internal cash flow growth, albeit with some competition as new operators enter the market to capture the benefits.

Although overbuilding has, at times, been a problem in the assisted and independent living sector such as in retirement villages, skilled nursing and hospital facilities are rarely in oversupply due to strict regulations surrounding permits to operate. A key attraction for investors is that health care REITs are generally recession resistant as demand for health services is relatively inelastic. Regardless of the state of the economy, people need medical attention, treatment and nursing facilities. Conversely, some specialized facilities, such as those providing cosmetic surgery or other non-critical procedures, are likely to be more vulnerable during periods of economic downturn. However, the health care industry is likely to continue to benefit from increasing demand for physicians, skilled nursing services and other medical services due to favourable population demographics.

3.2.3 Self-storage REITs

Self-Storage REITs own and manage properties comprising individual storage units, ranging from 1.5 x 1.5 metres to 6 x 6 metres, that are rented out often on a monthly basis (Willis, 2003). Many storage facilities are located in close proximity to major cities or in suburbs near industrial parks. The facilities allow individual renters to use the space for storing personal items such as furniture, clothing, books, documents, recreational vehicles and even boats.

Individual demand is often driven by changes in lifestyle, such as downsizing during retirement, relocation, marriage, divorce or the death of loved ones, or simply driven by the need to store items for future use. Business demand also arises, including the use of such space to store work documents and files by companies that lease expensive office space but are looking to limit their floor space requirement to save rent.

Sustained increasing demand for self-storage space has kept the sector profitable for fund managers, however the sector is susceptible to the usual risks borne by any commercial property investment, including overbuilding, then recession and then boom cycles. The self-storage sector appears to deliver above-average operating income growth and also offers some protection during periods of recession as individuals displaced or downsizing often require temporary storage space. However, the self-storage sector is not recession proof with periods of recession dampening consumer spending habits. A key risk to the sector is high renter turnover with operators needing to offer frequent rent discounting and/or rent free periods to protect cash flow during periods of market decline.

3.2.4 Hotels and lodging

Hotel and lodging REITs comprise properties offering a wide range of accommodation for leisure and business travellers, conferences and business meetings. Normally, a five star hotel typifies luxury across all areas of operation, with excellent design quality and attention to detail where guests enjoy highly personalized service. Accordingly, such hotels charge a higher rate than non-five star hotels. However, such hotels are expensive to build due to high land costs, high building materials and finishing costs and long construction periods. Less than five star rated hotels offer more limited services, amenities and facilities to guests without compromising guest security or cleanliness and with additional fee-based services such as Wi-Fi and other business facilities available upon request.

There are number of factors that affect hotel and lodging REIT performance including the quality of the property (rating), changing market conditions, discretionary income, business conditions, weather seasons, disease outbreaks, terrorist attacks, currency exchange rates, geopolitical shocks disrupting travel and room supply. The industry is well known for overbuilding, mainly in the luxury hotel sub-sector, where it may take several years between planning and completion and over which period the economy may have retreated (Block, 2011). However, when market conditions are favourable, owners are able to charge premium rates.

Given the cyclical nature of the business, hotel and lodging REITs are generally less suitable for risk-averse investors. There are few long-term traditional leases in the hotel and lodging sector, therefore cash flow protection may be limited, with operating costs becoming a major drain on income during low occupancy periods. To counter this frequent volatility in earnings, hotel and lodging REITs generally negotiate a master lease agreement with a hotel operator which could provide a minimum fixed income plus an opportunity to access the hotel operator's revenue and gross profits.

Capital expenditure is a major drain on profits as hotels and lodgings need to refurbish properties regularly to keep up with competition, especially newer properties in the market. So, unlike commercial properties which offer a high return on asset enhancement, hotels upgrade assets just to remain competitive. Therefore, it is important that REIT investors have a clear understanding of capital expenditure items covered under the master lease arrangement and those borne by the REIT.

In recognition of the nature of the risks inherent in the underlying business, while earnings may be good during peak periods compensate for, the overall cash flow position may just break-even as the higher room and occupancy rates during such peaks periods compensates for the decline in business during downturns. Accordingly, investors in the hotel and lodging sector must be prepared for a roller-coaster ride relative to other sectors.

3.2.5 Timber REITs

A timber REIT may be classified as a REIT in which more than 50% of the fund's total asset value is real property subject to the trade or business of producing timber. Timber REITs may manage many thousands of acres of timberland and produce softwood products, panels, particleboard and fibre products. Timber REITs may also extract minerals and receive royalties from coal bed methane, natural gas and oil production on their properties.

There have been significant shifts in institutional forest ownership globally as the demand-supply gap for wood is widening due to increased population. Construction for new housing and commercial properties has increased while forestry areas have decreased due to ever expanding city boundaries.

In the United States, timber REITs have evolved since the late 1990s when most corporations that traded in forestry either sold their timberlands or securitized their holdings. It is estimated that REITs and timber investment management firms control approximately 4.2–4.3% of all timberlands in the United States (Hood et al., 2015) with one group, Weyerhaeuser, holding 13 million acres of industrial grade timberland across the United States.

Unlike other REIT sectors where the value of assets generally depreciates over time, timber REITs experience value appreciation as trees keep growing. Timber REITs are among the least sensitive to a market downturn due to their unique nature, being a long-term biological asset. As such, timber REITs can, to some extent, schedule harvesting to suit market conditions. Thus the yield function of timberland can be independent from normal business cycles. Timber REITs often adopt timber harvesting contracts and long-term leases to limit investor's risks. However, similar to any biological asset, the key risk faced by the timber industry is natural disasters such as fires, flooding and earthquakes and longer periods of extreme climate events which can impact harvesting targets and affect cash flows negatively. Accordingly, this sector mainly suits REIT investors with a long-term view.

3.2.6 *Infrastructure REITs*

Infrastructure REITs buy, develop, own and manage industrial and technology infrastructure properties and collect rent from the tenants thereof. The infrastructure sector includes a growing number of REITs focused on serving emergent demand for mobile communication and information technology by fast-growing industries such as cloud services, gaming, entertainment, analytics and social media.

Infrastructure REIT assets may include telecommunications towers, wireless infrastructure solutions including towers, distributed antenna system (DAS) networks, backup power systems, electric transmission and distribution utility assets, outdoor advertising, renewable power generation assets and midstream and downstream energy infrastructure assets that perform utility-like functions, such as pipelines, storage terminals and transmission and distribution assets.

The real estate infrastructure assets are long-term assets similar to commercial properties. For example, fibre assets are usually buried underground and have a useful life of 30–50 years. Infrastructure REITs triple-net leases typically run for periods of 15 years thus offering investors steady cash flows and returns (Keenan, 2016). Infrastructure REITs are deemed to be relatively recession proof, due to the steady demand for real estate infrastructure by ever increasing numbers of internet and mobile phone companies reflecting demand for internet, wireless telecom, cloud computing and mobile solutions from businesses and individuals.

3.2.7 *Data centre REITs*

Data centre REITs own, develop, operate and/or manage large portfolios of wholesale and retail data centres which are designed to protect and secure the information technology (IT) infrastructure of customers that use them via cloud and other interconnection specific services. As such, the product sits in-between real estate and technology. However, compared to accessing specific technology stocks such as IBM, Microsoft, Google or Amazon, REIT investors gain access to a portfolio of real estate assets occupied by a range of such technology companies.

The performance of data centre REITs is dependent on the volume of worldwide internet traffic. Every time individual users search Google, listen to music online, stream videos on Netflix or share photos on Instagram, they add to the internet traffic volume and, consequently, the data centre REIT's earnings. The success of data centre REITs stems from their ability to accommodate a wide range of data needs. Although cloud vendors like Microsoft and Amazon sell space on their servers, companies generally are not able to manage systems independently and directly, allowing data centre REITs to offer such companies a mix of the use of their own services and those from other vendors (also known as hybrid cloud). The largest data centre REITs in the sector are the US based Equinix and Digital Reality that manage some 140 data centres in over 30 markets.

A key challenge faced by the data centre industry is managing and sustaining current growth levels in the face of increased demand for global data and secular drivers including cloud computing, big data, streaming media and other worldwide, internet and wireless services. As data centre REITs are the property companies that house, power and cool the computer servers stored therein by internet companies, they require purpose-built structures that meet data centre's needs. Re-modelling existing properties for such operations is possible but can be extremely difficult. At the same time, new data centres are costly to build and can take several years to complete from conception.

Unlike traditional REITs, data centre REIT performance is not heavily influenced by employment, consumer sentiment and GDP movements. This defensive nature means data centre REITs are in high demand despite periods of market instability. A good example is Equinix, the largest global data centre REIT, which derives almost half of its revenue from international markets and grew substantially during the recent "Brexit" market commotion when the company acquired London-based data centre TelecityGroup.

Accordingly, despite the economic situation, data centre REITs provide a cost-effective alternative for businesses as they continue to expand their internet technology operations. However, data centre REITs are not completely resistant to market downturns as performance can be impacted by broader market sell-offs, particularly in the overall technology stocks sector of the equities market.

3.2.8 Other specialty REITs

While there are a range of other sector specific REITs of interest, three of the smaller but significant sectors for investigation include agriculture, child care and petrol filling stations.

3.2.8.1 Agriculture REITs

Agriculture REITs, also known as farmland REITs, are REITs that own a diversified portfolio of agriculture assets that are leased to experienced agricultural operators/farmers mainly on long-term leases. Agriculture REITs offer a steady stream of income backed by rising global demand for food while farmland supply remains relatively fixed.

Agricultural REITs are a relatively recent development. Bulgarian REITs, known as Special Purpose Investment Companies (SPICs), were the first globally to introduce agricultural REITs in early 2000 (Fairbairn, 2014). Unlike commercial REITs, agricultural REITs have a very specific approach to the types and locations of farmland they own. For example, the Australian-based Rural Fund Group (RFF) specializes in the ownership of farmlands operating poultry, tree nut, orchards, vineyards, cotton and cattle assets. In the United States, Gladstone Land Corp. (LAND) focuses on fruit, vegetable and berry operations, Farmland Partners Inc. (FPI) on crops such as corn, wheat, soybeans, rice and cotton and American Farmland Co. (AFCO) mainly on grapes, nuts and specialty crops such as citrus.

Similar to timberland REITs, these REITs are relatively uncorrelated to economic cycles. Even during times of economic recession, basic food is a non-discretionary item as people don't stop eating. However, REIT investors need to understand that earnings can be impacted by food price cycles which are impacted by weather conditions and pest and disease invasions.

3.2.8.2 Child care REITs

Child care REITs are REITs that invest in properties providing early learning services. In Australia, two funds specialize in this sector: Folkestone Education Trust (FET) and Arena REIT. Folkestone is the largest Australian child care REIT with a portfolio of 388 properties valued at approximately A$829 million (ASX, 2017).

Child care REITs focus on providing investors with steady earnings from operators tied to long-term leases that are reviewed annually. The lease structure ensures tenant responsibility for the majority of outgoings, maintenance and capital expenditures. Both REITs have performed well on the back of a booming child care sector and increased government subsidies. In the future, limited land availability, mainly in inner city regions, means supply is likely to be restrained, forcing many new mixed-use developments to include child care centres to meet growing demand, supported by favourable demography and economic trends, which augurs well for the sector.

3.2.8.3 Petrol filling station REITs

A more recent introduction to the REIT industry has been REITs that specialize in petrol filling stations. In Australia, Viva Energy REIT owns a portfolio of 425 service station sites that are leased to Viva Energy, a fuel supplier that is the exclusive licensee for Shell products in Australia.

This sector is gaining traction with APN Property Group launching an A$106 million unlisted fund, comprising 23 Puma Energy service stations, in December 2016 (ASX, 2017). The leases are of a triple-net structure meaning the tenant, not the REIT, is responsible for all operating expenses including taxes, insurance and maintenance. Further, REIT unitholders are also protected under the lease in the event of site contamination which significantly reduces risk.

3.2.9 Summary

A common feature of specialist sector REITs is their tendency to focus on a particular industry or business, placing the structure of the lease agreement between the REIT and the occupier/operator at the centre of the risk/return equation. The REIT's requirement for a steady income stream with some potential for upside from the business conducted therein needs to be balanced against the requirement of the occupier/operator who has to be able to operate the business profitably if it is to be sustainable.

Specialist sector REITs have recently gained greater representation in institutional property portfolios. Newell and Peng (2006b) found that the growth of specialist sector REITs is driven by an increased appetite for property investment by pension funds, acceptance of higher risk levels by many investors (for example, value-added and opportunistic funds) and demographic changes favouring the retirement and health care property sectors. In addition, there is also the mismatch between available funds and available good quality core property assets both nationally and internationally.

While residential, health care, self-storage, timber, hotel and lodging, infrastructure and data centres are now becoming more widely accepted as REIT investment sectors, increasing demand for REITs by investors suggests that these specialist sector REITs may be expected to continue to grow. Further, with agricultural, child care and petrol filling stations emerging as assets around which a REIT may be formed, the range of property sectors suitable for investment by a REIT would appear to be limited only by the ability to structure a sustainable lease agreement between the REIT and the occupier/operator that meets the risk/return requirements of each.

3.3 Case study: US sector-specific REITs

The previous section investigated such sectors as residential, health care, self-storage, timber, infrastructure and data centres generally; this section now considers each specifically through a case study of US REITs.

The US market is one of the deepest and most mature REIT markets in the world and so, not surprisingly, is the home to a wide range of sector-specific REITs. A key motivating factor for fund managers seeking exposure to sector-specific REITs has been the need for new product diversity. Figure 3.2 provides an example of this diversity within US REITs market using data compiled by the National Association of Real Estate Investment Trusts (NAREIT).

As illustrated, investors have a choice of allocating capital to specific REIT sub-sectors or a combination of all sectors to achieve portfolio diversity. However, effective asset selection requires careful consideration of the basic characteristics and performance attributes of the various REIT sub-sectors, even by the most advanced investors.

The US, the most developed REIT market in the world, accounts for 65% of global REIT investment (EPRA, 2016a) and has a very broad sector specific investment universe. An entity will be qualified as a REIT in the US (NAREIT, 2016b), if it:

- invests at least 75% of its total assets in real estate;
- derives at least 75% of its gross income from rents from real property, interest on mortgages financing real property or from sales of real estate;

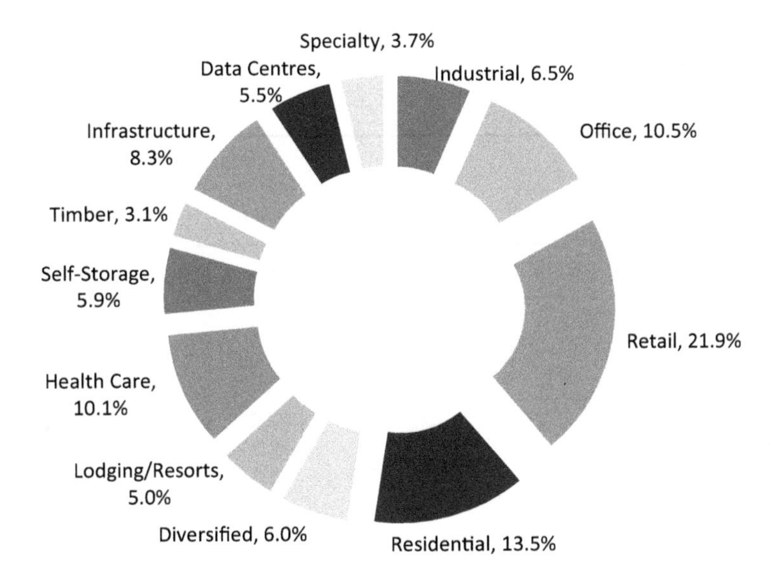

Figure 3.2 US Sub-Sector REITs by Market Cap: 2016

Total Market Cap (Equity REITs only): USD 914.80B

As of 30 November 2016

Source: Author's Compilation From NAREIT (2016a)

- pays at least 90% of its taxable income in the form of shareholder dividends each year;
- is an entity that is taxable as a corporation;
- is managed by a board of directors or trustees;
- has a minimum of 100 shareholders; and
- has no more than 50% of its shares held by five or fewer individuals.

To be classified as a sector-specific REIT, an entity must invest 75% or more of its gross assets in a particular sector, otherwise the entity will be classified as diversified (FTSE, 2016a).

Table 3.2 illustrates the US REIT sector distribution and their total returns. Almost all sectors made strong returns in 2014, ranging from 8.6% (timber REITs) to 69.7% (infrastructure REITs), but 2016 saw different outcomes between the sectors. For example, the industrial (24.98%), data centres (18.41%), lodging/resort (14.44%) and timber (10.09%) sectors achieved positive returns in 2016, but self-storage, residential and the retail sectors recorded -15.22%, -2.33% and -0.20% returns, respectively, in 2016. Also, infrastructure and health care REITs achieved 69.7% and 33.3% returns, respectively, in 2014 but only had 5.39% and 1.53% returns, respectively, in 2016, indicating the significant fluctuations in the US REIT market over the last two years and being an obvious indicator that a sector specific strategy could provide increased fund diversification opportunities.

Figure 3.3 shows the number of REITs in the US and its market capitalization over 1971–2016. There were 226 REITs in 1994 and the market cap was only US$44.3 billion at that time. The number of REITs in the US declined to 136 in 2008 during the global financial crisis but recovered to 216 in 2014, which is very close to the previous peak (226) in 1994. Conversely, the market cap of US REITs has grown significantly over the last two decades with the market cap in 2015 becoming US$973.9 billion, including mortgage REITs, being a 2,098% increase from the 1994 figure. Notably, after the global financial crisis in 2008, the market cap has grown more than fourfold (408%) and also more than twofold from the pre-global financial crisis in 2006 when the market cap was US$438 billion.

Table 3.2 US REITs Return and Market Cap by Sector: 2014–2016

	No of REITs Nov 2016	*Total Return (%)*		*Market Cap (USD B) Nov 2016*
		2014	*2016*	
FTSE NAREIT All Equity REITs	167	28	3.99	914.80
Industrial	11	21	24.98	59.24
Office	25	25.9	9.04	95.78
Retail	31	27.6	-0.20	200.70
Residential	21	40	-2.33	123.82
Diversified	15	27.2	4.04	54.54
Lodging/Resorts	17	32.5	14.44	45.80
Health Care	19	33.3	1.53	92.72
Self -Storage	5	31.4	-15.22	53.80
Timber	4	8.6	10.09	28.41
Infrastructure	5	69.7	5.39	76.36
Data Centres	6	-	18.41	50.04
Specialty	8	-	16.69	33.59
Total	**167**			**914.80**

Source: Author's Compilation From NAREIT (2015, 2016a)

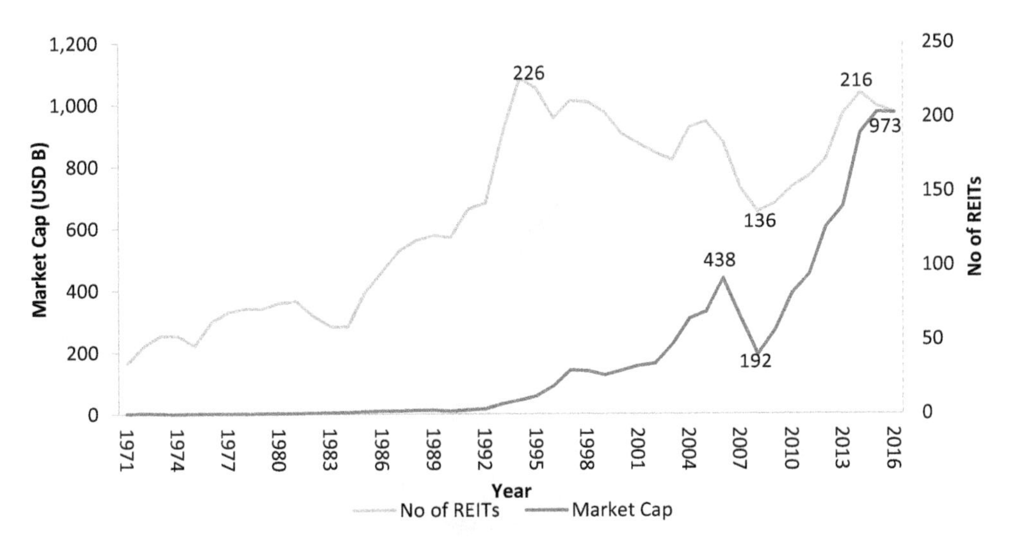

Figure 3.3 Number of US REITs and Market Cap: 1971–2014

Source: Author's Compilation From NAREIT (2016c)

Figures 3.2 and 3.4 illustrate sub-sector REIT markets in the US. As of November 2016, the retail sector (18%, 31 REITs) had the largest market share in the US REITs market, based on the number of REITs, followed by office (15%, 25 REITs) and residential (13%, 21 REITs), but it is also notable that non-traditional sectors, such as health care (11%, 19 REITs), specialty (5%, 8 REITs), data centres (4%, 6 REITs) and self-storage (3%, 5 REITs) also had good market shares and number of REITs. In terms of market cap, the retail sector (21.9%, US$200.7 billion) had the largest market share, followed by residential (13.5%, US$123.8 billion) and office (10.5%, US$95.8 billion).

Interestingly, non-traditional sectors, such as health care (10.1%, US$92.7 billion) and infrastructure (8.3%, US$76.4 billion) are larger than the industrial sector (6.5%, US$59.2 billion), self-storage (5.9%, US$53.8 billion), data centres (5.5%, US$50.0 billion) and lodging/resort (5.0%, US$45.8 billion). Timber and other specialty REIT sectors also have more than 3% market share while diversified REITs only have 6% market share. Accordingly, non-traditional sector REITs, such as infrastructure, timber, self-storage and data centres, which can produce regular incomes from the underlying assets, may be suitable investment vehicles for various investment strategies. This may also provide a useful guideline for many developing countries where REIT markets are evolving and are currently more focused on diversified REITs.

Table 3.3 and Figure 3.5 depict the development of sub-sector REITs market cap in the US over 1999–2016. Over the period from 1999 to 2016, the size of the REIT market in the US increased by 542%, in terms of market capitalization, from US$142 billion to US$915 billion. In the same period, other specialty REITs, health care REITs and self-storage REITs increased by 1,795%, 1,367% and 969% respectively and recorded the highest growth across the period. However, during the global financial crisis, health care REITs, self-storage REITs and other specialty REITs dropped 13%, 31% and 40%, respectively, being the smallest falls compared to other sectors.

The period shows that non-conventional sub-sectors have been growing significantly and were less affected by the global financial crisis because the underlying assets for those

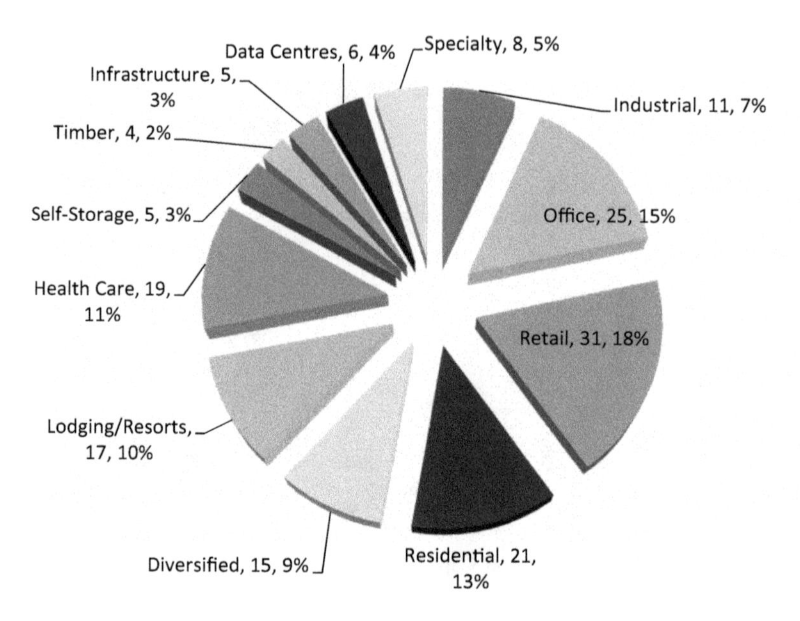

Figure 3.4 US Sub-Sector REITs by Number of REITs: 2016

Source: Author's Compilation From NAREIT (2016a)

Table 3.3 Sub-Sector REITs in the US: 1999–2016

Sector	*1999*	*2006*	*2008*	*2009*	*2012*	*2016*	*1999–2016*	*2006–2008*
FTSE NAREIT All Equity	(USD Billion)						542%	– 63%
REITs	142	439	164	242	539	915		
Industrial/Office	42	117	31	54	90	155	269%	– 73%
Retail	30	119	39	61	150	201	566%	– 68%
Residential	28	76	28	35	77	124	337%	– 63%
Diversified	12	34	13	16	37	55	346%	– 63%
Lodging/Resorts	8	30	7	13	27	46	439%	– 77%
Self-Storage	5	20	14	16	31	54	**969%**	**– 31%**
Health Care	6	23	20	32	68	93	**1,367%**	**– 13%**
Other Specialty*	10	21	13	15	59	188	**1,795%**	**– 40%**

* Including infrastructure, timberland, data centers

Source: Author's Compilation From NAREIT (1999–2016)

non-conventional REITs sectors were less correlated with other investment assets and had their own risk-return characteristics, being attractive for property fund managers and investors seeking diversified investment strategies.

With the previous section investigating such sectors as residential, health care, self-storage, timber, infrastructure and data centres generally, this section will now consider each specifically through a case study of US REITs:

3.3.1 Residential REITs in the US
3.3.2 Health care REITs in the US

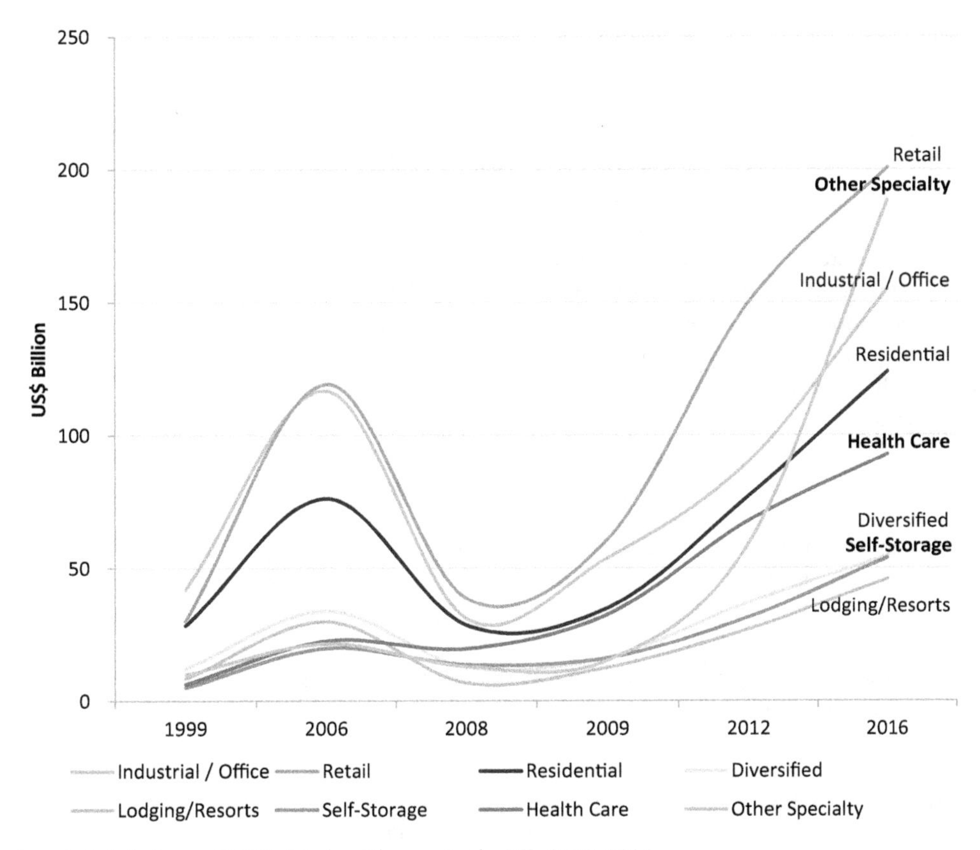

Figure 3.5 Sub-Sector REITs Market Changes in the US: 1999–2016

Source: Author's Compilation From NAREIT (1999–2016)

3.3.3 Self-storage REITs in the US
3.3.4 Timber REITs in the US
3.3.5 Infrastructure REITs in the US
3.3.6 Data centre REITs in the US
3.3.7 Other specialty REITs in the US

3.3.1 Residential REITs in the US

In addition to the traditional apartment market, US residential REIT investors now also have access to funds that specialize in mobile "manufactured" home parks, single-family homes and student housing.

3.3.1.1 Mobile 'manufactured' home parks

There are an estimated 200,000 manufactured home communities currently in North America, with the number growing (Equity Lifestyle Communities, 2017). A manufactured home is constructed on-site using traditional building techniques which meet local council and other planning requirements. Manufactured homes are different from mobile homes as they never move from the site. Similar to site-build homes, manufactured homes are available with unique floor plans, designs, fit-out and amenities. Residents enjoy amenities similar to single-family homes with an attractive main entrance, parking, pool, entertainment/play centres, gyms, tennis court and laundry facilities.

Manufactured home parks provide owners with a stable cash flow at modest capital expense compared to the requirements of financing apartments and traditional single-family homes. In the United States, Equity LifeStyle Properties, Inc., Sun Communities, Inc. and UMH Properties, Inc. operate in this sector-specific market.

The future of the manufactured home sector appears positive, mainly because supply has remained modest in the face of high demand as cities experience housing affordability issues. Increasing house prices have seen many individuals priced out of the traditional home markets. The steady rental return and low cost of borrowing has enticed investors to the manufactured home sector.

3.3.1.2 Single-family homes

Single-family home REITs acquire, renovate and lease houses to individuals and families. The REIT manages the properties, deriving weekly/monthly rental income and long-term capital appreciation.

Single-family home REITs were a product of the weak housing market during the 2007/08 global financial crisis period, when investors swooped on the huge pool of foreclosed homes, picking up properties at bargain prices and renting them to families that desired a traditional house without the burden of ownership and mortgage repayments.

There are three single-family home REITs listed in United States comprising American Homes 4 Rent with 48,000 homes, Colony Starwood Homes with 31,100 homes and Silver Bay Realty Trust Invitation Homes with 50,000 homes. These REITs manage a wide variety of houses in various cities and states which presents management and operational challenges. Each house has a different layout, fit-outs, appliances, water and energy services. This is one of the key reasons why single-family home REITs mainly focus only on areas where they already invest when purchasing new properties.

The single-family home REIT is a fast-growing sector, as the number of people renting houses, particularly in the younger demographic, has increased significantly in the face of rising house prices. However, it remains a relatively untested market and there are concerns about its long-term viability if the economic recovery results in greater home ownership in the future.

3.3.1.3 Student housing

Student housing is another specialized residential REIT sector where REIT managers offer to lease residential space to university/college students on a periodic basis. Usually student housing community spaces are leased at $300 to $700 per bed, with capacity ranging from several hundred to 1,000 students (Block 2011).

In the US, American Campus Communities, Education Realty Trust and Campus Crest Communities specialize in these types of properties. Generally, these REITs operate on the premise of satisfying the need for student housing for universities lacking the capability and resources to develop and manage such large housing facilities.

The risk of operating such facilities mainly arises from the seasonal nature of the university business where enrolments can vary within different teaching semesters. The turnover is normally high in student housing and periods of poor leasing can negatively impact revenue and profits. As such, property management can be quite intensive and require unique skills to manage the higher operating expenses. The upside is that there can be as many as three students occupying a single room at any point, generating higher income than that of an average apartment unit. For optimal success, REIT manager preferences would be leasing space annually on campus or within walking proximity of campus. As with other property, location is a key success factor for student housing.

3.3.2 Health care REITs in the US

Health care REITs invest in and manage a variety of health-related properties, such as skilled nursing facilities, (on-campus) medical office buildings, hospitals, acute care and rehabilitation hospitals, purpose-built health care facilities and senior housing communities.

As of November 2016, health care REITs accounted for 10.14% (US$92.7 billion) of the FTSE NAREIT All Equity REITs index and recorded 17.42% average annual returns for the last three years (2014–2016). There are 23 health care REITs in the US, including 18 publically listed on a stock exchange, four public non-listed REITs and one private REIT, with more details presented in Table 3.4.

3.3.3 Self-storage REITs in the US

Self-storage REITs invest in and manage warehouse and self-storage facilities, focusing on the ownership, operation, acquisition, development and redevelopment of such facilities. As of November 2016, self-storage REITs accounted for 5.88% (US$53.8 billion) of the FTSE NAREIT All Equity REITs index and recorded 18.94% average annual returns for the last three years (2014–2016).

There are six self-storage REITs in the US, including five listed on NYSE and one listed on NASDAQ, with more details presented in Table 3.5.

3.3.4 Timber REITs in the US

Timber REITs invest in and manage a variety of timber-related properties. Timber REITs grow trees, sell timber and manufacture solid wood products, paying dividends derived from harvesting portfolios of timberlands. As of November 2016, timber REITs accounted for 3.11% (US$28.4 billion) of the FTSE NAREIT All Equity REITs index and recorded 9.35% average annual return for the last three years (2014–2016).

There are four timber REITs in the US, including three listed on NYSE and one listed on NASDAQ, with more details presented in Table 3.6.

3.3.5 Infrastructure REITs in the US

Infrastructure REITs invest in and manage public and private infrastructure properties, such as communication towers, distributed antenna systems, pipelines, storage terminals, outdoor advertising facilities and renewable power generators. As of November 2016, infrastructure REITs accounted for 8.35% (US$76.4 billion) of the FTSE NAREIT All Equity REITs index and recorded 26.28% average annual return for the last three years (2014–2016).

There are six infrastructure REITs in the US, including four listed on NYSE and two listed on NASDAQ, with more details presented in Table 3.7.

Table 3.4 Health Care REITs in the US

No	*Company name (REIT)*	*Main underlying assets*	*Listing status*	*Stock Exchange*
1	Care Capital Properties, Inc.	Skilled nursing facilities	P-L	NYSE MK
2	CareTrust REIT, Inc.	Skilled nursing facilities	P-L	NASDAQ
3	Community Health Care Trust	Medical office buildings	P-L	NYSE MK
4	Global Medical REIT	Purpose-built health care facilities	P-L	NYSE MK
5	HCP, Inc.	Senior housing	P-L	NYSE
6	Health care Realty Trust	On-campus medical buildings	P-L	NYSE
7	Health care Trust of America, Inc.	Medical office buildings	P-L	NYSE
8	LTC Properties, Inc.	Seniors housing	P-L	NYSE
9	Medical Properties Trust Inc.	Acute care and rehabilitation hospitals	P-L	NYSE
10	National Health Investors, Inc.	Senior housing	P-L	NYSE
11	New Senior Investment Group	Senior housing	P-L	NYSE MK
12	NorthWest Health Care Properties REIT	Medical office buildings	P-L	TSX
13	Omega Health Care Investors, Inc.	Skilled nursing facilities	P-L	NYSE
14	Physicians Realty Trust	On-campus medical buildings	P-L	NYSE
15	Sabra Health Care REIT, Inc.	Skilled Nursing and Transitional care facilities	P-L	NASDAQ
16	Senior Housing Properties Trust	Senior living communities	P-L	NYSE
17	Ventas, Inc.	Senior housing communities	P-L	NYSE
18	Welltower Inc.	Senior housing	P-L	NYSE
19	CNL Health Care Properties Inc.	Senior housing	P-NL	N/A
20	Griffin-American Health Care REIT III	Medical office buildings	P-NL	N/A
21	Health care Trust, Inc.	medical office buildings	P-NL	N/A
22	Northstar Health Care Income, Inc.	Senior housing	P-NL	N/A
23	MedEquities Realty Trust, Inc.	Acute and post-acute facilities	PR	N/A

- P-L: Public Listed/P-NL: Public Non-Listed/PR: Private

Source: Author's Compilation From NAREIT and Websites of the REITs

Table 3.5 Self-Storage REITs in the US

No	*Company name (REIT)*	*Main underlying assets*	*Listing status*	*Stock Exchange*
1	CubeSmart	Self-storage facilities	P-L	NYSE
2	Extra Space Storage, Inc.	Self-storage facilities	P-L	NYSE
3	Global Self Storage, Inc.	Self-storage facilities	P-L	NASDAQ
4	Life Storage, Inc.	Self-storage facilities	P-L	NYSE
5	National Storage Affiliates	Self-storage facilities	P-L	NYSE
6	Public Storage	Self-storage facilities	P-L	NYSE

- P-L: Public Listed

Source: Author's Compilation From NAREIT and Websites of the REITs

Table 3.6 Timber REITs in the US

No	Company name (REIT)	Main underlying assets	Listing status	Stock Exchange
1	CatchMark Timber Trust, Inc.	Timberlands	P-L	NYSE
2	Potlatch Corporation	Timberlands	P-L	NASDAQ
3	Rayonier Inc.	Timberlands	P-L	NYSE
4	Weyerhaeuser	Timberlands	P-L	NYSE

- P-L: Public Listed

Source: Author's Compilation From NAREIT and Websites of the REITs

Table 3.7 Infrastructure REITs in the US

No	Company name	Main underlying asset	Listing status	Stock Exchange
1	American Tower Corporation	Wireless infrastructure solutions, towers, distributed antenna System (DAS) networks, backup power systems	P-L	NYSE
2	Communications Sales & Leasing, Inc.	Wireless infrastructure solutions, fibre strand miles, towers	P-L	NASDAQ
3	CorEnergy Infrastructure Trust	Pipelines, storage terminals, and transmission and distribution assets	P-L	NYSE
4	Crown Castle International Corp.	Shared wireless infrastructure, towers	P-L	NYSE
5	InfraREIT, Inc.	Electric transmission and distribution utility assets	P-L	NYSE
6	Landmark Infrastructure Partners LP	Wireless communication, outdoor advertising facilities, renewable power generations	P-L	NASDAQ

- P-L: Public Listed

Source: Author's Compilation From NAREIT and Websites of the REITs

3.3.6 Data centre REITs in the US

Data centre REITs invest in and manage information technology (IT)–related facilities that safely store data. Data centre REITs develop unique systems and solutions to power, cool and protect servers and IT assets, collecting rent from the users. As of November 2016, data centre REITs accounted for 5.47% (US$50.0 billion) of the FTSE NAREIT All Equity REITs index and recorded 9.97% average annual return for the last two years (2015–2016). (Data centre REITs only have return data from 2015.)

There are six data centre REITs in the US, including four listed on NYSE, two listed on NASDAQ and one public non-listed REIT, with more details presented in Table 3.8.

3.3.7 Other specialty REITs in the US

Other specialty REITs invest in and manage a mix of different property types which generate income, including prisons, jails, movie theatres, farms, casinos, outdoor adverting facilities and education properties. As of November 2016, such specialty REITs accounted for 3.67% (US$33.6 billion) of the FTSE NAREIT All Equity REITs index and recorded 9.19% average annual return for the last two years (2015–2016). (Specialty REITS only have data from 2015.)

Table 3.8 Data Centre REITs in the US

No	Company name	Main underlying asset	Listing status	Stock Exchange
1	Carter Validus Mission Critical REIT II, Inc.	Vital data centres and health care properties	P-NL	AMEX
2	CoreSite Realty Corporation	Data centres	P-L	NYSE
3	CyrusOne Inc.	Data centres	P-L	NASDAQ
4	Digital Realty	Data centres, cloud service platforms	P-L	NYSE
5	DuPont Fabros Technology, Inc.	Wholesale data centres	P-L	NYSE
6	Equinix, Inc.	Data centres, data connections	P-L	NASDAQ
7	QTS Realty Trust, Inc.	Data centres, cloud service platforms	P-L	NYSE

- P-L: Public Listed/P-NL: Public Non-Listed

Source: Author's Compilation From NAREIT and Websites of the REITs

Table 3.9 Specialty REITs in the US

No	Company name (REIT)	Main underlying assets	Listing status	Stock Exchange
1	CoreCivic	Prisons, jails and detention facilities	P-L	NYSE
2	EPR Properties	Entertainment, recreation and education properties	P-L	NYSE
3	Farmland Partners Inc.	Farmlands	P-L	NYSE
4	Gaming and Leisure Properties, Inc.	Gaming (casino) and related facilities	P-L	NASDAQ
5	Gladstone Land Corporation	Farms and farm-related properties	P-L	NASDAQ
6	Iron Mountain	Records and information storage facility	P-L	NYSE
7	Lamar Advertising Company	Outdoor advertising display facilities	P-L	NASDAQ
8	OUTFRONT Media Inc.	Outdoor digital and static displays	P-L	NYSE
9	The GEO Group	Rehabilitation facilities	P-L	NYSE

- P-L: Public Listed

Source: Author's compilation from NAREITs and websites of the REITs

There are nine specialty REITs in the US, including six listed on NYSE and three listed on NASDAQ, with more details presented in Table 3.9.

3.3.8 Summary

The US case study of sector specific REITs shows that, as with the traditional commercial sector broadening to include CBD and suburban offices, the sector specific REITs may also broaden within the specific sector, such as residential sector–specific REITs broadening to include manufactured homes, single-family homes and student housing.

As noted previously, demand for REITs by investors suggests that sector-specific REITs may be expected to continue to grow with US prisons, movie theatres and casinos now available as sector-specific REIT investments reinforcing that the extent of the range of suitable assets for REIT investment may only be limited by the ability to structure a sustainable lease agreement between the REIT and the occupier/operator that meets the risk/return requirements of each.

Given the size of the US REIT market, as time passes, greater amounts of return data will become available for sector-specific REIT performance, allowing a much clearer understanding of the risk/return profile of the different sectors, both relative to each other and to the traditional commercial, retail and industrial sectors.

3.4 Summary

Chapter 1 sought to identify critical contextual issues in international REITs, including the defining characteristics of REITs, the global evolution of REITs and the risk-return characteristics of real estate investment through REITs relative to non-REIT real estate investment with Chapter 2, the *Post-Modern REIT Era*, examining the evolution of the global REIT market, what is deemed to be best current market practice and how the market is expected to change going forward.

This chapter, *Emerging Sector REITs*, investigated such sectors as residential, health care, self-storage, timber, infrastructure and data centres generally and then specifically through a case study of US REITs.

A common feature of sector-specific REITs is their tendency to focus on a particular industry or business, placing the structure of the lease agreement between the REIT and the occupier/operator at the centre of the risk/return equation. The REIT's requirement for a steady income stream with some potential for upside from the business conducted therein needs to be balanced against the requirement of the occupier/operator who needs to be able to operate the business profitably if it is to be sustainable.

With sector-specific REITs having gained greater representation in institutional property portfolios due to an increased appetite for property investment by pension funds, acceptance of higher risk levels by many investors and demographic changes favouring the retirement and health care property sectors, demand for REITs by investors suggests that sector-specific REITs may be expected to continue to grow. This may be evidenced by the acceptability of US prisons, movie theatres and casinos as sector-specific REIT investments, reinforcing that the extent of the range of suitable assets for REIT investment may be limited only by the ability to structure a sustainable lease agreement between the REIT and the occupier/operator that meets the risk/return requirements of each.

Part I continues with chapters each reviewing a current theme of REIT evolution through the lens of contemporary research. Chapter 4, *Sustainable REITs*, analyses REIT environmental performance and the cost of equity with Chapter 5, *Islamic REITs*, examining the evolution of global Islamic REITs through a study of the Malaysian market. Chapter 6, *Behavioural Risk in REITs*, addresses the management of behavioural risk in global REITs and Part I concludes with Chapter 7 which reviews recent research into *REIT Asset Allocation*.

Part II then includes six chapters each analysing REITs in a region of the world, using case studies from a developed, developing and emerging REIT sector in the region, concluding with Chapter 14 which considers *Directions for the Future of International REITs*.

References

Australian Securities Exchange 2017, *Managed funds: Australian real estate investment trusts (A-REITs)*, ASX. Available at www.asx.com.au/products/managed-funds/areits.htm, 22 February 2017.

Block, R.L. 2011, *Investing in REITs*, John Wiley & Sons, Incorporated, Hoboken. Available at ProQuest Ebook Central, 22 February 2017.

CBRE 2015, *Global capital markets: Will new sources of capital extend the cycle?*, CBRE, Los Angeles.

EPRA 2016a, *Global REIT survey 2016*, European Public Real Estate Association, Brussels.

Equity Lifestyle Communities 2017, *Our MH communities portfolio*, Equity LifeStyle Properties (ELS). Available at https://equitylifestyleproperties.com/, 22 February 2017.

European Public Real Estate Association (EPRA) 2016b, *Global REIT Survey 2015: USA-US-REIT*, EPRA, Brussels.

Fairbairn, M. 2014, 'Like gold with yield: Evolving intersections between farmland and finance', *The Journal of Peasant Studies*, Vol. 41, No. 5, pp. 777–795.

FTSE 2016a, *Factsheet FTSE EPRA/NAREIT developed sector indices*, FTSE Group, London.

FTSE 2016b, *Factsheet FTSE EPRA/NAREIT emerging indices*, FTSE Group, London.

Hood, H., Harris, T., Siry, J., Baldwin, S. and Smith, J. 2015, *US timberland markets: Transactions: values and market research 2000 to mid-2015*, Timber Mart-South, Athens, GA.

Keenan, C. 2016, 'Infrastructure REITs are powering the growth of cloud computing', *REIT Magazine*, March/April 2016. Available at www.reit.com, 22 February 2017.

NAREIT 1999–2016, *NAREIT REIT Watch 1999–2016*, NAREIT, Washington.

NAREIT 2015, *NAREIT REIT Watch January 2015*, NAREIT, Washington.

NAREIT 2016a, *NAREIT REIT Watch December 2016*, NAREIT, Washington.

NAREIT 2016b, *What is a REIT?*, NAREIT, Washington. www.reit.com/investing/reit-basics/what-reit, 20 December 2016.

NAREIT 2016c, *FTSE Nareit Real Estate Index Historical Market Capitalization (1972–2015)*, NAREIT, Washington, www.reit.com/data-research/reit-market-data/us-reit-industry-equity-market-cap, 20 December 2016.

National Association of Real Estate Investment Trusts (NAREIT) 2017, *REIT sectors*, REIT.com. Available at www.reit.com, 22 February 2017.

Newell, G. and Peng, H.W. 2006b, 'The significance of emerging property sectors in property portfolios', *Pacific Rim Property Research Journal*, Vol. 12, No. 2, pp. 177–197.

Further reading

Benefield, J.D., Anderson, R.I. and Zumpano, L.V. 2009, 'Performance differences in property-type diversified versus specialized real estate investment trusts (REITs)', *Review of Financial Economics*, Vol. 18, No. 2, pp. 70–79.

Jackson, L.A. 2009, 'Lodging REIT performance and comparison with other equity REIT returns', *International Journal of Hospitality & Tourism Administration*, Vol. 10, No. 4, pp. 96–325.

Jayaraman, B. 2012, *Building wealth through REITs*, Marshall Cavendish, Singapore. Available at ProQuest Ebook Central, 22 February 2017.

Korhonen, J., Zhang, Y. and Toppinen, A. 2016, 'Examining timberland ownership and control strategies in the global forest sector', *Forest Policy and Economics*, Vol. 70, No. C, pp. 39–46.

Newell, G. and Fischer, F. 2009, 'The role of residential REITs in REIT portfolios', *Journal of Real Estate Portfolio Management*, Vol. 15, No. 2, pp. 129–140.

Newell, G. and Peng, H.W. 2006a, 'The role of non-traditional real estate sectors in REIT portfolios', *Journal of Real Estate Portfolio Management*, Vol. 12, No. 2, pp. 155–166.

Ngee, H.S.T.F.S. 2016, *Singapore's real estate: 50 years of transformation*, World Scientific Publishing Company, Singapore. Available at ProQuest Ebook Central, 22 February 2017.

Peterson, P. and Kuethe, T. 2015, 'Understanding farmland REITs', *farmdoc daily*, Vol. 5, p. 200, Department of Agricultural and Consumer Economics, University of Illinois, Urbana-Champaign, 28 October 2015.

Piao, X., Mei, B. and Xue, Y. 2016, 'Comparing the financial performance of timber REITs and other REITs', *Forest Policy and Economics*, Vol. 72, No. C, pp. 115–121.

Pruitt, A.D. 2014, 'Single-family rentals: Home sweet home?', *REIT Magazine*, March/April 2014 issue. Available at www.reit.com, 22 February 2017.

Stoller, B. 2016, *What data center investors should know*, DataCenter Knowledge, West Chester, OH. Available at www.datacenterknowledge.com, 22 February 2017.

Willis, G. 2003, *The smartmoney guide to real estate investing*, John Wiley & Sons, Hoboken. Available at EBSCOhost, 22 February 2017.

4

SUSTAINABLE REITS

REIT environmental performance and the cost of equity

Piet Eichholtz, Peter Barron and Erkan Yönder

4.1 Introduction

This book aims to identify key areas for research in the REIT discipline for the next five to ten years by surveying the current state of the REIT discipline around the world and identifying emerging and cutting edge research areas through a thematic review of current contextual issues and a regional analysis based on case studies.

This book comprises two parts, the first part being a thematic review of emerging and cutting edge global research into current contextual issues in REITs internationally and the second part being a regional analysis of REITs around the world, each written by authoritative academic authors from the world's leading Universities and REIT industry experts.

Part I includes six chapters each reviewing a current theme of REIT evolution through the lens of contemporary research. Chapter 1 focused on critical contextual issues in international REITs while Chapter 2, the *Post-Modern REIT Era,* examined the evolution of the global REIT market, what is deemed to be best current market practice and how the market is expected to change going forward, with Chapter 3, *Emerging Sector REITs*, investigating such sectors as timber and data centres.This chapter, *Sustainable REITs*, analyses REIT environmental performance and the cost of equity by investigating the relationship between sustainability and the cost of equity capital for 211 listed US equity REITs. Sustainability is measured using the two leading energy labelling programs, LEED and Energy Star, with the cost of equity assessed using four individual implied cost of equity models based on earnings per share forecasts, from which an average is taken with cost of equity estimates then regressed on the sustainability measures and relevant control variables.

Chapter 5, *Islamic REITs*, then examines the evolution of global Islamic REITs through a study of the Malaysian market. Chapter 6, *Behavioural Risk in REITs*, addresses the management of behavioural risk in global REITs and Part I concludes with Chapter 7 which reviews recent research into *REIT Asset Allocation.*

Part II includes six chapters each analysing REITs in a region of the world, using case studies from a developed, developing and emerging REIT sector in the region:

Chapter	Region	Developed **REIT Sector**	Developing **REIT Sector**	Emerging **REIT Sector**
8	North America	USA	Canada	Mexico
9	Latin America	Brazil	Argentina	Uruguay
10	Europe	UK	Spain	Poland
11	South East Asia	Singapore	Malaysia	Thailand
12	North Asia	Japan	Hong Kong	China
13	Oceania	Australia	South Africa	India

concluding with Chapter 14 which considers *Directions for the Future of International REITs.*

4.2 Background

Properties rated for energy efficiency command higher rents and asset values than otherwise equal properties without such a rating and have a better and more stable occupancy rate (Eichholtz, Kok and Quigley, 2010; Fuerst and McAllister, 2011; Wiley, Benefield and Johnson, 2010). A large part of the rent effect is driven by cost reductions. Eichholtz, Kok and Quigley (2013) show that a one-dollar decrease in energy costs is associated with a 95 cents higher rent and that the sustainability premium is robust to the US real estate crisis of 2008/09.

If the present value of these occupancy and rent premiums exceed that of the initial cost premium of sustainable real estate projects, then it can be said that sustainable properties, *ceteris paribus*, offer greater returns. While not much is known about these cost premiums, the market at least perceives significant costs associated with "going green". A survey by Building Design and Construction (2007) finds that 41% of respondents estimated that cost premiums were 11% or higher. (For more details, please see "LEED for new construction" by US Green Building Council (2007).) The World Economic Forum (2016) comes to the consensus that the perception of cost premiums lies in the range of 13% to 18%.

Most research on value and cost premiums are at the building level. While these results are interesting, it should be remembered that institutional investors do not invest primarily in individual buildings but in private and/or listed real estate funds and companies which manage portfolios of properties. These companies are the focus of this study.

Eichholtz, Kok and Yönder (2012) find strong evidence that REITs owning higher proportions of green property have higher operating performance – better return on equity and assets, higher cash flow – and a lower market beta, due to a higher and more stable occupancy at the asset level. Interestingly, that financial performance is the net result of costs and benefits, so at least, for REITs, the benefits of sustainability investments seem to outweigh the costs. Crucially, the authors do not report a statistically significant difference in the alpha of REITs with a higher proportion of environmentally certified real estate. In other words, energy efficiency seems to be priced in the stock market already, even during a period of market downturn.

But while the link between real estate sustainability and economic performance has been convincingly shown at both the asset and firm level, not much is known about the specific link between real estate sustainability and the cost of the capital to finance it. Eichholtz et al. (2017) make a first attempt to investigate the cost of debt capital for sustainable REITs, finding that an increase in the proportion of sustainable property in a REIT is associated with lower corporate bond spreads. For mortgages, more sustainable collateral leads to lower spreads as well. The link with the cost of equity capital remains uncharted territory and the aim of this study is to change that.

Buildings use large amounts of energy with heating, lighting and cooling systems having a huge influence on energy consumption and greenhouse gas emissions, representing almost 50% of US commercial building energy use (Kelso, 2012) In total, commercial buildings accounted for 18.4% of energy consumption in the US in 2012, with less than 1% of this coming from renewables. (For more details, see www.eia.gov.) Understanding how environmentally efficient solutions can tackle carbon emissions whilst having a net positive impact on investment performance is crucial for investors, regulators and policymakers. A lower cost of equity capital for sustainable REITs would be a strong incentive for corporate action on this matter.

This issue has already received some research attention outside of real estate. Recently, Dhaliwal et al. (2011) find that firms with superior corporate social responsibility (CSR) performance see a reduction in their cost of equity capital in the following year. Similarly, El Ghoul et al. (2011) show that firms with higher CSR scores exhibit a lower cost of equity capital. For real estate, this relationship is unclear.

In this study, analyst forecasts of REIT financial performance measures are obtained in order to build implied cost of equity estimates for analysis and the link to portfolio sustainability is investigated. A statistically significant relationship is found between the sustainability of REITs and the cost of equity at the firm level. Specifically, switching to a portfolio of assets which are all environmentally certified is found to be associated with a 38bps reduction in the cost of equity capital compared to an identical yet environmentally uncertified portfolio of properties. Contributing to the literature surrounding sustainability and financial performance, this study demonstrates how sustainable property features are "priced in" by the equity capital market. In other words, investors are taking into account the value of environmental performance.

The following section discusses the literature concerning the relationship between sustainability and financial performance and then the specific link between sustainability and the cost of capital and how the cost of equity capital can be estimated. The section thereafter discusses the dataset, with the following section detailing the methodology and the section thereafter presenting the results. The study ends with a discussion of the results and their implications.

4.3 Literature

There is increasing evidence that investors are aware of the advantages that sustainable property features offer for tenants, so they price in those features. Wiley, Benefield and Johnson (2010) control for differences in markets and find that Energy Star labelled properties achieve premium rents of 7.3% to 8.9% relative to otherwise identical properties. Pivo and Fisher (2010) show that Energy Star certification is associated with rental and sale price premiums of 5.2% and 8.5%, respectively. Eichholtz, Kok and Quigley (2013) find that "green" premiums are robust to the real estate crisis of 2008/09 and they show that more sustainability implies higher premiums. They find a positive yet nonlinear relationship between the rental increment and the LEED score of a property.

There is also a growing consensus around stranded assets and "brown discounts" where buildings that are not green sell for less or not at all. As green buildings increase their diffusion throughout the market, buildings will compete on sustainability metrics and energy efficiency thus making it harder to market non-green buildings. This may deter the negative externalities that unsustainable buildings are producing. However, the realization that sustainable property investment may simply be good business is not yet uniform. Marchettini et al. (2014) find that only 26% of Asian property companies implement sustainability elements.

While there is an academic consensus that sustainability in real estate is associated with higher market values and better economic performance, not much is known about whether that translates into better financing terms. For example, the lower (occupancy) risk of sustainable buildings would logically imply a lower cost of funding, but this has hardly been investigated empirically for real estate.

For corporations, an early paper investigating the topic by Healy and Palepu (2001) shows that certain financial performance measures can inform investors about the potential sustainability of future cash flows and therefore lead to a lower cost of capital at the firm level. Bassen, Holz and Schlange (2006) find that investors are becoming more sensitive to sustainability issues from a risk perspective. Investors expect considerably higher returns from stocks of firms that are significant emitters of toxic chemicals, firms with hazardous waste concerns and those with climate change concerns. Shareholders worry about the revoking of the operating licenses if the business fails to adhere to environmental regulations. Sharfman and Fernando (2008) find that US firms with better environmental risk management have a lower cost of capital Using the CAPM, they decompose the relationship to find lower systematic risk and less volatile financial performance, which the market rewards with lower equity financing costs.

El Ghoul et al. (2011) examine the relationship between sustainability and the cost of equity by using CSR scores as a proxy for sustainable business practices and implied cost of equity models. The negative association they observe between CSR and the cost of equity financing is due to an increased investor base (risk sharing) and lower perceived risk. Notably, improvements in employee relations, environmental policies and product strategies contribute to a lowering of firms' cost of equity.

Dhaliwal et al. (2011) report a negative relationship between the introduction of CSR programs and the cost of equity capital. After beginning these CSR programs, firms obtain more analyst coverage so reducing information asymmetry. Likewise, Li et al. (2013) report that firms in North America and Europe can reduce their cost of equity capital by implementing CSR strategies. Chava (2014) finds that investors demand higher returns from "sin" firms which emit significant toxic chemicals or contribute to climate change in a large way.

The only paper that looks at funding costs for sustainable real estate is Eichholtz et al. (2017), but the authors look at the cost of debt rather than the cost of equity. Using a sample of US REITs, they find a positive relationship between the degree to which REITs invest in sustainable real estate and their credit rating. They calculate the dynamic share of environmentally certified buildings within the REITs using the ratio of certified square footage of property to the total square footage of property in the REIT. In doing so, they find a negative association between the sustainability of the real estate portfolio and the interest rate spread on REIT bonds and mortgages. The economic link between sustainable property features and the cost of the equity to finance it is, as yet, unknown. Herein lies the contribution of this study.

4.4 Method and hypothesis development

4.4.1 Cost of capital

The implied cost of capital approach is used to measure REIT cost of equity. This approach attempts to define the cost of equity capital as an internal rate of return that equates the price of the stock today to the value of the sum of future expected dividends. Four implied cost of capital models are applied, following the literature. Based on each of the four models proposed in the following sub-sections, an average implied cost of capital measure is calculated.

4.4.1.1 Model 1: Gebhardt, Lee and Swaminathan (2001)

Gebhardt, Lee and Swaminathan (2001) use a model which calculates the cost of equity as a function of the return on equity. The model uses a version of a discounted cash flow model called the residual income model, where residual income refers to excess income of a firm after taking into account its cost of capital.

For practical use, a specific forecast period is identified. Analyst forecasts are used to estimate earnings for the first three years. Beyond this year, the ROE reverts to the median industry ROE in the final period by linear interpolation. This study forecasts up to 12 years after which a terminal value is used. The authors document little change in the estimates when the forecasting period is varied between 6 and 21 years.

Thus, the final model by Gebhardt, Lee and Swaminathan (2001) is as follows:

$$P_t = B_t + \frac{FROE_{t+1} - r_e}{(1+r_e)} B_t + \frac{FROE_{t+2} - r_e}{(1+r_e)^2} B_{t+1} + TV$$

where:

$$TV = \sum_{i=3}^{T-1} \frac{FROE_{t+i} - r_e}{(1+r_e)^i} B_{t+i-1} + \frac{FROE_{t+i} - r_e}{r_e(1+r_e)^{T-1}} B_{t+T-1}$$

and:

$$B_{t+i} = B_{t+i} + FEPS_{t+1}(1 - DPR_{t+i})$$

$$B_t = \textit{Book value at time } t$$

$$r_e = \textit{Cost of equity capital}$$

$$DPR_t = \textit{Expected dividend payout ratio in year } t$$

$$FROE_t = \textit{Forecasted return on equity for year } t$$

4.4.1.2 Model 2: Claus and Thomas (2001)

The Claus and Thomas (2001) model is a variation on the residual income or "abnormal earnings" model that sets a forecast horizon of five years, after which earnings grow at a constant long-term rate. Analyst earnings forecasts from the I/B/E/S database are not reported beyond year five, thus the model incorporates a terminal value to capture the net present value of abnormal earnings after year five.

$$P_0 = B_0 + \frac{ae_1}{(1+r_e)} + \frac{ae_2}{(1+r_e)^2} + \frac{ae_3}{(1+r_e)^3} + \frac{ae_4}{(1+r_e)^4} + \frac{ae_5}{(1+r_e)^5} + \left[\frac{ae_5(1+g_{ae})}{(r_e - g_{ae})(1+r_e)^5}\right]$$

where:

$$ae_{t+i} = FEPS_{t+i} - r_e B_{t+i-1}$$

$$B_{t+i} = B_{t+i} + FEPS_{t+1}(1 - DPR_{t+i})$$

$$g_{ae} = r_f - 0.03$$

The Claus and Thomas (2001) model has several advantages. First, a large proportion of the cost of equity estimate is formed using currently available data, lowering the influence of assumptions on the estimate. Second, the rate at which earnings can grow in perpetuity is not abstract and is easy to arrive at using economic logic about long-term economic fundamentals. This rate, g_{ae}, is provided by analysts on I/B/E/S. (With coverage from over 900 contributors and 18,000 individual analysts, I/B/E/S is generally regarded as the leading source for analysts' information regarding listed companies in the US. It offers mean estimates taken from individual analyst estimates of REIT financials.) Third, the abnormal earnings approach of this model allows the estimation of future price-to-book ratios, price-to-earnings ratios and returns on equity.

4.4.1.3 Model 3: Ohlson and Juettner-Nauroth (2005)

The model proposed by Ohlson and Juettner-Nauroth (2005) is a generalization of the dividend growth model. The model uses both a short-term and perpetual growth rate. The short-term growth rate is simply the average of the change in one- and two-year forecasts and the long-term analyst forecasted growth rate. The perpetual growth rate is assumed to be the expected rate of inflation. By focusing on next year's earnings and short/long-run abnormal earnings growth expectations, the model shows how the stock price depends on forward earnings per share and how they grow over time, modelled by two measures of growth that are independent of any dividend policy. Thus, the model focuses on the prediction of earnings growth rather than how these are distributed in the future.

$$r_{oj} = A + \sqrt{A^2 + \frac{FEPS_{t+1}}{P_t}\left(g_2 - (y-1)\right)}$$

where: $A = \frac{1}{2}\left((y-1) + \frac{DPS_{t+1}}{P_t}\right)$

$$DPS_{t+1} = DPS_0$$

$$g_2 = \frac{STG + LTG}{2}..$$

$$STG = \frac{FEPS_{t+2} - FEPS_{t+1}}{FEPS_{t+1}}..$$

$$(y-1) = r_f - 0.03$$

4.4.1.4 Model 4: Easton (2004)

The model proposed by Easton (2004) generalizes the price-earnings-growth model (which is the price-earnings ratio divided by the short-term earnings growth) and uses a two-year forecasting horizon. After this, earnings grow in perpetuity. It relies on the fact that future abnormal growth in earnings adjusts for the difference between next year's accounting and economic earnings.

The model uses current prices and future earnings estimates to obtain a price to earnings ratio in a scenario where abnormal growth in earnings is constant, which Easton (2004) shows is a reasonable assumption to make.

$$P_0 = \frac{EPS_2 + r_e * DPS_2 - EPS_1}{r_e^2}$$

where:

$$DPS_{t+1} = DPS_0$$

In order to analyse the relationship between the sustainability of REITs and their cost of equity, the cost of equity estimates are analysed first using the implied (ex-ante) approach. The estimation technique uses the economic relationships between observed prices, book values and forecasted earnings in order to build an implied cost of equity estimate. Again, the necessary model inputs are obtained from I/B/E/S.

The four models developed in the literature by Gebhardt, Lee and Swaminathan (2001), Claus and Thomas (2001), Ohlson and Juettner-Nauroth (2005) and Easton (2004) will be used to obtain estimates of $r_{GLS,}$ $r_{CT,}$ r_{OJ} and r_{ES} respectively. r_{OJ} is estimated in closed form. For r_{CT}, r_{GLS} and r_{OJ} the calculations are more involved and numerical techniques are used to find the estimates, whilst ensuring that the solutions are restricted to fall between 0 and 1. The literature does not offer a consensus as to which model is best. Therefore, following the methodologies of such studies as El Ghoul et al. (2011) and Hail and Leuz (2006), the average of the four cost of equity estimates is taken for the hypothesis testing and for the estimate r_{AVG}. Robustness analysis is performed using each of the four measures separately.

4.4.2 Measuring the sustainability of REITs

To obtain sustainability estimates for the REITs, the methodology used by Eichholtz, Kok and Yönder (2012) is employed. First, an estimate of the sustainability of the individual properties within the REITs is obtained, which can then be combined in order to form one sustainability estimate per REIT per year.

A dynamic REIT sustainability measure is calculated, at the portfolio level, by using the ratio of the certified square footage of the REIT (LEED or Energy Star). Using this method, the level of sustainability of a REIT is approximated by the proportion of sustainable property that the REIT portfolio is comprised of:

$$Green\ Share_{it}^{g} = \frac{\sum_{l} Sqft\ of\ Certified\ Property_{ilt}^{g}}{\sum_{l} Sqft\ of\ Property_{ilt}^{g}} \times 100$$

where: *i* is the REIT
t is the year
l is the property
g is the environmental certification (either Energy Star, LEED, or a combination of the two).

4.4.3 Hypothesis development

Following Dhaliwal et al. (2011) and El Ghoul et al. (2011), a negative association is predicted between REIT portfolio sustainability and the cost of equity capital. As outlined previously, this may be due to more stable property cash flows and thus a lower overall risk level as perceived by analysts. Furthermore, it could be due to improved reputation and improved investor expectations about future cash flows. In summary, it may be expected that REITs with a higher proportion of environmentally certified property will have a lower cost of equity capital, on average.

Regarding the control variables, the literature finds that firm size and market-to-book have a negative impact on the cost of capital, while higher leverage and beta imply increased systematic risk and therefore an increase in the cost of equity capital.

A linear regression model is estimated in order to test the hypothesis that a REIT's cost of equity depends on its portfolio green share as well as a set of control variables.

$$Cost\,of\,equity_{it} = \beta_0 + \beta_1 Green\,Share_{it} + \beta_2 Beta_{it} + \beta_3 Log\left(Size\right)_{it} + \beta_4 Leverage_{it} + \beta_5 Market - to - Book_{it} + u_{it}$$

where:

u_{it} is the random error term for firm *i* at time *t*.

4.5 Data

4.5.1 Leadership in Environmental

Leadership in Environmental and Energy Design (LEED) is a sustainable building rating scheme, developed by the US Green Building Council (USGBC) and designed to encourage the diffusion of sustainable property and green building practices throughout the market. LEED is used as a benchmark not only in the operation of green buildings, but also in their design and construction.

A defining component of LEED certification is that it takes into account six different aspects of sustainability. LEED defines its certification based on the categories of sustainable sites, water efficiency, energy and atmosphere, materials and resources, indoor environmental quality and, finally, innovation in design. This contrasts with the other leading US certification scheme – Energy Star – which only uses energy performance as a measure of sustainability.

The six aspects of LEED are aimed to reduce the negative environmental impact that buildings can have, to inhibit unsustainable construction methods and to improve financial performance of buildings through reduced energy usage, lower operating costs and improved wellbeing of occupants. LEED is an attempt to measure how a building affects the occupants, the owner and the stakeholders in every measure of sustainability.

Developed by the US Environmental Protection Agency (EPA) and the Department of Energy (DOE), Energy Star is a voluntary labelling system for commercial buildings that began in 1999. Certification can only be awarded to the top 25% of buildings, graded on energy efficiency, being awarded based on a benchmark level of energy use. This benchmark is based on the predicted energy use of the building, using national data to compute the average energy usage of an identical building. Energy Star aims to help organizations measure, track and improve energy performance. It offers a measure of building-level energy performance on a 1 to 100 score.

4.5.2 REITs

The dataset for this study includes information on 211 US REITs, which are tracked from 2003 to 2015. The data source is SNL Real Estate and only includes REITs in the sample for which SNL offers a complete set of control variables and building-level information. The total dataset comprises 2,743 firm-year observations. LEED and Energy Star–certified buildings in the REITs are identified by matching addresses of the REIT-owned properties with certification data provided by the Environmental Protection Agency (EPA) and the US Green Building Council (USGBC).

Table 4.1 shows the descriptive statistics. LEED Share and Energy Star Share refer to the proportion of property space in the REIT that is certified by LEED and Energy Star, respectively. Green Share is a measure of the total proportion of REIT property space that is certified by LEED, Energy Star or both. The mean of the total proportion of green space per REIT is 1.39%. This is much lower than the actual proportion of total certified space in the entire sample, because of the fact that larger REITs, on average, have higher proportions of certified properties. Table 4.1 also shows that Energy Star certification is slightly more prevalent than LEED certification over the sample period.

To avoid selection bias, REITs are included containing no environmentally certified properties over the entire period, as in Eichholtz, Kok and Yönder (2012). Out of the 2,743 total observations, 1,665 have a Green Share of zero, and out of the 211 REITs, 92 had no environmentally certified properties in their portfolio over the complete sample period.

The highest LEED and Energy Star Shares observed are 1, or 100%. This is due to one observation, Gramercy Property Trust Inc., which was launched in 2012 and which initially acquired a small number of properties, all of which were environmentally certified.

Figure 4.1 depicts the average proportion of environmentally certified floor space per REIT over time. As documented by Holtermans, Kok and Pogue (2015) and Eichholtz et al. (2017), the diffusion of sustainability in the REITs in the sample was positive, yet extremely slow between 2003 and 2006. This is especially true for LEED certification, which can be explained by the slow evolution of this programme into a system which accurately represents and incorporates emerging sustainability technologies. Furthermore, LEED's focus is on new constructions, which take time to appear in the market (Kok, McGraw and Quigley, 2011). Since 2010, the growth rate of the combined share of Energy-Star and LEED certified floor space has remained relatively stable, reaching 4.8% of all floor space in the REIT sample by 2015. Figure 4.1 shows that the growth towards more sustainable investment practices in US REITs continues.

Figure 4.2 shows the proportion of REITs in the sample which hold at least one green property in their portfolio. This figure exhibits the S-Shape pattern of innovation theorized by Griliches (1957). Because innovation in sustainability certification comes from outside the traditional real estate investing mindset, investors are likely to be skeptical or unaware of the benefits of sustainability. As theorized by Rogers (2003), the uptake begins to accelerate when

Table 4.1 Descriptive Statistics (2003–2015)

Variables	*Mean*	*Std. Dev.*	*Min*	*Max*	*Obs.*
Green Share	1.39%	5.40%	0%	100%	2743
LEED Share	0.72%	3.61%	0%	100%	2743
Energy Star Share	1.02%	5.06%	0%	100%	2743
Beta	1.003	0.822	-0.035	2.890	2743
Book value	1163.7	1284.6	7.4	4861.7	2743
Leverage	189.9	173.0	4.0	729.5	2743
Market-to-Book	1.91	1.22	0.41	5.50	2743

Notes: The table shows the descriptive statistics. The energy efficiency variables Green share, LEED share and Energy Star represent the proportion of certified property in the REIT portfolio.

Source: Authors

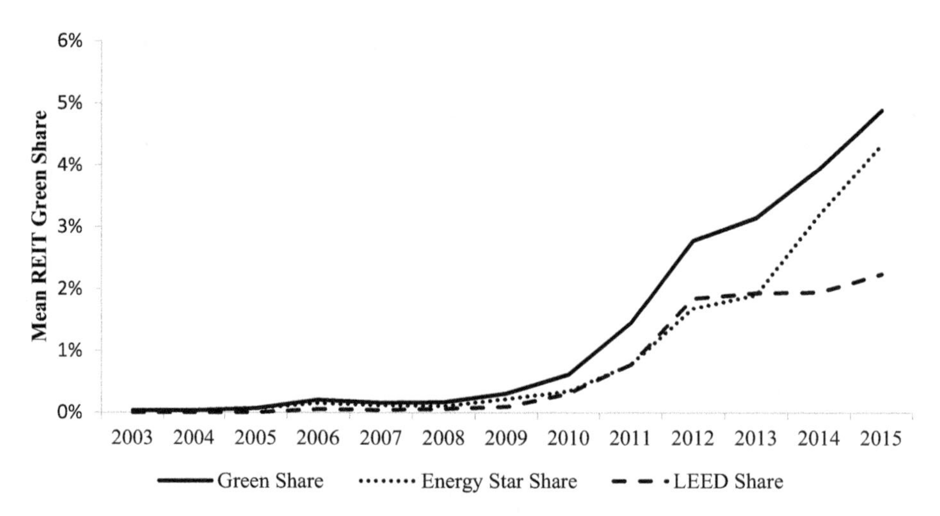

Figure 4.1 Uptake of Environmental Certification in US REITs (2003–2015)

Source: Authors

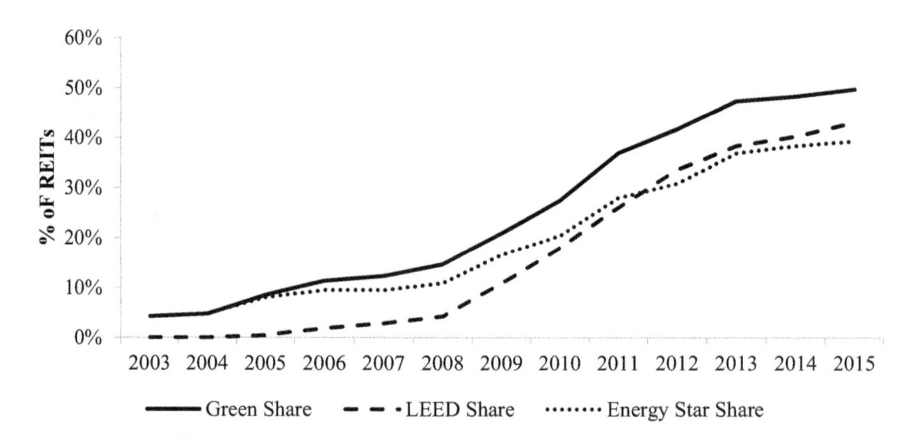

Figure 4.2 REIT Level Diffusion of Environmental Certification (2003–2015)

Source: Authors

between 10% and 20% of firms have invested to some extent in environmentally certified property. Despite the fact that the growth in the number of REITs adopting sustainability investment practices seems to have slowed down, growth is still positive.

The REITs are categorized into six sectors by property type: industrial, office, residential, retail, mixed and specialty. Figure 4.3 shows the breakdown of the sample by property type. Of the four core property types, retail makes up the largest proportion of the sample at 21%. There is also a significant number of mixed REITs, forming 18% of the sample. The majority of these firms either contain a mixture of office and industrial assets or a split between office, industrial and retail property.

Following the methodology of Ambrose and Linneman (2001), a REIT is classified as mixed as long as the most abundant property type in the REIT comprises less than 75% of the properties in that REIT. This classification is not without its problems. If a REIT in the sample has

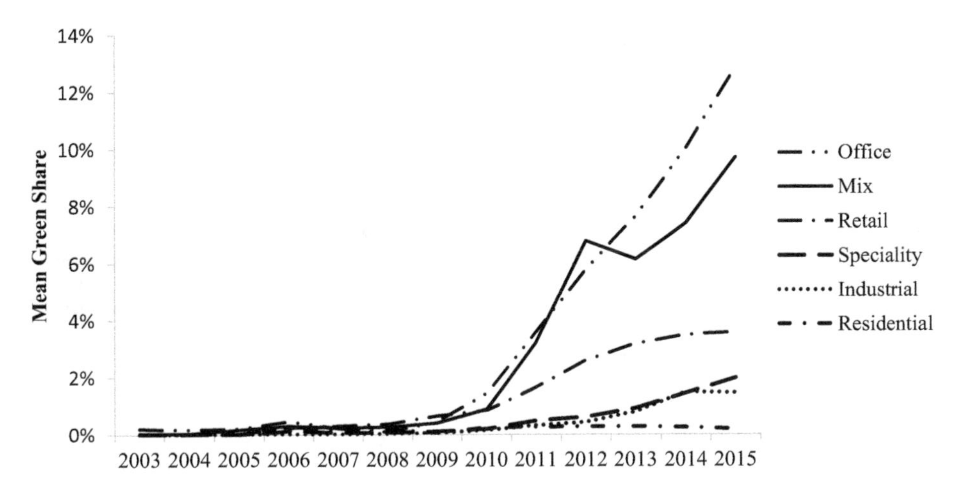

Figure 4.3 Average Green Share by REIT Type

Source: Authors

75% of one property type but then 5% of five other property types, then this classification does not take into account the costs associated with managing six property types. This is different to the NAREIT definition of a diversified REIT, which is simply a "catch all" definition categorizing REITs as diversified if they undertake a strategy which does not focus on one of the 13 property types identified by NAREIT, regardless of whether the strategy incorporates multiple property types.

Using the adopted categorization of REITs, the diffusion of environmentally certified properties over time, by property type, can be analysed. Figure 4.3 shows that, in 2015, every category of REIT had a higher mean allocation to environmentally certified property than at the start of the study period in 2003.

Very strong differences are found in sustainable real estate diffusion across REIT property types. Office REITs see the largest uptake in green property, with a mean proportion of just 2.1% in 2003, growing to 12.9% of floor space in 2015. In fact, this number underestimates the true diffusion of sustainable office space in REIT portfolios, since the next-highest category in Figure 4.3 is mixed real estate which is also predominantly office space. The current body of literature evaluating sustainable commercial property performance tends to concentrate on offices, showing improved rents, asset values and vacancy rates in sustainable offices. Sustainability adoption is far lower in the other REIT property types, with only about 3.5% in retail REITs, 1.6% in industrial and 0.2% in residential REITs. The data do not indicate a slowdown in sustainable office uptake.

4.6 Results

First, the cost of equity measures are estimated using the four models described in detail previously. Table 4.2 shows summary statistics for the implied cost of equity estimates from 2003 to 2015. The average of the four individual model estimates over the sample is 7.44%. The results from the ES and OJ models produce higher cost of equity estimates with means of 9.27% and 8.57%, respectively, whereas the GLS and CT models produce estimates of 5.04%

Table 4.2 Implied Cost of Equity Estimates

Variables	*Mean*	*Std. Dev.*	*Min*	*Max*	*Obs.*
r_{CT}	6.87%	4.20%	1.04%	28.42%	487
r_{GLS}	5.04%	8.14%	0.40%	94.87%	487
r_{OJ}	8.57%	2.96%	2.23%	23.54%	487
r_{ES}	9.27%	4.14%	1.05%	33.04%	487
r_{AVG}	7.44%	4.02%	2.51%	39.39%	487

Notes: The table presents the descriptive statistics of the cost of equity estimates. We obtain the estimates of r_{GLS}, r_{CT}, r_{OJ} and r_{ES} using the models by Gebhardt, Lee and Swaminathan (2001), Claus and Thomas (2001), Ohlson and Juettner (2005) and Easton (2004), respectively.

Source: Authors

and 6.87%, respectively. The findings indicate that it is the GLS model that produces the lowest bound. The GLS model also produces the widest range of estimates, from 0.4% all the way up to 94.9%.

These four estimates concur with previous estimates on REIT cost of equity. For example, Damodaran (2015) calculates yearly cost of equity for REITs and estimates a mean of 8.3% from 2003 to 2015. (For more details, see http://pages.stern.nyu.edu/~adamodar/.) So, whilst the cost of equity is a theoretical concept that is not directly observable, the estimates seem in line with the consensus in the literature. As expected, the estimates are positively correlated with each other, with the highest correlation observed between the OJ and ES models (0.95). The lower correlations of r_{GLS} with the other estimates are consistent with previous studies.

To illustrate the relationship between the cost of equity and portfolio sustainability, scatter diagrams are used to plot the cost of equity estimate, $r_{AVG,}$ on the y-axis and Green Share, LEED Share and Energy Star Share on the x-axis in Figure 4.4. All observations with no certified properties are excluded for better illustration. The bivariate analysis shows a negative association between the core cost of equity estimate and measures of greenness. It may be noted that, in particular, many firms with very high shares of environmentally certified properties also have lower than average cost of equity capital estimates.

Table 4.3 presents the results of the linear regression equation described in the methodology section. All regression models include time dummies and property type controls. It may be noted that the REIT share of environmentally certified property is negatively associated with the average implied cost of equity estimate for all three share measures. In unreported regressions, this key result is found to be robust to using the cost of equity estimates from each model proposed instead of using the average cost of equity estimate of the four models, except for r_{GLS}. The analysis is also robust to using analyst dispersion as a control variable. In a final unreported robustness analysis, in order to rule out whether the findings are driven by the most optimist analysts, the four models are estimated excluding the top 5%, 10%, 25% and 50% of observations with the highest growth forecasts, respectively. The findings are robust to each of the subsample analysis.

The findings on the control variables are in line with expectations and the literature. The coefficient on beta is positive and statistically significant, meaning that firms with higher systematic risk also incur higher costs of raising equity capital and REITs that are more highly leveraged have higher costs of equity, again most likely channelled through the higher risk that these firms exhibit. Larger firms incur a lower cost of capital and an increase in the

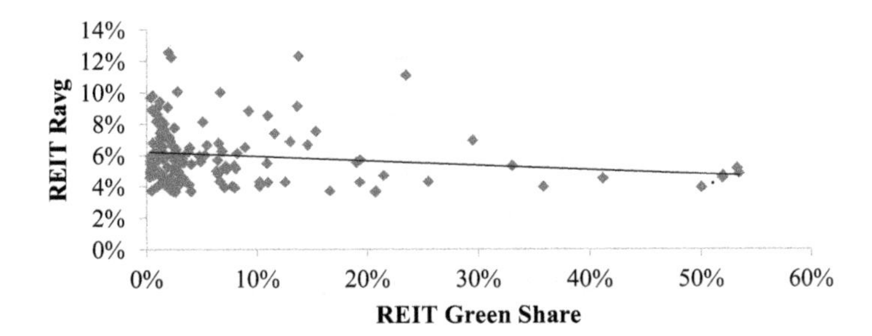

Panel A. Green Share and the cost of equity

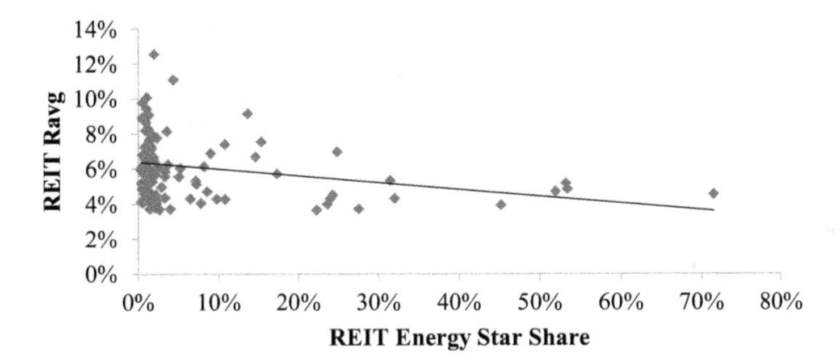

Panel B. Energy Star share and the cost of equity

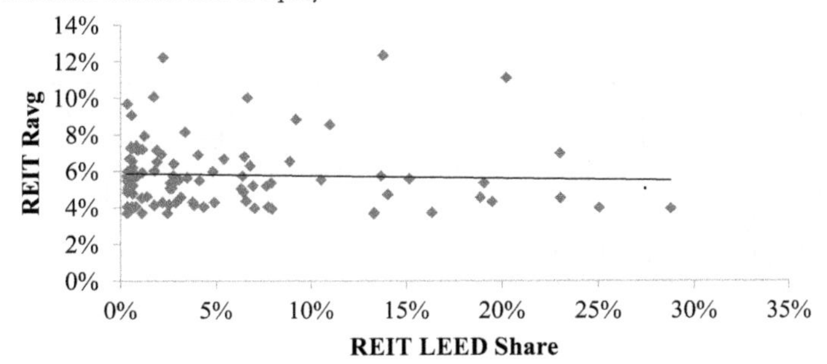

Panel C. LEED Share and the cost of equity

Figure 4.4 REIT Cost of Equity (r_{AVG}) *vs Green Share*

Source: Authors

market-to-book value also leads to a lower cost of equity, perhaps due to increased growth opportunities.

With this study focused on the coefficients regarding the different portfolio sustainability indicators, the coefficients are interpreted in several ways. First, if a REIT that had no certified properties were to switch to a 100% allocation of otherwise identical yet certified properties, it would enjoy a reduction in the cost of equity of 37.7bps on average, holding all other factors constant. This is a 5% decrease over the mean firm cost of equity estimate of 7.44%.

Table 4.3 Environmental Performance and the Cost of Equity

Variables	*(1)*	*(2)*	*(3)*
	r_{AVG}	r_{AVG}	r_{AVG}
Green Share	-0.0378***		
	(0.0134)		
LEED Share		-0.0736**	
		(0.0355)	
Energy Star Share			-0.0283**
			(0.0112)
Beta	0.0045**	0.0044**	0.0045**
	(0.0022)	(0.0022)	(0.0022)
log(Book Value)	-0.0065***	-0.0067***	-0.0065***
	(0.0014)	(0.0013)	(0.0014)
Leverage	0.0034***	0.0033***	0.0035***
	(0.0011)	(0.0011)	(0.0011)
Market-to-Book	-0.0068***	-0.0066***	-0.0067***
	(0.0013)	(0.0013)	(0.0013)
Constant	0.1362***	0.1358***	0.1353***
	(0.0093)	(0.0091)	(0.0096)
Year-Fixed Effects			
Property Type-Fixed Effects			
Observations	487	487	487
R-squared	0.378	0.378	0.374

This table shows the regression results of LEED/Energy Star certification and firm characteristics on the REIT cost of equity. The energy efficiency variables Green Share, LEED Share and Energy Star Share represent the proportion of certified property in the REIT portfolios. The average cost of equity from the four cost of equity models is the dependent variable. Firm characteristics include beta, the logarithm of book value, leverage and the market to book value. The regressions include year dummies and property type dummies. All regressions include heteroscedasticity-robust standard errors that are clustered by REIT. *, **, and *** represent statistical significance at the 10%, 5%, and 1% level, respectively.
Source: Authors

As with previous literature, which details LEED's increased ability to explain financial performance over Energy Star, the coefficient on LEED is found to be more negative than that of Energy Star at -0.0736 compared to -0.0283 for Energy Star. This means that a REIT with a 0% allocation to certified properties would see a reduction in its cost of equity of 73.6 bps or 28.3bps by switching to an allocation of properties certified by LEED and Energy Star, respectively. These two results are significant at the 5% level.

Second, the coefficients are interpreted in terms of smaller, yet still economically significant, changes in certified property allocation. It is found that a one-standard deviation increase in the proportion of environmentally certified property in a REIT is associated with a 5.40% × -0.0378 = 2.04bps reduction in the cost of equity capital on average at the 1% significance level. A one-standard deviation increase in the LEED share of property is associated with a 3.61% × -0.0736 = 2.66bps reduction in the cost compared to a 5.06%−0.0283 = 1.43bps reduction in the cost of equity for a one-standard deviation increase in the Energy Star share of certified property.

Third, the results are compared to previous cost of capital studies to evaluate economic significance. El Ghoul et al. (2011) use qualitative measures to show that a 10-point increase in CSR score based on six dimensions of sustainability leads to a 45bps reduction in the cost of equity capital The results of Chava (2014) are perhaps more easily interpreted. The study finds that firms exhibit a cost of equity that is 138bps higher for a firm that has environmental concerns in all four qualitative dimensions of products, pollution prevention, clean energy and environmental reporting. For REITs, Eichholtz et al. (2017) find that, as the LEED share increases by one-standard deviation, the bond spread decreases by 8–19bps and overall better bond ratings.

Fourth, the magnitude of the findings from Table 4.3 are carefully interpreted in terms of their relative magnitude. One-standard deviation (5.40%) of the share of environmentally certified property translates to a more than tripling of the average share of certified property (1.39%). Thus, a one-standard deviation increase in an average REIT's green share, for example, would be a very large increase in its current allocation to sustainable property. This gives some insight into the economic significance of the findings. Whilst a 2.04bps reduction in the cost of equity would be in itself economically small, this effect would only be associated with a very significant increase in the current portfolio share of sustainable property.

Based on this logic, the results are viewed as more economically relevant in the context of REITs which are either looking to greatly expand their current portfolio or are newly formed and building a portfolio of assets. Real estate is, in general, an asset class needing a long-term investment strategy. Rather that these results incentivizing investment managers to trade unsustainable properties for sustainable ones (which, incidentally, would involve significant transaction costs), this result is more useful for firms constructing a portfolio and thus benefiting from the average 37.7bps reduction in the cost of equity capital for a fully certified portfolio vs an identical yet unsustainable portfolio.

Hence, as in Eichholtz et al. (2017), who find a small economic significance of sustainability on the cost of debt at a portfolio level, the association of sustainability with the cost of equity appears to be small in magnitude, yet statistically and economically significant. In summary, the results from models (1) to (3) from Table 4.3 indicate that real estate investment firms with a higher proportion of environmentally certified property enjoy lower equity financing costs, on average.

4.7 Discussion

Real estate has a significant role to play in curbing global carbon emissions and unrenewable energy usage. In 2015, residential and commercial building energy usage was around 40% of total US energy consumption according to the US Energy Information Administration. (For more details, see www.eia.gov.) Energy usage from buildings is forecasted to increase 25% by 2025 from 2005 levels, leading to total energy costs from buildings in the US increasing $130 billion to $430 billion over that 20-year period.

The link between real estate's sustainability and its economic performance measures – including risk indicators such as occupancy volatility and beta – is well-established. It would, therefore, seem safe to conject that the cost of capital to finance sustainable real estate is lower than it is for conventional real estate. However, this is uncharted terrain and this study investigates the impact of energy efficiency and environmental performance on firms' cost of equity capital. Focusing on publicly listed real estate investment trusts, this chapter is an addition to Eichholtz et al.

(2017) who address the link between sustainability and the cost of debt for REIT mortgages and bonds.

Using the four leading different model specifications suggested in the literature, the cost of equity capital at the firm level is assessed and how these estimates are associated with portfolio greenness is investigated. As is common in cross sectional non-experimental studies, endogeneity cannot be ruled out and thus the direction of the relationship between greenness and the cost of equity is not identified.

Taking into account these concerns, it is found that an increase in allocation to environmentally certified properties from 0% to 100% of the portfolio is associated with significant reductions in the cost of equity for LEED and Energy Star–certified portfolios of 73.6bps and 28.3bps, respectively. Robustness tests validate this result.

These findings have several implications for participants in the real estate market. First, the evidence in this chapter could be yet another incentive for real estate investors and developers to consider "going green" and integrating sustainable practices into their projects. The evidence presented in this study indicates that this does not just contribute to society in a net positive way, but that it may also benefit the firm in the form of lower equity financing costs. Second, as in Brounen and Kok (2011), the study contributes to literature which suggests that environmental certification, as a standardized measure of energy performance and other sustainability measures, is conveying information in a meaningful way and fulfilling its role as a strong market signal. Third, given that the results suggest that information on portfolio sustainability is value-relevant, it would be beneficial for REITs and their – potential – investors that REITs are more transparent regarding this issue.

4.8 Summary

Chapter 1 sought to identify critical contextual issues in international REITs, including the defining characteristics of REITs, the global evolution of REITs and the risk-return characteristics of real estate investment through REITs relative to non-REIT real estate investment with Chapter 2, the *Post-Modern REIT Era*, examining the evolution of the global REIT market, what is deemed to be best current market practice and how the market is expected to change going forward and Chapter 3, *Emerging Sector REITs*, investigating such sub-sectors as timber and data centres.

This chapter, *Sustainable REITs*, analysed REIT environmental performance and the cost of equity by investigating the relationship between sustainability and the cost of equity capital for 211 listed US equity REITs. Sustainability is measured using the two leading energy labelling programs, LEED and Energy Star, finding that the sustainability of REITs' portfolios is strongly property driven, with office REITs being most sustainable. The cost of equity is assessed using four individual implied cost of equity models based on earnings per share forecasts, from which an average is taken. Cost of equity estimates are regressed on the sustainability measures and relevant control variables, finding a statistically significant relationship between the cost of equity capital and REIT greenness. Overall, a 38bps average reduction in the cost of equity is documented when a REIT would have a 100% certified portfolio, as compared to a completely uncertified one, with the results being robust to different specifications and approaches.

Part I continues with chapters each reviewing a current theme of REIT evolution through the lens of contemporary research. Chapter 5, *Islamic REITs*, examines the evolution of global Islamic REITs through a study of the Malaysian market. Chapter 6, *Behavioural Risk in REITs*, addresses the management of behavioural risk in global REITs and Part I concludes with Chapter 7 which reviews recent research into *REIT Asset Allocation*.

Part II then includes six chapters each analysing REITs in a region of the world, using case studies from a developed, developing and emerging REIT sector in the region, concluding with Chapter 14 which considers *Directions for the Future of International REITs*.

References

Ambrose, B.W. and Linneman, P. 2001, 'REIT organizational structure and operating characteristics', *Journal of Real Estate Research* Vol. 21, pp. 141–162.

Bassen, A., Holz, H.-M. and Schlange, J. 2006, *The influence of corporate responsibility on the cost of capital*, University of Hamburg, Hamburg.

Brounen, D. and Kok, N. 2011, 'On the economics of energy labels in the housing market', *Journal of Environmental Economics and Management*, Vol. 62, pp. 166–179.

Building Design and Construction 2007, *Green buildings research white paper*, US Green Building Council.

Chava, S. 2014, 'Environmental externalities and cost of capital', *Management Science*, Vol. 60, pp. 2223–2247.

Claus, J. and Thomas, J. 2001, 'Equity premia as low as three? Evidence from analysts' earnings forecasts for domestic and international stock markets', *Journal of Finance*, Vol. 56, pp. 1629–1666.

Damodaran, A. 2015, 'Cost of capital by sector (US)', available at people.stern.nyu.edu/adamodar/New_Home_Page/datafile/wacc.htm, 1 January 2017.

Dhaliwal, D.S., Oz Li, Tsang, A. and Yang, Y.G. 2011, 'Voluntary nonfinancial disclosure and the cost of equity capital: The initiation of corporate social responsibility Rreporting', *Accounting Review*, Vol. 86, pp. 59–100.

Easton, P.D. 2004, 'PE ratios, PEG ratios, and estimating the implied expected rate of return on equity capital', *Accounting Review*, Vol. 79, pp. 73–95.

Eichholtz, P., Holtermans, R., Kok, N. and Yönder, E. 2017, *Environmental performance and the cost of capital: Evidence from commercial mortgages and REIT bonds*, Working Paper, Maastricht University, Maastricht.

Eichholtz, P., Kok, N. and Quigley, J.M. 2010, 'Doing well by doing good: Green office buildings', *American Economic Review*, Vol. 100, pp. 2494–511.

Eichholtz, P., Kok, N. and Quigley, J.M. 2013, 'The economics of green building', *Review of Economics and Statistics*, Vol. 95, pp. 60–63.

Eichholtz, P., Kok, N. and Yönder, E. 2012, 'Portfolio greenness and the financial performance of REITs', *Journal of International Money and Finance*, Vol. 31, pp. 1911–1929.

El Ghoul, S., Guedhami, O., Kwok, C.C.Y. and Mishra, D.R. 2011, 'Does corporate social responsibility affect the cost of capital?', *Journal of Banking and Finance*, Vol. 35, p. 2388–2406.

Fuerst, F. and McAllister, P. 2011, 'Green noise or green value? Measuring the effects of environmental certification on office values', *Real Estate Economics*, Vol. 39, pp. 45–69.

Gebhardt, W.R., Lee, C. and Swaminathan, B. 2001, 'Toward an implied cost of capital', *Journal of Accounting Research*, Vol. 39, pp. 135–176.

Griliches, Z. 1957, 'Hybrid corn: An exploration in the economics of technological change', *Econometrica*, Vol. 25, pp. 501–522.

Hail, L. and Leuz, C. 2006, 'International differences in the cost of equity capital: Do legal institutions and securities regulation matter?', *Journal of Accounting Research*, Vol. 44, pp. 485–531.

Healy, P.M. and Palepu, K.G. 2001, 'Information asymmetry, corporate disclosure, and the capital markets: A review of the empirical disclosure literature', *Journal of Accounting and Economics*, Vol. 31, pp. 405–440.

Holtermans, R., Kok, N. and Pogue, D. 2015, *National green building adoption index*, CBRE, Los Angeles.

Kelso, J.D. 2012, *Buildings energy data book*, Department of Energy.

Kok, N., McGraw, M. and Quigley, J.M. 2011, 'The diffusion of energy efficiency in building', *American Economic Review: Papers and Proceedings*, Vol. 101, pp. 77–82.

Li, Q., Luo, W., Wang, Y. and Wu, L. 2013, 'Firm performance, corporate ownership, and corporate social responsibility disclosure in China', *Business Ethics: A European Review*, Vol. 22, pp. 159–173.

Marchettini, N., Brebbia, C., Bastianoni, S. and Pulselli, R. (eds) 2014, *The sustainable city IX: urban regeneration and sustainability*, WIT Press, Sienna.

Ohlson, J.A. and Juettner-Nauroth, B.E. 2005, 'Expected EPS and EPS growth as determinants of value', *Review of Accounting Studies*, Vol. 10, pp. 349–365.

Pivo, G. and Fisher, J.D. 2010, 'Income, value, and returns in socially responsible office properties', *Journal of Property Research*, Vol. 32, pp. 243–270.

Rogers, E.M. 2003, *Diffusion of innovations*, Free Press Simon & Schuster, New York.

Sharfman, M.P. and Fernando, C.S. 2008, 'Environmental risk management and the cost of capital', *Strategic Management Journal*, Vol. 29, pp. 569–592.

US Green Building Council 2007, *LEED for new construction version 2.2 reference guide*. US Green Building Council.

Wiley, J.A., Benefield, J.D. and Johnson, K.H. 2010, 'Green design and the market for commercial office space', *Journal of Real Estate Finance and Economics*, Vol. 41, pp. 228–243.

World Economic Forum 2016, *Environmental sustainability principles for the real estate industry*, World Economic Forum, Geneva.

5

ISLAMIC REITS

A study of the Malaysian market

Muhammad Najib Razali and Tien Foo Sing

5.1 Introduction

This book aims to identify key areas for research in the REIT discipline for the next five to ten years by surveying the current state of the REIT discipline around the world and identifying emerging and cutting edge research areas through a thematic review of current contextual issues and a regional analysis based on case studies.

This book comprises two parts, the first part being a thematic review of emerging and cutting edge global research into current contextual issues in REITs internationally and the second part being a regional analysis of REITs around the world, each written by authoritative academic authors from the world's leading Universities and REIT industry experts.

Part I includes six chapters each reviewing a current theme of REIT evolution through the lens of contemporary research. Chapter 1 focused on critical contextual issues in international REITs while Chapter 2, the *Post-Modern REIT Era,* examined the evolution of the global REIT market, what is deemed to be best current market practice and how the market is expected to change going forward, with Chapter 3, *Emerging Sector REITs*, investigating such sectors as timber and data centres.

Chapter 4, *Sustainable REITs*, analysed REIT environmental performance and the cost of equity with this chapter, *Islamic REITs*, examining the evolution of global Islamic REITs (IREITs) through a study of the Malaysian market. The requirement for Shariah compliance is fundamental to IREITs, impacting IREITs in many diverse ways including strategic asset allocation and stock selection. However, with the Muslim population constituting 24% of the world's population, the potential for growth in Islamic financial products and fund management services is enormous.

Chapter 6, *Behavioural Risk in REITs*, then addresses the management of behavioural risk in global REITs and Part I concludes with Chapter 7 which reviews recent research into *REIT Asset Allocation*.

Part II includes six chapters each analysing REITs in a region of the world, using case studies from a developed, developing and emerging REIT sector in the region:

Chapter	Region	Developed REIT Sector	Developing REIT Sector	Emerging REIT Sector
8	North America	USA	Canada	Mexico
9	Latin America	Brazil	Argentina	Uruguay
10	Europe	UK	Spain	Poland
11	South East Asia	Singapore	Malaysia	Thailand
12	North Asia	Japan	Hong Kong	China
13	Oceania	Australia	South Africa	India

concluding with Chapter 14 which considers *Directions for the Future of International REITs*.

5.2 Islamic financial markets

The Muslim population of 1.5 billion, which constitutes 24% of the world's 6.3 billion population, creates enormous potential for Islamic financial products and fund management services. Based on the 2007–2008 World Islamic Banking Competitive Report published by McKinsey, global Islamic banking assets under management were estimated at US$750 billion and expected to grow to US$1 trillion by 2010 (McKinsey, 2006). The Asian Banker Research Group predicted that the global Islamic finance industry will grow at a steady rate of 15% to 20%pa. Islamic banking has grown from a near zero base in 1970 into a sizeable market with approximately US$2 trillion in value by 2014 (Hoggarth, 2016).

Currently, there are more than 250 Islamic financial institutions operating in 75 countries and 100 Islamic equity funds managing assets in excess of US$5 billion. As of 31 December 2014, Islamic funds estimated at US$1.8 billion comprise 36% of the total listed equity funds globally (Bank Negara Malaysia, 2014a). The enormous potential in Islamic financial markets has attracted keen interest from both Islamic and conventional financial institutions in creating various Islamic financial products (Ibrahim, Ong and Parsa, 2009). Many financial institutions in non-Muslim countries in the West (such as the United Kingdom, Luxembourg, France, the US and others) have taken active initiatives to tap into the growing Islamic financial markets. For instance, Luxembourg Central Bank is the first central bank in Europe to join the Islamic Financial Services Board, London has operated an Islamic finance scheme for over 30 years and France has appointed several Islamic finance specialists to accelerate the development of its Islamic finance system. Furthermore, Dow Jones have established two separate Islamic Market Titans Indices for the European and US markets, respectively.

Countries that embrace and promote an Islamic finance system need appropriate regulatory frameworks for risk management and information disclosure. Governments of these countries have developed rules and regulations for their Islamic finance system to ensure compliance with the Islamic finance system requirements and with Shariah requirements during the implementation stage. They supervise investment activities of Islamic institutions operating in their countries. Furthermore, several international bodies have been established to monitor and standardize rules and processes, which are important in supporting sound and stable Islamic financial market activities across the world. These key bodies include, among others, the Islamic Financial Services Board (IFSB), the Accounting and Auditing Organization for Islamic Financial Institutions (AAOIFI), the Islamic Fiqh Academy (IFA) and the International Islamic Rating Agency (IIRA).

The rapid growth in Islamic financial markets has attracted keen interest from researchers in studying the sustainability and potential of the Islamic financial system. This chapter aims to provide a broad analysis of the development of Islamic REITs and to discuss relevant Shariah rules that govern the investments and business activities of Islamic REITs.

The next section of this chapter gives an overview of the development of the Islamic financial market globally, followed by a discussion of key features underlying the Islamic financing system. The section thereafter compares the structures of conventional and Islamic REITs, followed by a description of the development of Islamic REIT markets in Malaysia and other countries and case studies of Islamic REITs to illustrate issues relating to structure, investment portfolios and management with conclusions then drawn.

5.3 Features of the Islamic financial system

Islamic finance relies on an asset-backed feature in the product design to protect investors against extreme market risk. It offers a relatively resilient alternative to the conventional financial system that has proven better able to withstand extreme volatility during crises. During the 2007–2008 global financial crisis, Islamic banks outperformed conventional banks with higher than average returns on assets and liquidity (Parashar and Venkatesh, 2010; Hoggarth, 2016).

Principles of Islamic law advocate fairness and equity in wealth distribution, with Shariah Compliance being the key feature of the Islamic financial system. Table 5.1 compares various features between conventional and Islamic financial systems.

Table 5.1 Differences Between Islamic Finance and Conventional Finance

Differences:

	Islamic Finance	*Conventional Finance*
Deposits	• Agreement of rewards is through Musharaka and Mudaraba • Risk born by the bank • Total reward belongs to banker after servicing the depositors at fixed rate	• Agreement of rewards is fixed and pre-determined • Risk and reward with depositors
Financing and investment	• Finance agreement on offering loan without interest • Overdrafts/Credit Cards • Facility in the form of Murabaha and debit card • Leasing • All risk are owned by Islamic finance • House Finance/Mortgages • Financing provided through diminishing Musharaka • Investment • Only "Halal" Investments are allowed	• Finance agreement on offering loan on a fixed reward • Overdrafts/Credit Cards • Usage of credit card and limitation of overdrawing issued by bank. • Leasing • All risk are owned by the owner not bank/leasing institution • House Finance/Mortgages • Loan is provided with interest • Investment • Consider "Halal" and "Haram" activities in investing

Similarities

- Deposits are collected from savers under both type of institutions
- Credit facilities to business and industry
- Leasing can be done through both financial systems
- Mortgages can be applied from both financial systems
- Both financial systems offer investment

Source: Razali and Sing (2015)

The Islamic financial system encourages mutual sharing of risk and profit between parties involved in transactions. In Islamic profit sharing contracts (equity-based), lenders (investors) and borrowers (entrepreneurs) agree to share profit based on their respective ownership shares in assets (Jobst, 2007). However, direct exchanges of money for debt are not allowed without asset transfers.

Islamic banks are also prohibited from money lending with interest earnings (Riba), however they could contribute interest-free capital into borrowers' businesses that generate income from permissible activities (Halal) (Jobst, 2007). Permissible business activities in an Islamic financing scheme must adhere strictly to a set of Shariah guidelines. Investment activities that are deemed to be Haram are prohibited. Such activities include savings, investment deposits, interest-yielding bonds and acquisition of shares of companies involved in the sale of tobacco, alcohol and pork, betting, gambling and pornography production (Wilson, 1997).

Islamic law encourages entrepreneurship, trade and commerce activities that bring societal benefits and development. They prohibit investments that are speculative in nature, such as derivative instruments, forward contracts and future agreements. Prohibiting the use of derivatives could restrict risk hedging options for Islamic investment schemes. The absence of a viable money market in the Islamic financial system could increase liquidity risk. Some researchers argue that Shariah Compliance imposes additional management costs on the Islamic financial system (Derigs and Marzban, 2009). However, others show that Islamic financial products do not underperform other financial products (Ibrahim and Ong, 2008; Ibrahim, Ong and Parsa, 2009).

With the advancement of financial engineering and innovation, complex agreements could be structured to replicate interest-bearing claims in Shariah compliant lending and thus overcome the strict rules that disallow interest as a form of compensation (Mirakhor and Zaidi, 1988). Islamic securitization is an example of such financial innovation that transforms Islamic risk-sharing contracts between borrowers and lenders into market-based financing (Jobst, 2007). The pass-through structure is a suitable mechanism for Islamic securitization, where a special purpose vehicle is established to facilitate the "off-balance-sheet" transfer of permissible assets (cash flows) from an originator's book. In the process of executing the transfer of ownership rights of assets, various categories of unsecured pass-through securities are also issued to investors. Jobst (2007) provides a useful discussion about Shariah Compliance and the various ways of creating Islamic securitization backed by Shariah permissible assets.

Halal financial products are popular among Muslim investors, who have a strong propensity to invest in financial products that adhere to Shariah principles. Shariah Compliance consists of a set of religious requirements, which is either based on the Holy Quran and Hadith (words, actions or habits of the Islamic prophet Muhammad) or the Ijtihad, which is based on the interpretation of established Muslim scholars. Islamic financial products appeal to Muslim investors who follow and abide by Islamic laws (Shariah) in their daily life (El Qorchi, 2005). Table 5.2 describes a range of Islamic financial instruments:

Muslim scholars rely on the Shariah principles in the Quran, Hadith and Ijtihad to assess if business activities are in adherence with Shariah Law (Binmahfouz, 2012). However, there have been many activities or misconduct that is not stated in the Quran and Hadith and/or not found within the standard guidance of the Muslim scholars. In some countries, when different views occur, a special council is established to provide independent interpretations and recommendations on a case by case basis. Discussions, debates and opinions derived by such a special council must be consistent with the principles in the Quran and Hadith (Saeed, 2011).

Table 5.2 Range of Islamic Financial Instruments

No.	*Type*	*Description*
1.	Debt instruments (Murabaha)	A purchase and resale contract in which a tangible asset is purchased by a bank at the request of its customer from a supplier, with the resale price determined based on cost plus profit markup; Salam, a purchase contract with deferred delivery of goods (opposite to Murabaha), which is mostly used on agricultural finance long-term projects; and Qard al-Hassan (benevolent loan), an interest-free contract that is usually collateralized.
2.	Quasi-debt instruments (Ijara)	A leasing contract whereby a party leases an asset for a specified rent and term. The owner of the asset (banks) bears all risks associated with ownership. The asset can be sold at a negotiated market price, effectively resulting in the sale of the Ijara contract. The Ijara contract can be structured as a lease-purchase contract whereby each lease payment includes a portion of the agreed asset price and can be made for a term covering the asset's expected life.
3.	Profit-and-loss-sharing instruments (Musharaka)	An equity participation contract under which a bank and its client contribute jointly to finance a project. Ownership is distributed according to each party's share in the financing. They include Mudaraba, a trustee-type finance contract under which one party provides the capital for a project and the other party provides the labour. Profit sharing is agreed between the two parties to the Mudaraba contract and the losses are borne by the provider of funds except in the case of misconduct, negligence or violation of the condition agreed upon by the bank.

Source: Hoggarth (2016

5.4 Global conventional REIT markets

Despite an early start in the 1960s, the US REIT market was transformed in the 1990s with a large influx of institutional investors and funds igniting explosive growth (Figure 5.1). The number of REIT listings has increased from 34 to 211 in 1997 and the market capitalization has expanded by 94 times. Significantly, 92 new REIT were listed between 1990 and 1997 and the market capitalization expanded by nearly 15 times during this period.

The US experience in transforming the REIT market has paved the way for the setting up of REIT markets in other parts of the world. Despite early inertia and inactivity in the 1990s (except for Australia), REITs started to emerge and expand rapidly in Asia and subsequently in Europe in the 2000s. Table 5.3 shows a list of countries outside the US that have established REITs as a securitized vehicle for investing in income generating commercial real estate.

5.5 Islamic REITs

Islamic property funds with Asia-focused property portfolios have attracted keen interest in recent years (Ibrahim and Ong, 2008). For example, established funds include Baitak Asia Real

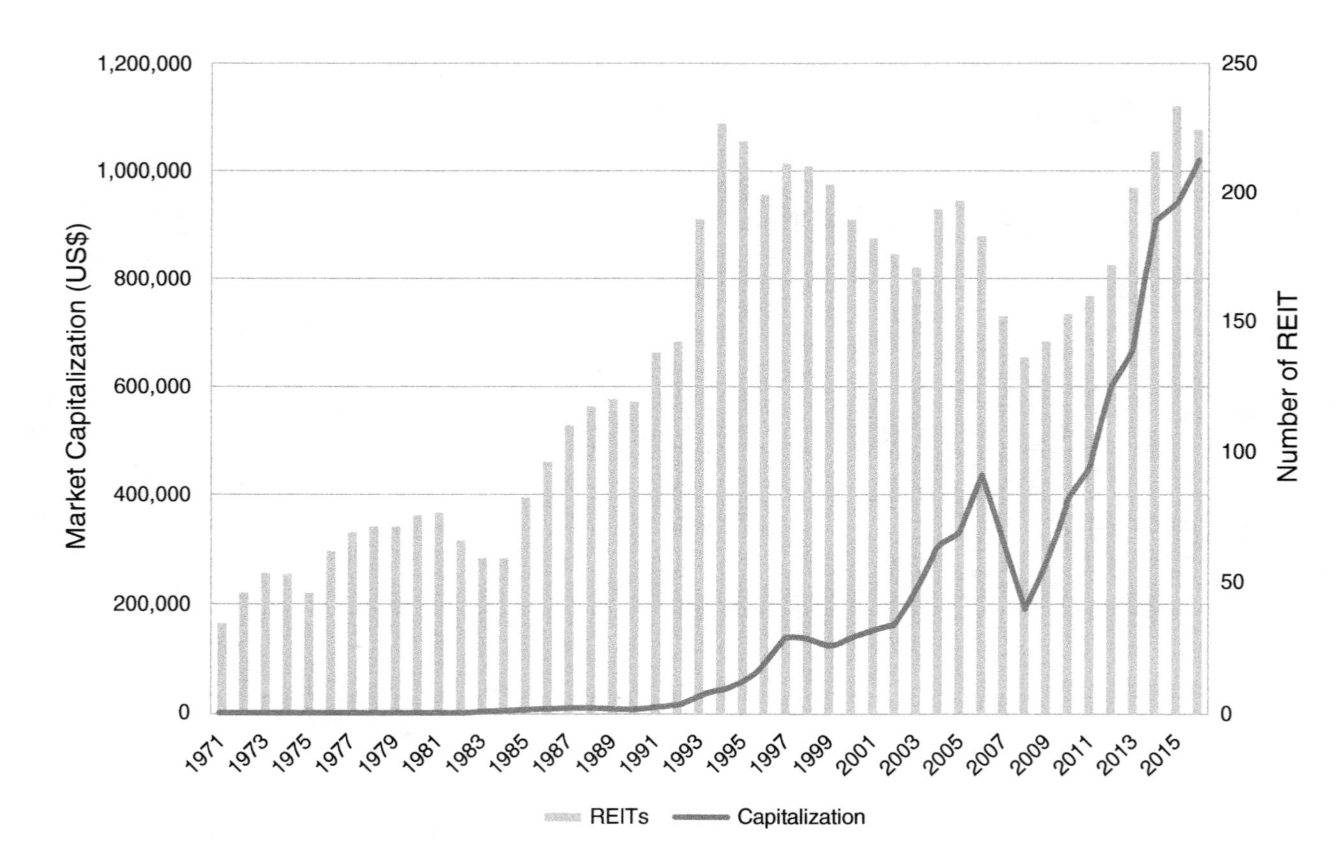

Figure 5.1 US REIT Market

Source: National Association of REITs (NAREIT) and Authors

Table 5.3 Global REIT Regimes

No	*Country*	*Continent*	*Year of Inception*	*Number of REITs*	*Market Capitalization (US$ in Bill.)*	
					As of 2015	*As at IPO#*
1	United States	America	1961	270	1,364.976	303.971
2	Australia	Oceania	1971	56	125.098	0.037
3	Japan	Asia	2001	52	124.501	3.108
4	United Kingdom	Europe	2007	38	109.194	37.099
5	France	Europe	2003	28	66.599	15.975
6	Singapore	Asia	2002	36	62.414	1.217
7	Canada	America	1994	47	59.568	4.722
8	Hong Kong SAR, China	Asia	2005	9	33.268	8.135
9	South Africa	Africa	2000	31	25.601	0.973
10	Mexico	America	2011	11	20.811	0.936
11	Belgium	Europe	1995	16	14.985	0.261
12	Spain	Europe	2013	13	10.178	0.216
13	Turkey	Asia	1995	24	8.739	0.798
14	Malaysia	Asia	2005	16	8.144	0.082
15	Netherlands	Europe	1969	3	5.325	0.072
16	Germany	Europe	2007	5	4.540	0.320
17	Ireland	Europe	2013	3	3.898	1.562
18	Italy	Europe	2007	3	3.807	1.671
19	Greece	Europe	2005	3	3.210	0.244
20	Taiwan	Asia	2005	5	3.088	1.022
21	Mauritius	Africa	2015	1	2.710	2.710
22	Israel	Africa	2006	3	2.433	0.010
23	New Zealand	Oceania	2007	2	2.111	0.198
24	Thailand	Asia	2014	16	1.823	0.609
25	Virgin Islands (US)	America	2012	1	0.985	0.870
26	Brazil	America	2011	1	0.643	0.980
27	United Arab Emirates	Asia	2014	1	0.535	0.516
28	China	Asia	2013	1	0.498	0.458
29	Bulgaria	Europe	2005	48	0.370	0.003
30	Pakistan	Asia	2015	1	0.322	0.322
31	Finland	Europe	2013	1	0.132	0.053
32	Korea, Rep.	Asia	2002	3	0.110	0.406
33	Botswana	Africa	2008	1	0.066	0.041
34	Kenya	Africa	2015	1	0.055	0.055
35	Jamaica	America	2008	1	0.006	0.006

Note: The information and statistics are compiled from data collected from Bloomberg. A list of 35 countries in the world is shown in the database to have at least one REIT listed on the respective bourse and the countries are arranged by the market capitalization as of 2005 in a descending order. The table also includes data on the continent, the year in which the first REIT IPO was listed and the number of REITs listed as of 2005. For the initial market capitalization figures, statistics as in 2000 are used for those countries with the first REIT listed before 2000 as the reference and for others the market capitalization in the year of the first REIT IPO is used in the estimation of the statistics.

Source: Seah et al. (2016)

Estate Fund, Islamic European Real Estate Fund, Al Islamic Far Eastern Real Estate Fund and Bumiputra Real Estate Trust Fund.

Islamic REITs (IREITs) are a form of ethical property investment that adhere to Islamic principles. Despite sharing the same framework, IREITs differ from conventional REITs because their investment activities are guided by a set of Islamic principles, which are referred to as Shariah Compliance (SC). SC principles prohibit any form of business activity, trade, service or exchange that is not Halal (religiously permissible) under the SC rules. These religiously non-permissible activities are deemed by Islamic laws to have a negative impact on society, institutions and/or culture (Osmadi, 2007).

Some non-Halal business activities include those that are related to Riba and Gharar and that involve unethical businesses, such as gambling, production or sale of non-Halal products, conventional insurance, entertainment activity, sale of tobacco-based products, stockbroking, hotels and/or resorts (Dusuki, 2007). Riba are loans where borrowers are contractually obliged to pay interest to lenders who provide the principal (capital) for the loan. Gharar means businesses with high uncertainty, including, for example, transactions conducted in dubious places, with ambiguous product types and/or sales that have been suspended or with future delivery.

However, IREITs may hold real estate assets in sectors such as residential, industrial, commercial, retail, office, car parks, hotels, institutions and health care which generate a steady stream of rental income from business activities that are permitted under the SC guidelines. In other words, only Halal or religiously ethical business activities are allowed in the properties held by IREITs (Dusuki, 2008).

Figure 5.2 presents the structure of a typical IREIT. A distinctive feature of this structure is the formation of the Shariah Committee, which is not found in the conventional REIT structure. The Shariah Committee, comprising Muslim scholars and/or persons with an Islamic education background, is formed to provide advice and guidance on SC matters relating to IREIT business activities, including investment, deposit, asset management, operation and financing.

IREITs pool resources by issuing SC securities to investors via the initial public offering (IPO) channel. Capital raised is deployed to fund acquisitions of a portfolio of Halal real estate assets. These assets are held through trustees who are entrusted with the responsibilities of managing properties in the portfolio and distributing rental income ("pass-through") back to beneficiaries (investors). In an externally advised model, IREIT trustees usually outsource asset, investment and capital management services to third-party asset managers and outsource routine maintenance and servicing works to professional property managers. The internally advised model, widely adopted in the US, is still relatively new and has not been used by IREITs in Asia. Unlike conventional REITs, the appointed Shariah Committee will advise IREITs against leasing out space in their properties to Non-Halal tenants, whose business activities are not consistent with Islamic principles.

At this stage of IREIT development, the need for a transparent regulatory framework and guidelines governing IREITs remains a significant challenge. While certain elements are strictly prohibited under SC rules, such as interest, gambling and trading in liquor, other elements, such as alcohol, Non-Halal food and entertainment (gambling) are deemed to be something that is a mix of the permissible and the prohibited. The mixed elements are likely to attract different interpretations by Shariah Committees in different jurisdictions. Abozaid (2008) and some Islamic scholars take the view that the mixed elements are not necessarily prohibited if their inclusion has no direct effect on permissible activities in the business as a whole.

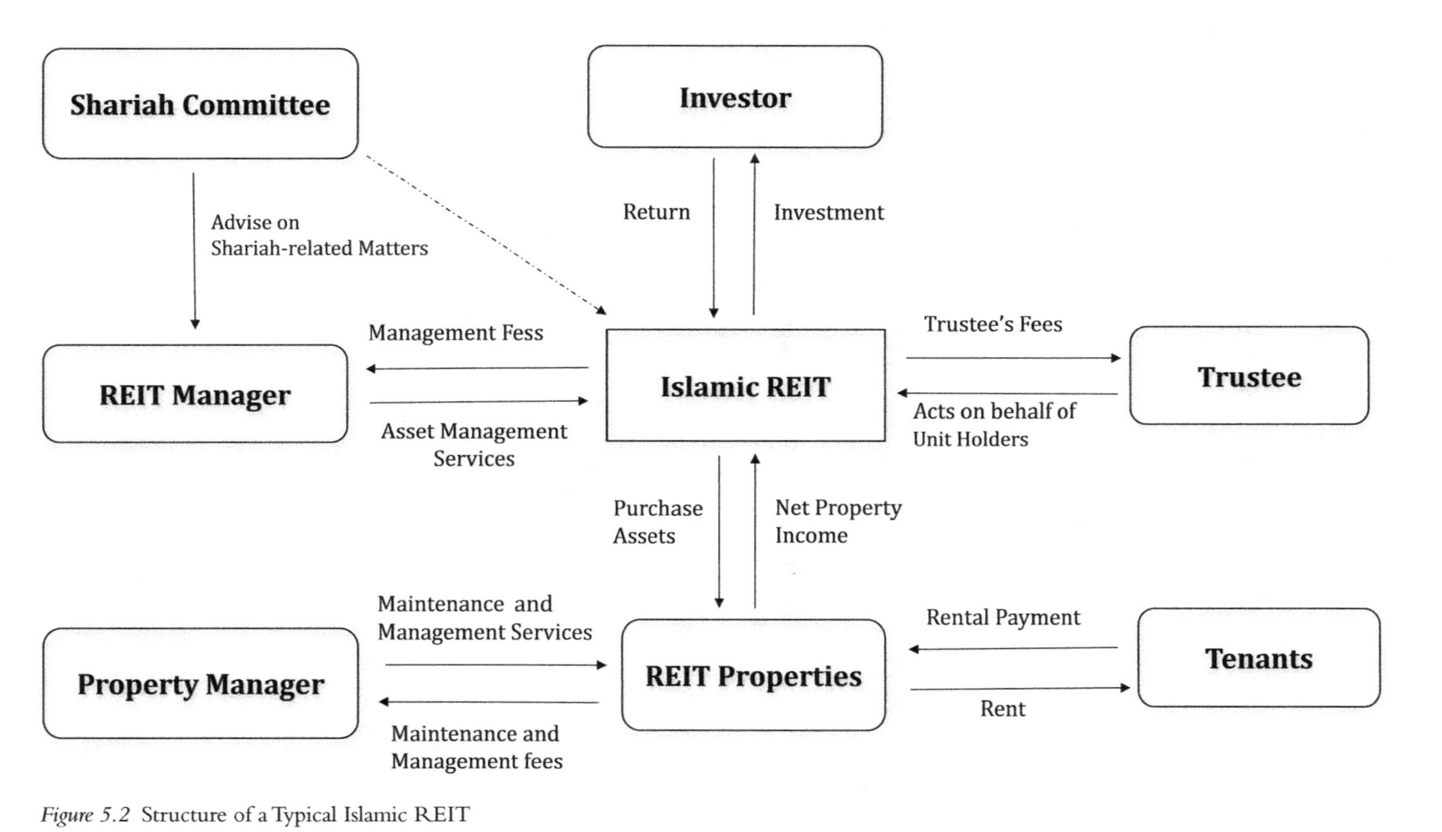

Figure 5.2 Structure of a Typical Islamic REIT

Source: Authors

Many Islamic scholars have debated the substantive issue of Haram (non-permissible) elements. Despite intense discussions between Shariah advisors and some IREITs' boards of directors, they are still divided with respect to the percentage of non-permissible elements that could render activities to be prohibited (Non-Halal). In countries such as Malaysia, the benchmark for non-permissible elements is set at 20%.

5.6 Malaysia's Islamic financial market

As a country with a predominantly Muslim population, Malaysia has a well-established Islamic financial system regulated by the Islamic Banking Act 1983. Bank Islam, the first fully fledged Islamic bank was established in 1983 under the Islamic Banking Act. Malaysia's Islamic finance industry has existed since the 1970s and has grown by an average rate of 18% to 20%pa, with Islamic banking assets estimated at US$65.6 billion in 2014 (Source: Bank Negara Malaysia, the Central Bank of Malaysia, 2014).

Economic-based Islamic institutions actively look for SC investment opportunities for the Islamic community. The Pilgrim Fund Board (known as Lembaga Tabung Haji), which was the first Islamic financial institution established in Malaysia in 1963 to manage pilgrimage savings, channels a substantial portion of its funds into SC investments. The Pilgrim Fund Board has expanded rapidly over the years to become the largest Islamic fund manager in Malaysia with more than US$14.46 billion (RM$55 billion) funds under management as of 2015, based on an exchange rate of US$1: RM$3.80 as in July 2015 (Tabung Haji, 2015).

The strong demand from both Muslim and non-Muslim investors, especially institutional investors, motivated the Malaysian government to launch its first Capital Market Master Plan in 2001 (Securities Commission of Malaysia, 2001). Dr Zeti Akhtar Aziz, the then Governor of the Central Bank of Malaysia, outlined ten strategies for the creation of a sustainable global Islamic financial centre:

- to identify the nature of the Islamic banking system, either a dual banking system or a single banking system, that is appropriate for the country;
- to ensure that the Islamic banks are well-capitalized and resilient;
- for countries opting for a dual banking system, where conventional banks are allowed to offer Islamic banking products, to ensure absolute separation of banking operations to avoid co-mingling between the Islamic and conventional funds;
- to build strong leadership in the management team that has a firm commitment toward performance and capacity enhancement;
- to put in place a comprehensive legal infrastructure including enactment of an Islamic banking law;
- to develop a comprehensive Shariah framework including the establishment of a Shariah advisory committee to ensure that all Islamic transactions, products and operations are Shariah compliant;
- to set up a robust and effective regulatory and supervisory framework;
- to build up research and development capability and incorporate core Islamic values into banking products and business models;
- to develop financial market infrastructure for sustainable Islamic banking; and
- to promote consumer education and awareness about Islamic banking (Aziz, 2005).

IREITs are one of the several Islamic financial products that have been actively promoted in Malaysia as a channel for investors to invest in SC property investments (Azhar and Saad, 2012;

Rozman et al., 2016). The creation of IREITs encourages investments in SC commercial properties with returns derived from capital appreciation and annual investment income from the leasing of accommodation in the portfolio (Ting and Noor, 2007; Dusuki, 2008; Ibrahim, Ong and Parsa, 2009; Rozman et al., 2016).

5.7 Emergence of the Islamic REIT market in Malaysia

Malaysia, as a country practicing moderate Islamic rules of life, is well-positioned to be a market leader for IREITs. Some flexibility, such as allowing 70% foreign shareholders in Islamic REITs as well as 20% of non-SC permissible business, facilitates the establishment of new IREITs and/or the conversion of existing REITs into IREITs.

Malaysian REIT market history dates back to 1989 following the listing of Arab Malaysian First Property Trust on the Bursa Malaysia (the Malaysian stock exchange, formerly known as the Kuala Lumpur Stock Exchange) in September 1989. The first Malaysian REIT was modelled on the Australian listed property trust (LPT) structure. Two other public listed Malaysian LPTs, First Malaysian Property Trust and Mayban Property Trust Fund One and one other property trust, Amanah Harta Tanah PNB, were subsequently launched on the stock exchange between 1989 and 1990. The Mayban Property Trust Management Berhad, which took over the management of Pelaburan Hartanah Nasional Berhad, launched Amanah Harta Tanah PNB 2 in November 2001. However, Malaysia's LPT market was relatively inactive with thin trading activities (Sing, Ho and Mak, 2002) and the First Malaysian Property Trust was subsequently delisted from the Bursa Malaysia in July 2002.

In 2005, the Malaysian LPT market underwent a major repositioning and modernization exercise resulting in the release of the new *Guidelines on Real Estate Investment Trusts (REITs)*. The new guidelines adopt the US-style REIT structure and the first modernized REIT, Axis Real Estate Investment Trust, was introduced on the Bursa Malaysia on 29 July 2005. The modernization of Malaysian REITs has gained traction, attracting a slew of new listings in the market including YTL Hospitality REIT (formerly known as Starhill REIT), the country's largest REIT, listed on 16 December 2005 and UOA Real Estate Investment Trust listed on 30 December 2005, among others (Hamzah et al., 2010). As of December 2014, 15 REITs including three Islamic REITs traded on the Bursa Malaysia with Table 5.4 showing a list of publicly traded REITs in Malaysia and information on their respective listing date and market capitalization.

In November 2005, the Malaysian Securities Commission issued Guidelines for IREITs with Malaysia being the first country in the world to have a formal regulatory framework for IREITs. The Malaysian Securities Commission recognizes IREITs as a collective investment scheme that invests in real estate. The Securities Commission appoint the Shariah Advisory Council (SAC) and entrust it with the responsibility for setting up and enforcing Shariah Compliance guidelines for IREITs. IREITs are also required under the guidelines to form a Shariah Committee internally to advice on permissible investment activities.

The first IREIT in Malaysia was established on 10 August 2006 with the IPO listing of Al-Aqar Health Care REIT. Its property portfolio consists of specialist hospitals and medical centres distributed across different parts of Malaysia. As of December 2014, its assets were valued at US$259 million (RM$831 million) with gearing of 46%. The second IREIT was a plantation REIT sponsored by Boustead Group and listed under the name Al-Hadharah Boustead REIT on Malaysia's Bursa Stock Exchange in January 2007. Al-Hadharah Boustead REIT was subsequently privatized and delisted from the Bursa Malaysia on 16 July, 2013.

Table 5.4 Profile of Malaysia REITs

No	*Real Estate Investment Trust (REITs)*	*Date of Listing on the Bursa Malaysia*	*Market Capitalization as at 2014 (RM$ million)*	*Property Type*
1.	Axis REIT*#	29 July 2005	RM 1,982 million	Office, Industrial
2.	Al-Aqar KPJ REIT*	10 August 2006	RM 960 million	Health care
3.	AmFirst REIT	21 December 2006	RM 641 million	Office
4.	Atrium REIT	02 April 2007	RM144 million	Industrial
5.	Amanahraya REIT	26 February 2007	RM475 million	Office, Retail, Industrial, Hotel
6.	Capitamalls REIT	16 July 2010	RM 2500 million	Retail
7.	KLCC REIT*	9 May 2013	RM 12100 million	Office, Retail, Hotel
8.	IGB REIT	21 September 2012	RM4515 million	Retail
9.	Pavilion REIT	7 December 2011	RM 4400 million	Retail, Office
10.	Sunway REIT	8 July 2010	RM 4220 million	Retail, Hotel, Office
11.	Hektar REIT	4 December 2006	RM 604 million	Retail
12.	Quill Capital REIT	8 January 2007	RM 475 million	Office
13.	Tower REIT	12 April 2006	RM 359 million	Office
14.	UOA REIT	30 December 2005	RM 672 million	Office, Retail
15.	YTL Hospitality REIT	16 December 2005	RM 1350 million	Hotel, Retail, Apartment
16.	Amanah Harta Tanah REIT	28 December 1990	RM 116 Million	Office

Note: The table provides a list of Malaysian REITs that are currently traded on the Bursa Malaysia (formerly the Kuala Lumpur Stock Exchange). * indicates that the REITs are listed as Islamic REITs (IREITs). Axis REIT was converted to an IREIT on 11 December 2008. The listing dates and the market capitalization as at 14 December 2014 are given in the table. Information on their property portfolio by sector is also included.

Source: Authors' Compilation from Datastream and Bursa Malaysia, 2015

The third IREIT was created through the conversion of Axis REIT into an IREIT on 11 December 2008. Axis REIT was originally listed as a conventional REIT on 3 August 2005 with a portfolio of office and industrial properties. Axis REIT acquired 33 properties valued at US$42.5 million (RM$136 million) and has gearing of 32%. KLCC REIT was the latest IREIT to be listed on the Bursa Malaysia in May 2013. As in 2014, there are three Islamic REITs in Malaysia with a market capitalization of US$3 billion (RM$15 billion), representing 42% of the overall REIT market in Malaysia by market capitalization.

IREITs invest mainly in SC properties with leases that only allow tenants to run business activities permitted under the Shariah guidelines (Dusuki, 2008). However, the rules give a buffer for tenants engaged in non-permissible activities. These non-permissible activities must not occupy more than 20% of the total floor area or rental income generated from these non-permissible activities must also not exceed 20% (Saeed, 2011). For non-space activities, decisions must be based on Ijtihad, the reasoning process applied by Islamic jurists.

A set of sector and financial screening mechanisms is used to evaluate non-permissible business activities and practices (Derigs and Marzban, 2009). The sector screen, which is also known as the business line screen, limits income from Haram activities such as arms and ammunition, alcohol, tobacco, gambling (maysir) and Non-Halal food products (e.g. pork) to not more than 5% of the total firm revenue. The financial screen forbids business activities in conventional financial services, such as investments in low gearing firms, conventional fixed income instruments (such as bonds), interest-based instruments/accounts and derivatives, as well as short selling. Thresholds on liquidity (accounts receivables, cash and short-term investments), interest income and gearing (total debt to total asset ratio) are usually adopted in the financial ratio screen.

IREIT managers must adhere to Shariah principles when entering into borrowing agreements with banks. IREITs are not allowed to use conventional loans, where banks charge Riba, which is interest on a loan. However, IREITs may use two types of financing contracts which include a short-term financing agreement, known as Muharabah, and a long-term financing agreement, known as Musharakah, under the Islamic Banking Act to raise new capital. (Bank Negara Malaysia, 2010a). The Shariah Guidelines forbid IREITs from participating in risky trades known as Gharar, such as derivatives contracts with uncertain terms for future delivery of underlying assets. Excessive Gharar is not allowed, but minor Gharar, if unavoidable, is tolerated.

5.8 Other IREIT markets

Malaysia's position as the Islamic financial hub in the South East Asia region has been duly recognized, with its experience in developing an IREIT market emulated by other countries in the Middle East region. Kuwait, Bahrain, United Arab Emirates (UAE) and Qatar are among the Gulf Cooperation Council (GCC) countries that have stepped up efforts in establishing IREIT markets. Examples of the newly established GCC's IREITs include Emirates REIT in Dubai, Inovest REIT in Bahrain, Al-Mehrab in Kuwait and Regency REIT in Qatar. Saudi Arabia, as the largest capital market in the region, set up a legal framework for IREITs via the Saudi's Capital Market Authority in 2006 but IREITs have not yet been established in that country.

Given the enormous extent of investment grade commercial real estate assets in the Middle East region, the potential of the GCC IREIT markets is immeasurable. However, the relatively slow pace of current development in GCC IREIT markets may be attributable to various obstacles, including a lack of regulatory incentives in terms of dividend distribution and tax transparency, strict laws on foreign real estate ownership and low levels of investor awareness.

Singapore, as a non-Muslim country, has also set up the first IREIT via the listing of Sabana REIT on the Singapore Stock Exchange in 2010. It holds a portfolio of SC industrial properties in Singapore, which was valued at US$507 million (S$936 million) in March 2015. Some countries in Europe have also relaxed regulatory frameworks to prepare for the future development of Islamic finance. The following sections give a brief overview of IREIT developments in the GCC countries and Singapore.

5.8.1 Kuwait

After the global financial crisis, Kuwait has relied heavily on real estate investment to pull its economy out of the crisis. Kuwait launched its first REIT, known as Al Mahrab Tower REIT, in 2007. It is a private REIT, which is not listed on the public stock market. Al Mahrab Hotel Tower in the holy city of Mecca, valued at about US$100 million in 2014, is the main income generating source for the Kuwaiti IREIT. As part of its effort to promote IREITs, the Kuwaiti government introduced the Capital Market Law (known as the Law No.7) in 2010 in contemplation of the establishment of equity IREITs.

5.8.2 United Arab Emirates (UAE)

Dubai Islamic Bank and Eiffel Management, a French asset management company, jointly established Emirates REIT, the first (private) IREIT in UAE based in the Dubai International Financial Centre (DIFC) in November 2010. Dubai Islamic Bank contributed seven commercial properties secured on long leases as the seed assets into the initial portfolio of Emirates REIT. The REIT manager, Emirates REIT Management (Private) Limited incorporated on 27 October 2010 in the DIFC, being a joint venture between DIB (25%) and Eiffel Management (75%).

On 8 April 2014, Emirates REIT became a publicly listed entity on NASDAQ Dubai through an initial public offer. Due to high interest from institutional investors, especially from the Middle Eastern countries and the UK, the Emirates REIT's IPO was 3.5 times oversubscribed, raising a total of US$175 million new capital. At the IPO date, Emirates REIT's portfolio consisted of ten properties with a combined net leasable area of 1.2 million square feet and valued at US$323.1 million. In March 2015, DIB sold its 25% stake in the REIT manager to Eiffel Management, which now owns 100% of the REIT Manager.

As of 2016, Emirates REIT's investment portfolio includes Index Tower (Dubai International Financial Centre), Office Park (Dubai Internet City), GEMS World Academy (School and Education Facilities), Loft Offices (Dubai Media City), Jabel Ali School (School and Education Facilities), Le Grande Community Hall (Dubai Marina), Building 24 (Dubai Internet City) and Indigo 7 Building (Commercial Property).

5.8.3 Bahrain

The Central Bank of Bahrain (CBB) granted approval for the establishment of the first IREIT in Bahrain in 2009, known as Innovest REIT, with a start-up capital of US$79.8 million. Bahrain's first IREIT was set up as a private REIT with a plan to list the fund on the Bahrain bourse. However, the listing of Innovest REIT has not materialized.

The first successful listing of an IREIT on the Bahrain bourse arose following the successful closure of the IPO of the Eskan Bank REIT in December 2016. The US$38.2 million Eskan Bank REIT IPO was reserved for Bahraini and GCC individual and intuitional investors only with the portfolio comprising two income-generating properties wholly owned by Bahrain Property Musharaka Trust, Danaat Al Madina in Isa Town and Segaya Plaza in Manama.

5.8.4 Singapore

As a non-Muslim country, the Singapore government takes a broad-based approach to developing its Islamic financial system. The Mendaki Growth Fund launched in 1991 was considered one of the earliest Shariah compliant funds worldwide (Lai and Samers, 2016). With a small domestic market for Islamic financial products, Singapore's focus has been on building infrastructure and intermediating capital flows between the Middle East region and Asia (Lai and Samers, 2016).

Sabana REIT was the first IREIT listed on the Singapore Exchange in November 2010. Sabana REIT's investment portfolio comprises 23 industrial buildings in Singapore, generating a gross revenue of US$72 million (S$100.8 million). With a total asset value estimated at US$0.79 billion (SG$1.1 billion) and a market capitalization of approximately US$337 million (S$523.7 million) in 2015, Sabana REIT is the largest IREIT in the world.

5.9 Malaysian IREITs: case studies

Case studies of three IREITs in Malaysia, namely Axis REIT, Al-Aqar Health Care REIT and KLCC (stapled) REIT are provided in this section through a description of the organizational structure of the IREIT, the portfolio composition and an analysis of key financial indicators and performance.

5.9.1 Axis REIT

Axis REIT was established and listed as a conventional REIT on 3 August 2005 with seven buildings in the portfolio including Axis Business Park, Menara Axis, Crystal Plaza, Infinite Centre, Axis Plaza, Complex Development and Kayangan Depot. On 11 December 2008, Axis REIT converted its status into an IREIT. By 2015 the number of properties owned increased to 35 and the assets under management increased from R$300 million to R$1.62 billion.

Axis REIT owns a diversified portfolio of properties including office, retail, light industrial, warehouse and logistics properties, distributed across several states in Malaysia including the Federal Territory (Klang Valley), Johor, Penang, Negeri Sembilan and Kedah. Table 5.5 shows the net rental income and market value of properties in the portfolio of Axis REIT.

Figure 5.3 shows the structure of Axis REIT. The Shariah advisor for Axis REIT is IBFIM, an industry-owned institute that is committed to provide services related to the Islamic financial industry. Axis REIT Managers Berhad is the asset manager of Axis REIT, which is entrusted with the responsibility for all administration-, management- and finance-related matters. Axis Property Services is an external property manager providing maintenance and operational management services for Axis's portfolio of properties. The trustee for Axis REIT is RHB Trustee Berhad.

5.9.2 Al-Aqar Health Care REIT

Al-Aqar Health Care REIT was listed on the Bursa Malaysia on 10 August 2006. The initial portfolio consists of six properties in the health care sector in Malaysia, comprising of the following:

- Ampang Puteri Specialist Hospital Building;
- Damansara Specialist Hospital Building;
- Selangor Medical Centre Building;
- Ipoh Specialist Hospital Building;
- Puteri Specialist Hospital Building; and
- Johor Specialist Hospital Building.

As of 2015, Al-Aqar REIT has expanded its portfolio to 25 assets which are geographically diversified and distributed across three countries, Malaysia, Indonesia and Australia. Al Aqar Health Care REIT's portfolio consists mainly of health care properties located in Malaysia and Indonesia. Table 5.6 shows the net rental income and market value of properties in the portfolio of Al-Aqar Health Care REIT.

Figure 5.4 shows the structure of Al-Aqar Health Care REIT. The Shariah Committee is mandated under the SC rules and comprises members who have a wide knowledge of Shariah principles, being responsible for advising the IREIT on Shariah-related matters.

Table 5.5 Net Property Income and Market Value of Property Portfolio of Axis REIT

Properties	*Net Property Income*		*Market Value**	
	RM$	*US$*	*RM$*	*US$*
Axis Business Park	6,586,489	1,529,892	119,000,000	27,640,992
Crystal Plaza	7,840,360	1,821,137	109,400,000	25,411,131
Menara Axis	8,374,418	1,945,187	120,000,000	27,873,270
Infinite Centre	2,643,703	614,072	42,000,000	9,755,644
Wisma Kemajuan	5,163,804	1,199,434	62,000,000	14,401,189
Axis Business Campus	(180,557)	(41,939)	73,200,000	17,002,694
Axis Shah Alam DC 1	1,739,905	404,140	27,500,000	6,387,624
Giant Hypermarket	3,217,538	747,361	41,000,000	9,523,367
FCI Senai	1,545,759	359,045	17,000,000	3,948,713
Fonterra HQ	1,043,434	242,366	15,000,000	3,484,159
Quattro West	3,890,144	903,592	55,800,000	12,961,070
Strateq Data Centre	4,829,463	1,121,774	53,200,000	12,357,149
Niro Warehouse	1,593,830	370,210	17,700,000	4,111,307
BMW Centre PTP	3,367,250	782,136	30,300,000	7,038,001
Delfi Warehouse	1,543,683	358,562	15,600,000	3,623,525
Axis Vista	2,936,528	682,089	56,000,000	13,007,526
Axis Steel Centre	5,308,326	1,233,003	70,000,000	16,259,407
Bukit Raja Distribution Centre	6,230,059	1,447,101	97,000,000	22,530,893
Seberang Prai Logistic Warehouse 1	1,514,434	351,769	20,200,000	4,692,000
Seberang Prai Logistic Warehouse 2	576,796	133,977	8,000,000	1,858,218
Tesco Bukit Indah	6,014,276	1,396,979	92,000,000	21,369,507
Axis PDI Centre	5,510,980	1,280,075	85,000,000	19,743,566
Axis Technology Centre	2,701,631	627,527	53,200,000	12,357,149
D8 Logistics Warehouse	3,478,201	807,907	32,500,000	7,549,010
Axis Eureka	2,338,343	543,144	54,000,000	12,542,971
Bayan Lepas Distribution Centre	4,397,507	1,021,441	51,500,000	11,962,278
Seberang Prai Logistic Warehouse 3	6,050,353	1,405,359	63,500,000	14,749,605
Emerson Industrial Facility Nilai	2,443,732	567,623	30,000,000	6,968,317
Wisma Academy Parcel	5,495,277	1,276,428	79,500,000	18,466,041
The Annex	542,287	125,961	18,000,000	4,180,990
Axis MRO Hub	3,829,826	889,581	53,000,000	12,310,694
Axis Shah Alam DC 3	13,192,372	3,064,288	183,390,000	42,597,324
Axis Steel Centre @ SiLC	11,057,487	2,568,403	155,500,000	36,119,112
Axis Shah Alam DC 2 ^	2,400,766	557,643	46,000,000	10,684,753
Proposed acquisition – Beyonics iPark Campus ^^	112,809	26,203	.na.	
Total	**139,331,213**	**32,363,470**	**2,046,990,000**	**475,469,200**

Note:
^^ Acquired on 31 March 2015.
^ Sharing of rental income upon payment of redemption sum. The proposed acquisition was completed on 28 January 2016.
*Market value based on valuation conducted by independent registered valuers approved by the SC in 2015.
Source: Annual Report of Axis REIT 2015

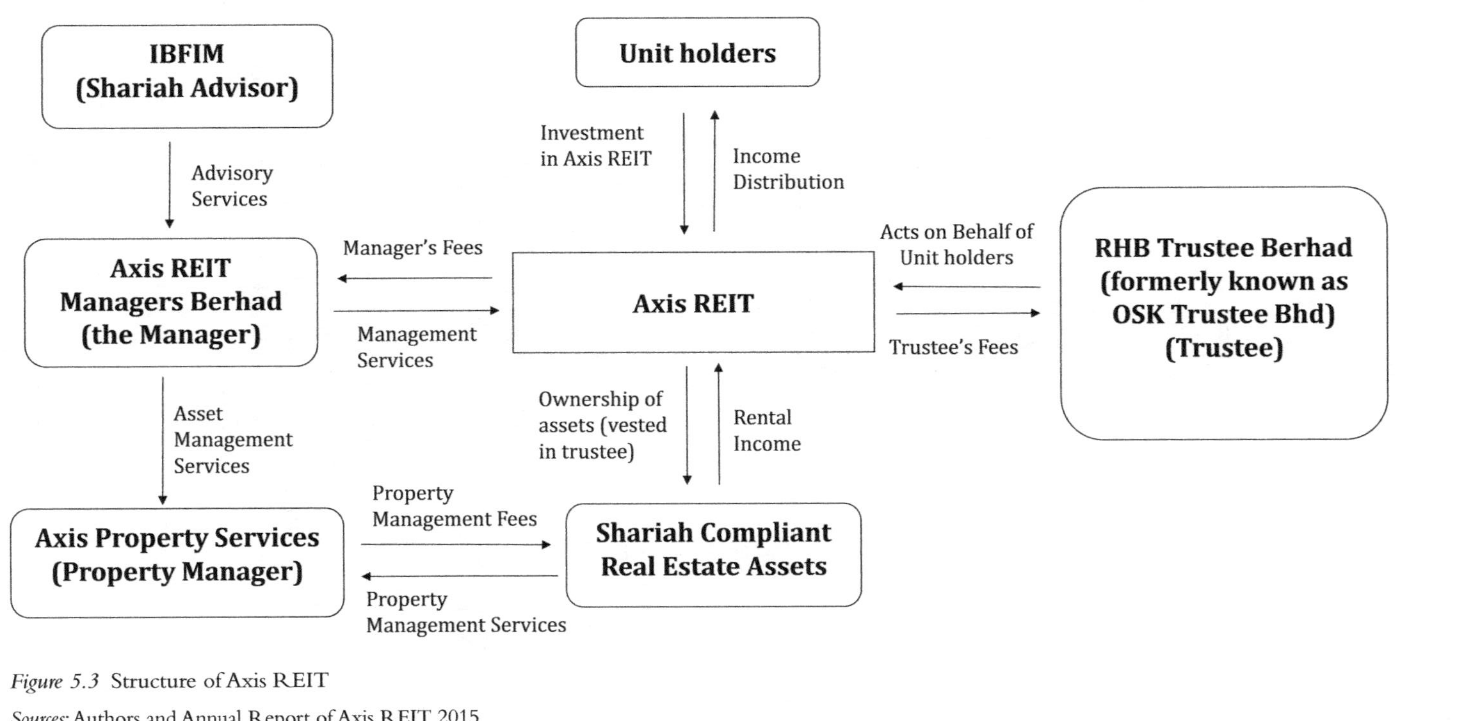

Figure 5.3 Structure of Axis REIT

Sources: Authors and Annual Report of Axis REIT 2015

Table 5.6 Net Property Income and Market Value of Property Portfolio of Al-Aqar Health Care REIT

Properties	*Rental Income*		*Fair Market Value*★	
	RM$	*US$*	*RM$*	*US$*
KPJ Ampang Puteri Specialist Hospital	9,602,892	2,230,533	132,000,000	30,661
KPJ Damansara Specialist Hospital	8,169,845	1,897,669	116,000,000	26,944
KPJ Johor Specialist Hospital	8,221,553	1,909,680	114,000,000	26,480
Selesa Tower	7,246,386	1,683,171	103,500,000	24,041
KPJ Ipoh Specialist Hospital	5,133,854	1,192,477	72,500,000	16,840
KPJ Selangor Specialist Hospital	4,616,775	1,072,372	64,500,000	14,982
KPJ Perdana Specialist Hospital	3,076,433	714,585	44,500,000	10,336
KPJ Penang Specialist Hospital	4,376,809	1,016,633	63,200,000	14,680
KPJ Tawakkal Specialist Hospital	8,592,156	1,995,762	123,000,000	28,570
KPJ Seremban Specialist Hospital★	4,267,266	991,189	66,000,000	15,330
Kedah Medical Centre	3,560,367	826,992	51,600,000	11,986
KPJ Kajang Specialist Hospital	3,159,393	733,855	46,100,000	10,708
Tawakal Health Centre	3,042,935	706,805	44,700,000	10,383
Puteri Specialist Hospital	3,093,091	718,455	43,000,000	9,988
Sentosa Medical Centre	1,914,993	444,809	28,700,000	6,666
Kuantan Specialist Hospital	1,472,540	342,038	21,700,000	5,040
KPJ Health Care University College, Nilai★	1,592,018	369,790	98,363,000	22,847
KPJ College, Bukit Mertajam	1,107,296	257,200	16,200,000	3,763
Kota Kinabalu Specialist Hospital	1,088,788	252,901	na	
Taiping Medical Centre	699,014	162,365	9,900,000	2,300
Kluang Utama Specialist Hospital	275,730	64,046	4,300,000	999
KPJ Klang Specialist Hospital	7,369,258	1,711,711	100,000,000	23,228
Rumah Sakit Bumi Serpong Damai	5,402,377	1,254,849	15,200,000	3,531
Rumah Sakit Medika Permata Hijau	2,271,854	527,700	na	
Jeta Gardens Aged Care Facility and Retirement Village	11,591,768	2,692,504	142,560,000	33,113
Total	110,945,391	25,770,090	1,521,523,000	353,415

Note: # Based on market value as in 2015. ★Inclusive of additional purchase during the year.
Source: Annual Report of Al-Aqar Health Care REIT

5.9.3 KLCC REIT

KLCC Property Holding Berhad (KLCCP), the sponsor of KLCC REIT, was listed on the main Board of the Bursa Malaysia on 18 August 2004. KLCCP underwent a corporate restructure on 9 May 2013 resulting in the creation of KLCC REIT. KLCC REIT is a stapled security product listed on the Bursa Malaysia, where KLCC REIT units are stapled with the ordinary shares of the sponsors (KLCCP).

Upon the completion of the restructuring exercise, the sponsor transferred three office properties (Petronas Twin Tower, Menara ExxonMobil and Menara 3 Petronas) into the KLCC REIT portfolio. As of 2014, the property portfolio of KLCC REIT was valued at RM$8.9 billion. The market capitalization was estimated RM$12.1 billion, which made KLCC REIT the biggest REIT in Malaysia. The revenue and the net property income for the KLCC REIT were estimated at RM$592.9 million and RM$564.7, respectively. Table 5.7 shows the rental income and market values of the assets in the portfolio of KLCC REIT in 2015.

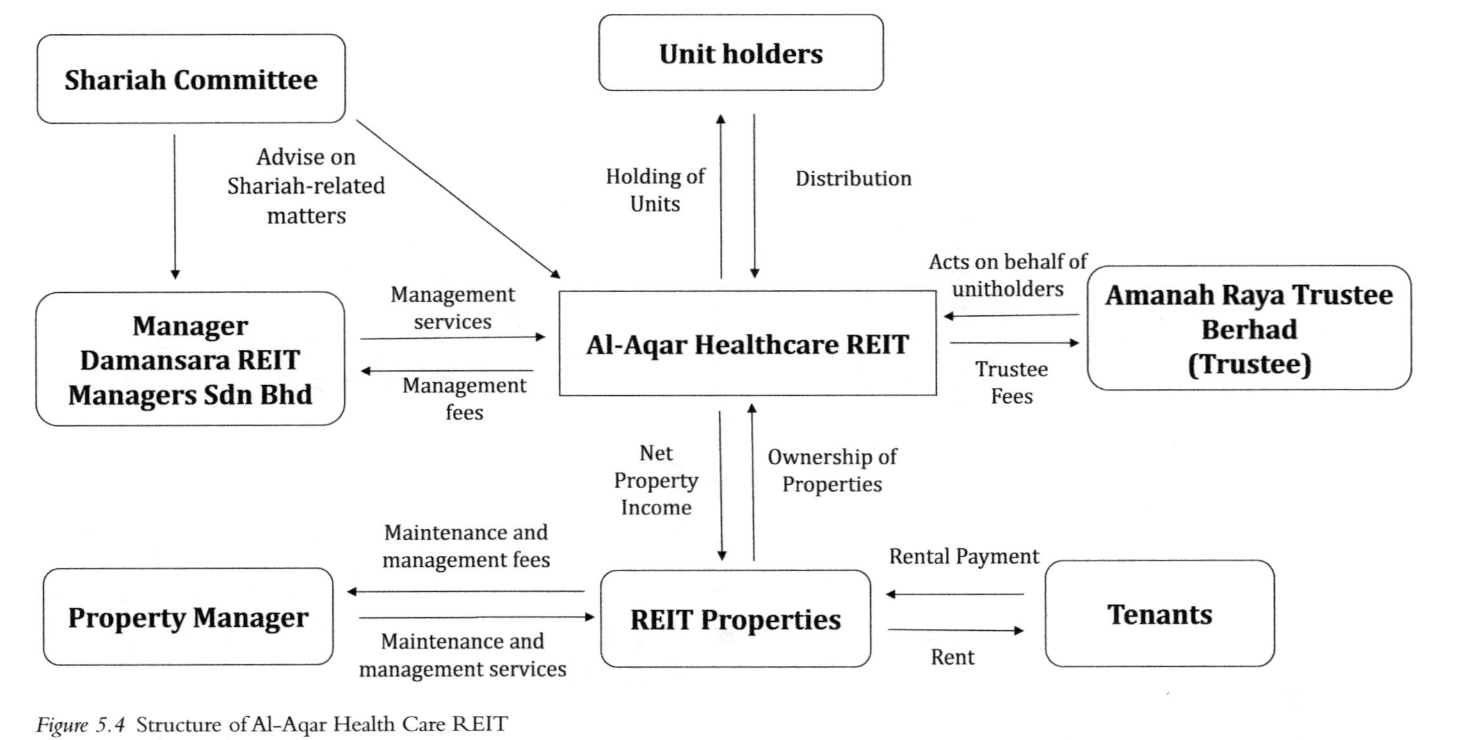

Figure 5.4 Structure of Al-Aqar Health Care REIT

Sources: Authors and Annual Report of Al-Aqar Health Care REIT

Table 5.7 Net Property Income and Market Value of Property Portfolio of KLCC REIT

Properties	*Net Property Income*		*Market Value*	
	RM$	*US$*	*RM$*	*US$*
Petronas Twin Towers	422,100,000	98,044,226	6,820,000,000	1,584,130,819
Menara ExxonMobil	29,400,000	6,828,951	500,000,000	116,138,623
Menara 3 Petronas (Office/Retail)	-	-	1,980,000,000	459,908,947
-Office Tower	87,900,000	20,417,170	-	-
-Retail Podium	30,300,000	7,038,001	-	-
Total	569,700,000	132,328,347	9,300,000,000	2,160,178,389

Note: The data is reported for the FY2015 at 31 December 2015.

Source: Annual Report of KLCC REIT 2015

Figure 5.5 shows the structure of KLCC REIT. The model is a typical externally advised REIT model, where KLCC REIT Management Sdn Bhd, a wholly owned subsidiary of the sponsor, is appointed as the asset manager of KLCC REIT to provide investment, asset and capital management services. Two wholly owned subsidiaries of the sponsor, which are KLCC Urusharta Sdn Bhd and KLCC Parking Management Sdn Bhd, are appointed to provide facility management services and car parking management services, respectively, for the property portfolio.

5.9.4 Financial analyses of Malaysian IREITs

Except for Al-Aqar Health Care REIT, Axis REIT and KLCC REIT were not originally listed as IREITs on the Bursa Malaysia. KLCC Property Holdings Berhad, which is backed by the national petroleum company, Nasional Petroliam Nasional Bhd (Petronas), was subsequently turned into an IREIT by stapling securities of the holding company with the newly created KLCC REIT securities on 9 May 2013. Axis REIT made the conversion from a conventional REIT into an IREIT on 11 December 2008. The listing and conversion dates are summarized in Table 5.8.

Figure 5.6 shows the daily stock prices of the three IREITs. All three IREIT stock prices show positive upward trends from both the dates of IPO listing and from the conversions. Al-Aqar Health Care REIT priced its IPO at RM$1 per unit and the stock price has since risen by about 58% to RM$1.58 as on 31 January 2017. The conversion into an IREIT has also pushed the share price of Axis REIT from a low of RM$0.57 (as on 9 December 2017) to RM$1.68 on 31 January 2017. The KLCC REIT price increased from the pre-stapling unit price of RM$7.25 on 7 May 2013 to RM$7.99 on 31 January 2017.

Figure 5.7 shows the market capitalization of the three IREITs. KLCC REIT saw a huge jump in market capitalization following the bundling of the three assets, which included Petronas Twin Towers, Menara ExxonMobil and Menara 3 Petronas, into the KLCC REIT Stapled portfolio of RM$15 billion. The current market capitalization of KLCC REIT is estimated at RM$14,424.61 million on 31 January 2017. Al-Aqar Health Care REIT has also expanded its IPO market capitalization from RM$104.21 million to RM$1,150.60 million in 31 January 2017. The market capitalization of the Axix REIT increased substantially from RM$112.92 million in the pre-IREIT period to

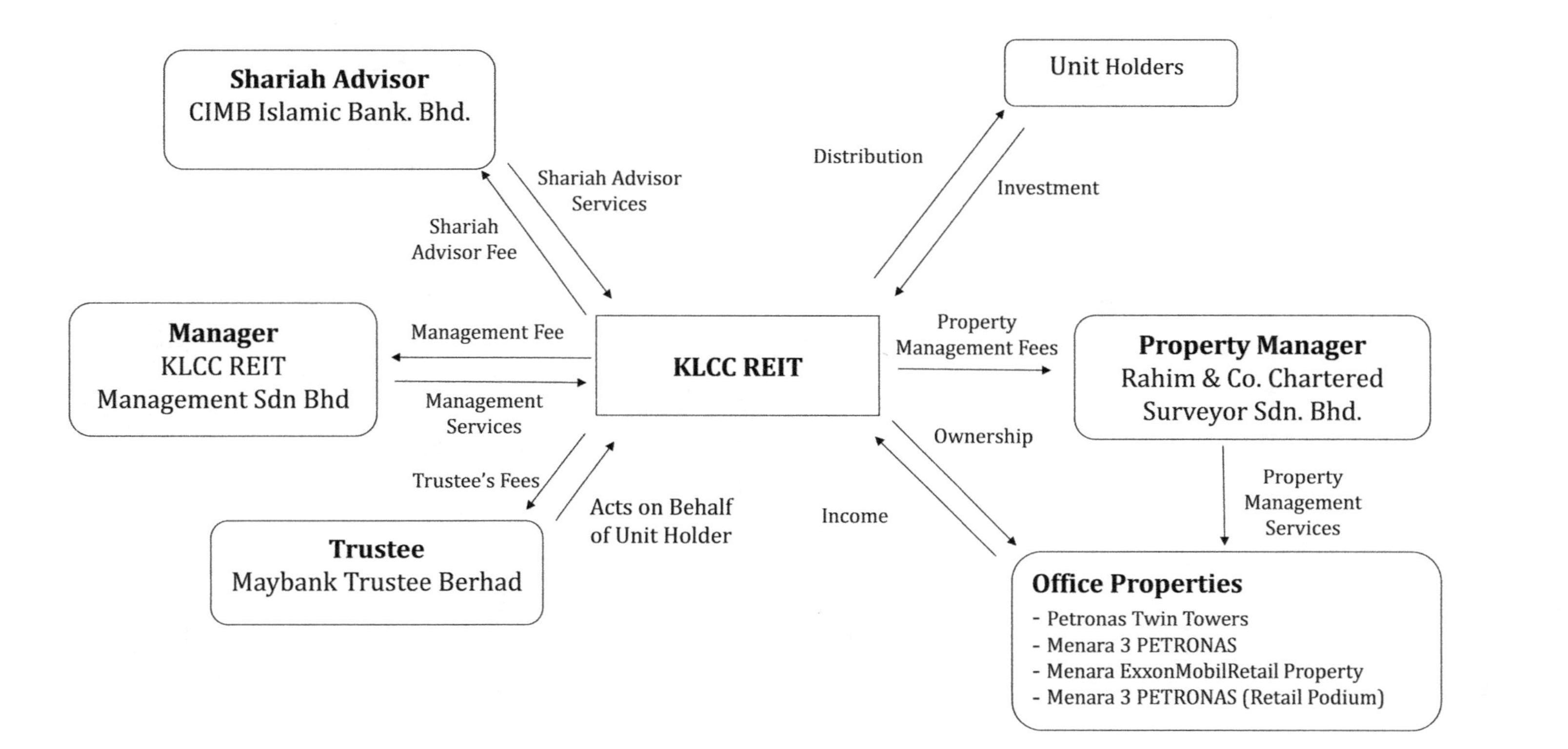

Figure 5.5 Structure of KLCC REIT

Sources: Authors and Annual Report of KLCC REIT 2015

Table 5.8 Financial Performance Statistics of Three IREITs

IREIT	*KLCC Stapled Securities*	*Axis REIT*	*Al-Aqar Health Care REIT*
Ticker	*KLCCSS MK Equity*	*AXRB MK Equity*	*AQAR MK Equity*
Public Listing Date	18-Aug-2004	3-Aug-2005	10-Aug-2006
Islamic Status Date (IPO/ Conversion)	9-May-2013	11-Dec-2008	10-Aug-2006
Date for Financial Statistics:	31-Jan-2017	31-Jan-2017	31-Jan-2017
A) Stock Price Indicators			
Stock price	7.99	1.68	1.58
Total Return Index	13.414	3.617	3.043
Current Market Capitalization (RM$ million)	14424.6117	1856.6916	1150.5978
Return on Common Equity (%)	2.6753	8.9087	8.2847
Return on Asset (%)	1.9199	5.5768	4.3386
Dividend Per Share	0.0985	0.021	0.0517
Dividend Yield (%)	4.668	5.238	5.316
B) Financial Ratio			
Price to Earnings (PE) Ratio	42.5575	15.1351	16.2384
Earnings per Share (EPS) Ratio	0.0782	0.022	0.0216
Debt to Equity Ratio	17.2718	56.0432	74.1787
Debt to Asset Ratio	14.3539	34.7847	40.197
Price to Book Ratio	1.1274	1.3329	1.3053

Sources: Authors' compilation and Bloomberg

RM$1,856.69 million in January 2017. Currently, there are 35 properties in its portfolio, which are distributed across Malaysia.

Table 5.8 provides financial performance statistics for the three IREITs. Axis REIT and Al-Aqar Health Care REIT, which own mainly industrial properties, show higher returns on common equity (ROE) of 8.91% and 8.28% and returns on assets (ROA) of 5.58% and 4.34%, respectively. However, KLCC REIT's portfolio consists mainly of prime grade-A office properties, showing a relatively lower ROE and ROA of 2.68% and 1.92%, respectively.

In terms of distributions per share, KLCC REIT has the highest dividend per share of 9.85 cents (in RM$) and Al-Aqar Health Care REIT has the highest dividend yield of 5.316% among the three IREITs. Panel B of Table 5.8 compares key financial ratios, based on data obtained from Bloomberg. KLCC REIT has the highest PE ratio of about 42.58, whereas the two industrial property IREITs have PE ratios of between 15.14 and 16.24.

The earnings per share of KLCC REIT is also relatively higher at 0.078 compared to 0.022 for the two industrial IREITs. In terms of debt, Al-Aqar Health Care REIT has relatively higher debt to equity and debt to asset ratios of 74.18% and 40.20%, respectively. In comparison, the debt level is the lowest for KLCC REIT, which is estimated at 17.27% and 14.35%, given as the ratios of the equity and asset values, respectively. The price to book ratio shows that the three IREITs are trading at premiums to their book value, ranging from 1.127 to 1.333.

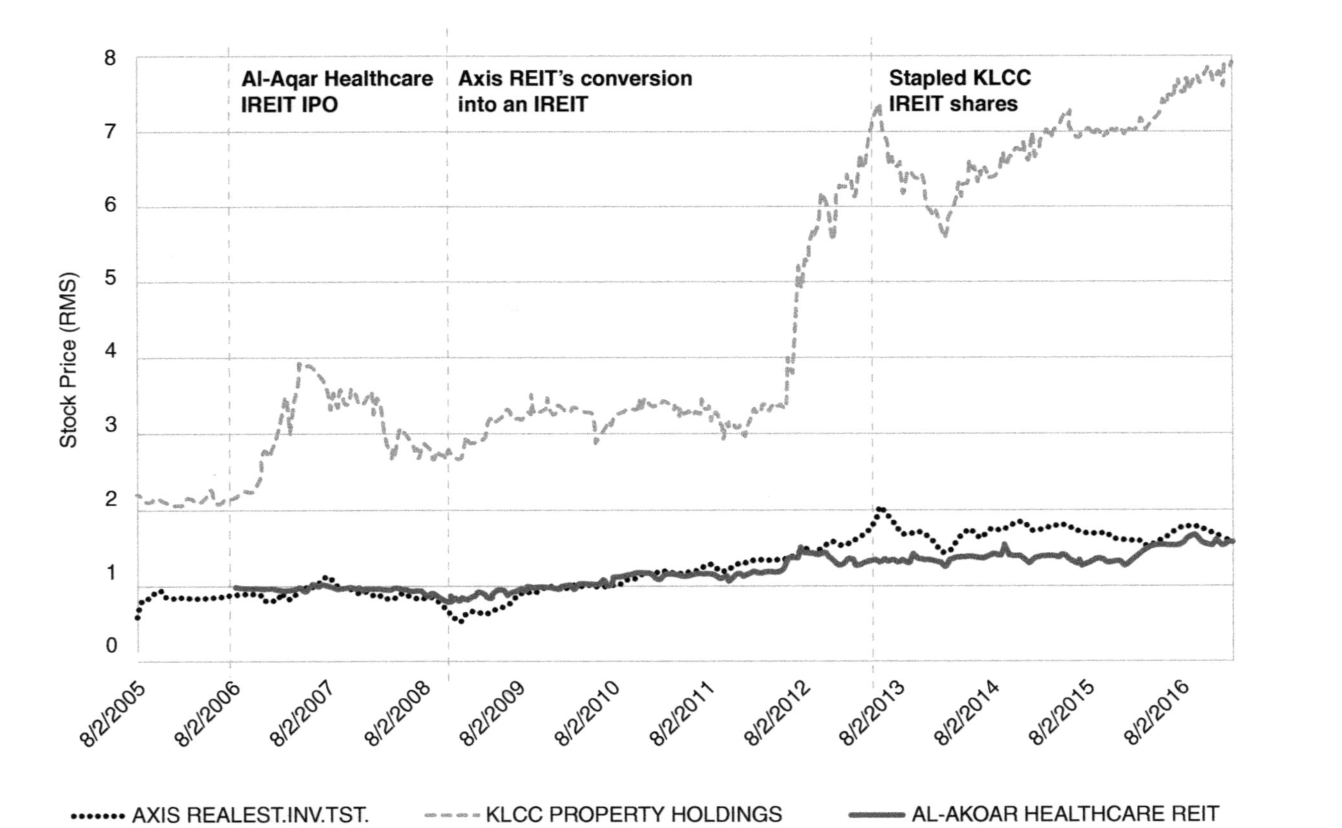

Figure 5.6 Daily Stock Prices of the Three Malaysian IREITs

Sources: Authors and Bloomberg

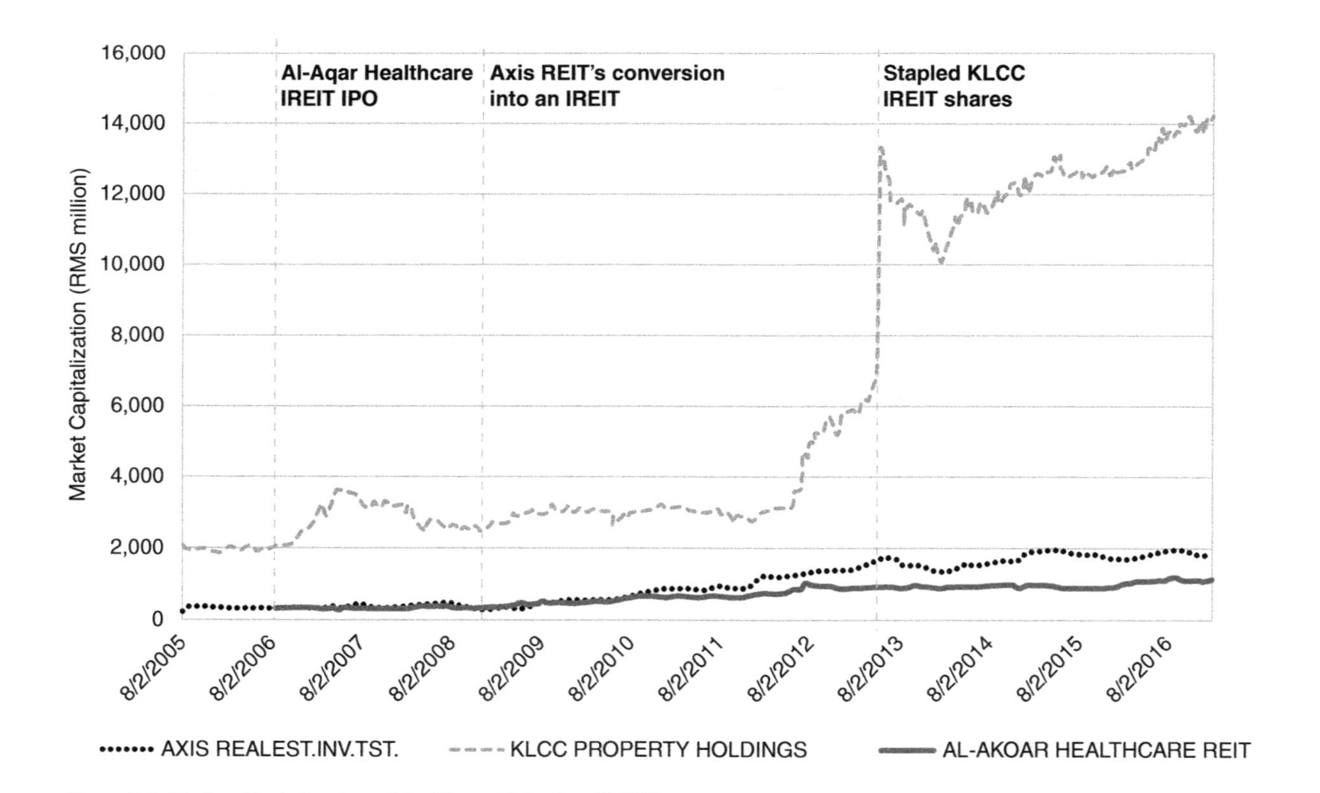

Figure 5.7 Market Capitalization of the Three Malaysian IREITs

Sources: Authors and Bloomberg

5.10 Impact of SC rules on IREITs

Governed by a set of Islamic principles, the SC rules limit IREITs' exposure to a subset of the investment universe. IREITs are only allowed to generate a steady stream of income from Halal real estate leasing businesses (Newell and Osmadi, 2009; Saeed, 2011; Razali and Sing, 2015; Wong, 2015). Several researchers have investigated the possible implications of the SC restrictions for IRETs compared to conventional REITs.

Derigs and Marzban (2009) argue that the SC requirements increase management costs. Sing and Loh (2014) show that SC risk is significantly priced into SC real estate funds. Ibrahim and Ong (2008) and Ibrahim, Ong and Parsa (2009) develop a synthetic SC-REIT portfolio and compare its performance with that of the US REIT index funds. They find no significant differences between the risk-adjusted returns of the two stock portfolios. Many researchers have used Malaysian REIT samples to study the performance between conventional REITs and IREITs. Newell and Osmadi (2009) show that IREITs have lower risk-adjusted returns relative to conventional REITs during the global financial crisis period. They argue that IREITs offer good diversification benefits to investors. Omokolade et al. (2014) find that idiosyncratic risks are strongly correlated with expected returns of IREITs. Morad and Masih (2015) find that IREITs underperform conventional REITs in their models which control for various macroeconomic risk factors. Rozman et al. (2016) find weak evidence of causality from IREIT returns to the returns of other mixed-asset portfolios.

Razali and Sing (2015) use the time-varying beta model to estimate the systematic risks of Axis REIT before and after the conversion to an IREIT (Figure 5.8) with the results showing that the before betas of Axis REIT are significantly higher than the betas after its conversion into an IREIT in 2008. The authors argue that the Shariah principles governing the investment and borrowing activities of IREITs reduce IREITs investors' exposure to stock market volatilities.

5.11 Discussion

The emergence of IREITs is not by accident, with the strong demand for new Islamic financial products by individual and institutional investors in both Muslim and non-Muslim countries providing strong impetus for financial institutions and for firms that own portfolios of SC properties to offer IREITs as alternatives to conventional REITs. The governments of Muslim countries have taken active steps to promote the Islamic REIT and other non-Muslim countries are also taking measures to prepare their respective markets to tap into the growing demand for Islamic financial investments including REITs.

Malaysia, as one of the leading Islamic financial hubs in the region, has the first mover advantage by introducing IREITs to the market. Its experience offers opportunities for SC firms in Malaysia to enhance regulatory frameworks, improve management expertise and also embrace new thinking in popularizing IREITs not just within the GCC and those regions with majority Muslim populations, but also to reach out to other non-Muslim markets that have not yet embraced IREITs.

Currently, IREIT development remains in its infancy and a relatively slow pace of development is expected. It is important for the financial regulators and central bankers of different countries to work closely to remove possible institutional obstacles that may stand in the way of IREIT development, including a lack of financial incentives for promoting IREITs, some restrictive SC rules that impede financing and investment risk hedging and inconsistent Shariah rules governing IREITs across different jurisdictions. IREITs, in the long term, could offer

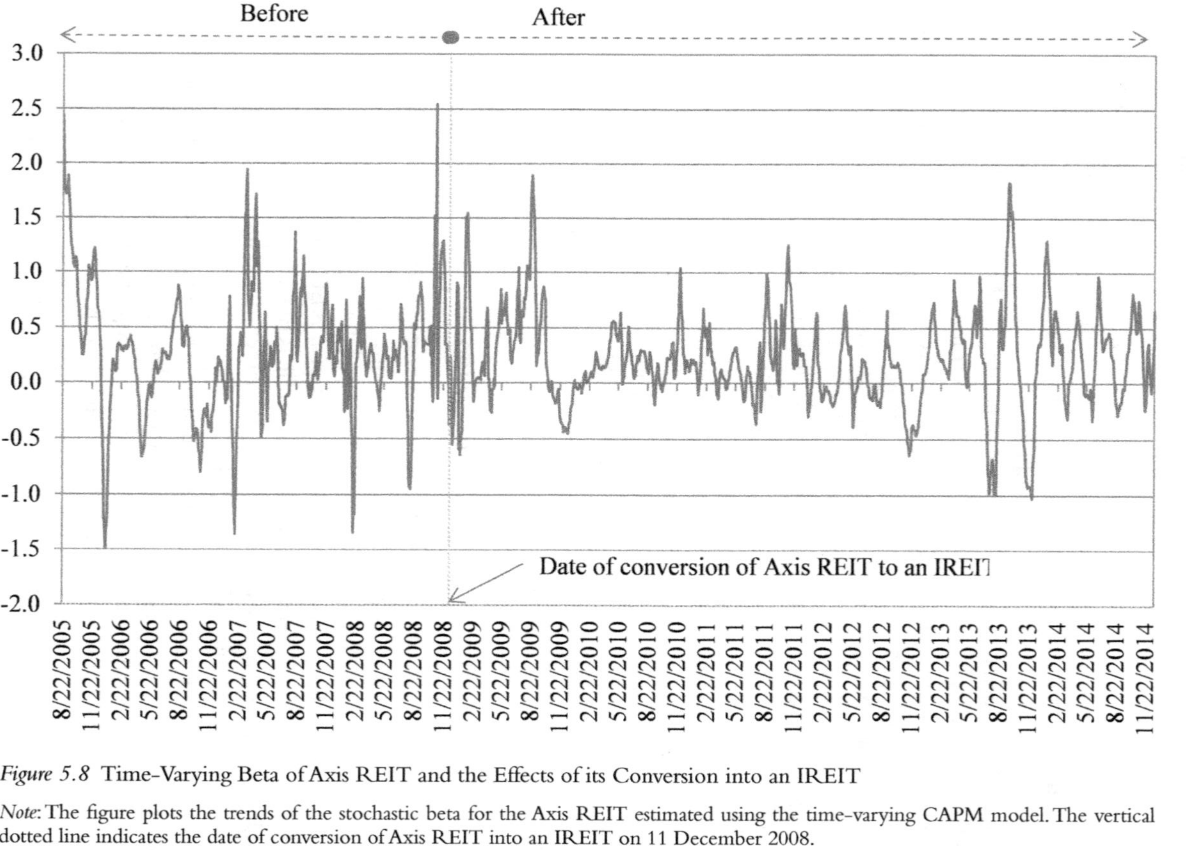

Figure 5.8 Time-Varying Beta of Axis REIT and the Effects of its Conversion into an IREIT

Note: The figure plots the trends of the stochastic beta for the Axis REIT estimated using the time-varying CAPM model. The vertical dotted line indicates the date of conversion of Axis REIT into an IREIT on 11 December 2008.

Source: Razali and Sing (2015)

investors alternative investments that are consistent with religious beliefs and core values in life. Integrating IREITs with the main-stream REIT markets globally is also essential to enhance price efficiency, capital flows and liquidity in the IREIT markets.

5.12 Summary

Chapter 1 sought to identify critical contextual issues in international REITs, including the defining characteristics of REITs, the global evolution of REITs and the risk-return characteristics of real estate investment through REITs relative to non-REIT real estate investment with Chapter 2, the *Post-Modern REIT Era*, examining the evolution of the global REIT market, what is deemed to be best current market practice and how the market is expected to change going forward. Chapter 3, *Emerging Sector REITs*, investigated such sectors as timber and data centres with Chapter 4, *Sustainable REITs*, analysing REIT environmental performance and the cost of equity.

This chapter, *Islamic REITs*, examined the evolution of global IREITs through a study of the Malaysian market. Reviewing the features of the Islamic financial system, the role of asset backing to protect investors from extreme market risk is identified together with the absence of interest and the encouragement of mutual sharing of risk and profit between lenders and borrowers. The role of Shariah Compliance and permissible activities is outlined with IREITs permitted to hold real estate assets in sectors such as residential, industrial, commercial, retail, office, car parks, hotels, institutions and health care which generate a steady stream of rental income from business activities that are permitted under the Shariah Compliance guidelines.

Shariah Compliance not only requires the involvement of a Shariah Committee in the property investment decision-making process of an IREIT but also impacts IREITs in many diverse ways including strategic asset allocation and stock selection. Effectively, only Halal or religiously ethical business activities are allowed in the properties held by IREITs. Despite such constraints, with the Muslim population constituting 24% of the world population, the potential for growth in Islamic financial products and fund management services is enormous with well-developed IREIT markets in Malaysia and emerging markets in Kuwait, UAE, Bahrain and Singapore.

Part I continues with chapters each reviewing a current theme of REIT evolution through the lens of contemporary research. Chapter 6, *Behavioural Risk in REITs*, addresses the management of behavioural risk in global REITs and Part I concludes with Chapter 7 which reviews recent research into *REIT Asset Allocation*.

Part II then includes six chapters each analysing REITs in a region of the world, using case studies from a developed, developing and emerging REIT sector in the region, concluding with Chapter 14 which considers *Directions for the Future of International REITs*.

References

Abozaid, A. 2008, 'Examining bay al-inah and its new applications in the Islamic financial institutions', *Journal of Al-Tamaddun*, Vol. 4.

Azhar, N.E. and Saad, N.M. 2012, 'Syariah REITs vis-a-vis conventional REITs: An analysis', *International Journal of Academic Research in Business and Social Sciences*, Vol. 2, No. 7, pp. 1–17.

Aziz, Z.A. 2005, *Building a Progressive Islamic Banking Sector – Charting the Way Forward*, Seminar on 10-Year Master Plan for Islamic Financial Services Industry, Putrajaya.

Bank Negara Malaysia 2014a, Guidelines on Musharakah and Mudharabah Contracts for Islamic Banking Institutions, Bank Negara Malaysia, Kuala Lumpur.

Bank Negara Malaysia 2014b, available at www.bnm.gov.my/index.php?ch=statistic&pg=stats_exchangerates, November 25, 2014.

Bank Negara Malaysia 2015, available at www.bnm.gov.my/index.php?ch=statistic&pg=stats_exchangerates, November 25, 2014.

Binmahfouz, S.S. 2012, *Investment characteristics of Islamic investment portfolios: Evidence from Saudi mutual funds and global indices*, Doctoral Thesis, Durham University.

Derigs, U. and Marzban, S. 2009, 'New strategies and a new paradigm for Shariah-compliant portfolio optimization', *Journal of Banking & Finance*, Vol. 33, pp. 1166–1176.

Dusuki, A.W. 2007, *Practice and prospect of Islamic real estate investment trusts (s-reits) in Malaysian Islamic capital market*, The International Conference on Islamic Capital Markets, Jakarta.

Dusuki, A.W. 2008, 'Understanding the objectives of Islamic banking: A survey of stakeholders' perspective', *International Journal of Islamic and Middle Eastern Finance and Management*, Vol. 1, No. 2, pp. 132–148.

El Qorchi, M. 2005, 'Islamic finance gears up', *Finance and Development*, Vol. 42, No. 4, p. 46.

Hamzah, A.H., Rozali, M.B. and Tahir, I.M. 2010, 'Empirical investigation on the performance of the Malaysian real estate investment trusts in pre-crisis, during crisis and post-crisis period', *International Journal of Economics and Finance*, Vol. 2, No. 2, p. 62.

Hoggarth, D. 2016, 'The rise of Islamic finance: Post-colonial market-building in central Asia and Russia', *International Affairs*, Vol. 92, No. 1, pp. 115–136.

Ibrahim, M. and Ong, S. 2008, 'Shariah compliance in real estate investment', *Journal of Real Estate Porfolio Management*, Vol. 14, No. 4, pp. 105–124.

Ibrahim, M., Ong, S. and Parsa, A. 2009, 'Shariah property investment in Asia', *Journal of Real Estate Literature*, Vol. 17, pp. 233–248.

Jobst, A.A. 2007, 'The economics of Islamic finance and securitization', *The Journal of Structured Finance*, Vol. 13, No. 1, pp. 6–27.

Lai, P.Y.K. and Samers, M. 2016, 'Conceptualizing Islamic banking and finance: A comparison of its development and governance in Malaysia and Singapore', *The Pacific Review*, pp. 1–20.

McKinsey, 2006, *The World Islamic Banking Competitiveness Report 2007–08, 'Capturing The Trillion Dollar Opportunity'*, McKinsey, London.

Mirakhor, A. and Zaidi, I. 1988, *Stabilization and growth in an open Islamic economy*, IMF Working Paper #22.

Morad, S.N. and Masih, M. 2015, *Islamic REIT response to macroeconomic factors: A Markov regime switching auto regressive approach* (No. 65237), University Library of Munich, Germany.

Newell, G. and Osmadi, A. 2009, 'The development and preliminary performance analysis of Islamic REITs in Malaysia', *Journal of Property Research*, Vol. 26, No. 4, pp. 329–347.

Omokolade, A., Ong, S.E., Ibrahim, M.F. and Newell, G. 2014, 'The idiosyncratic risks of a Shariah compliant REIT investor', *Journal of Property Research*, Vol. 31, No. 3, pp. 211–243.

Osmadi, A.B. 2007, *REITs: A new property dimension to Islamic finance*, 13th Pacific-Rim Real Estate Society Conference, Fremantle, Western Australia.

Parashar, S. and Venkatesh, J. 2010, 'How did Islamic banks do during global financial crisis?' *Banks and Bank System*, Vol. 4, pp. 54–62.

Razali, M.N. and Sing, T.F. 2015, 'Systematic risk of Islamic REITs and conventional REITs in Malaysia', *Journal of Real Estate Portfolio Management*, Vol. 21, No. 1, pp. 77–92.

Rozman, A.T., Razali, M.N., Azmi, N.A. and Ali, H.M. 2016, 'The dynamic of linkages of Islamic REITs in mixed-asset portfolios in Malaysia', *Pacific Rim Property Research Journal*, Vol. 22, No. 3, pp. 245–265.

Saeed, M. 2011, *The outlook for Islamic REITs as an investment vehicle*, Working Paper, Department of Economics, Management School, Lancaster University.

Seah, K.Y., Shilling, J.D., Sing, T.F. and Wang, L. 2016, '*REITarization' effects on global real estate investments*, Working Paper, Institute of Real Estate Studies, National University of Singapore.

Securities Commission of Malaysia 2001, available at www.sc.com.my/wp-content/uploads/eng/html/cmp/EXECUTIV.PDF, July 2015.

Sing, T.F., Ho, D.K.H. and Mak, M.F.E. 2002, 'Real estate market reaction to public listings and acquisition news of Malaysia REITs', *Journal of Real Estate Portfolio Management*, Vol. 8, No. 3, pp. 209–228.

Sing, T.F. and Loh, K.W. 2014, 'Predictability of Shariah-compliant stock and real estate investments', *International Real Estate Review*, Vol. 17, No. 1, pp. 23–46.

Tabung Haji 2015, available at www.tabunghaji.gov.my/background, July 2015.

Ting, K.H. and Noor, A.R.M. 2007, *Islamic REITs: A Syariah-compliant investment option*, Asian Real Estate Society Annual Conference, 9 July 2007, Macau.
Wilson, R. 1997, 'Islamic finance and ethical investment', *International Journal of Social Economics*, Vol. 24, No. 11, pp. 1325–1342.
Wong, Y.M. 2015, *Malaysia REIT: First decade development and returns characteristics*, 21st Annual Pacific-Rim Real Estate Society Conference, 18–21 January 2015, Kuala Lumpur, Malaysia.

6
BEHAVIOURAL RISK IN REITS
Managing behavioural risk in global REITs

Larry Wofford and Michael Troilo

6.1 Introduction

This book aims to identify key areas for research in the REIT discipline for the next five to ten years by surveying the current state of the REIT discipline around the world and identifying emerging and cutting edge research areas through a thematic review of current contextual issues and a regional analysis based on case studies.

This book comprises two parts, the first part being a thematic review of emerging and cutting edge global research into current contextual issues in REITs internationally and the second part being a regional analysis of REITs around the world, each written by authoritative academic authors from the world's leading Universities and REIT industry experts.

Part I includes six chapters each reviewing a current theme of REIT evolution through the lens of contemporary research. Chapter 1 focused on critical contextual issues in international REITs while Chapter 2, the *Post-Modern REIT Era,* examined the evolution of the global REIT market, what is deemed to be best current market practice and how the market is expected to change going forward with Chapter 3, *Emerging Sector REITs*, investigating such sectors as timber and data centres. Chapter 4, *Sustainable REITs*, analysed REIT environmental performance and the cost of equity with Chapter 5, *Islamic REITs*, examining the evolution of global Islamic REITs through a study of the Malaysian market.

This chapter, *Behavioural Risk in REITs*, seeks to develop a cognitive risk management framework for REIT entrepreneurs and executives. Beginning with a brief overview of the cognitive and behavioural landscape and the notion of cognitive risk, attention is then turned to how to effectively manage cognitive risk in an enterprise setting. Building on the fundamental elements contributing to cognitive risk, the power of positive alignment between these fundamental elements is contended to be critical to cognitive risk management. The power of mindfulness, the need for reflective practitioners and the concept of management as a practice are identified as crucial elements of cognitive risk management implementation.

Part I then concludes with Chapter 7 which reviews recent research into *REIT Asset Allocation*.

Part II includes six chapters each analysing REITs in a region of the world, using case studies from a developed, developing and emerging REIT sector in the region:

Chapter	Region	Developed **REIT Sector**	Developing **REIT Sector**	Emerging **REIT Sector**
8	North America	USA	Canada	Mexico
9	Latin America	Brazil	Argentina	Uruguay
10	Europe	UK	Spain	Poland
11	South East Asia	Singapore	Malaysia	Thailand
12	North Asia	Japan	Hong Kong	China
13	Oceania	Australia	South Africa	India

concluding with Chapter 14 which considers *Directions for the Future of International REITs*.

6.2 Human cognition, behavioural finance and real estate

Human cognition and its limits became a fruitful area of inquiry among US behavioural scientists in the post-war era. Herbert Simon, one of the most influential pioneers, formulated the difference between behavioural and behaviouristic research with a key point of his discussion being the acknowledgement of constraints to human cognition (Simon, 1955). George Miller, 1956, in his seminal paper, *The Magical Number Seven Plus or Minus Two: Some Limits on Our Capacity for Processing Information*, expanded the conversation by specifying the tendency of the human mind to remember between five and nine facts/data points at any given time. From these beginnings, Simon (1979, 1980) developed the well-known theory of bounded rationality whereby human beings can and will think rationally in general, but the ability to find optimal solutions is conditioned upon the difficulty of the task at hand, the amount of time to arrive at a solution and the person's own cognitive limitations.

Cognitive psychology, in treating the broad issue of how a limited human brain handles seemingly infinite information, discovered that people employ a variety of filters and time-saving devices to cope. Tversky and Kahneman (1971, 1972, 1973, 1974, 1979) identified and categorized these mental shortcuts as various biases and heuristics. The trade-offs embodied in biases and heuristics involve time and simplicity on the one hand and lost information and possible cognitive errors on the other. Biases and heuristics make the world more intelligible to the human mind, but possible options and opportunities may be ignored including those that are better for solving the current problem and may deviate from economic rationality. The departure point for most research in economics, finance and real estate is the assumption that individuals will act in ways to maximize their utility, with this assumption dating from the 18th-century economists Francois Quesnay and Adam Smith (Stigler, 1987) and being the foundation for the Efficient Markets Hypothesis (Fama, 1970).

Beyond examining deviations from economic rationality, cognitive research also addresses whether such deviations are systematic or random. If deviations from rationality are systematic, then formulating strategies for correcting or eliminating them is more likely to produce more significant results than if such deviations are random. Research indicates that deviations from rationality frequently exhibit a systematic component (Wofford, Troilo and Dorchester, 2010, 2011a).

In the 1970s and 1980s, behavioural research radiated from cognitive psychology exclusively to other disciplines, including economics, finance and real estate. The first effort in real estate was Ratcliff's (1972) book examining the behaviour of market participants in the 1960s, but a coherent research programme did not develop for two more decades. Wofford (1985) built on Ratcliff's work by highlighting the impact of cognitive processes on real estate investment decisions. Diaz (1990) initiated an empirical behavioural real estate research programme with his efforts carrying into the millennium as the field of real estate grappled with "establishing the boundaries" of cognitive research in its domain (Black et al., 2003).

As mentioned, utility maximization is the crux of economic rationality with behavioural research in economics and finance evaluating the reasons and extent to which decisions deviate from rational choice theory. Hardin (1999) and Ritter (2003), building on the seminal work of Tversky and Kahneman (1971, 1972, 1973, 1974, 1979), articulated several of the common heuristics and biases with regard to investment decisions which may be briefly summarized as the following:

- **representative heuristic** – this heuristic leads people to extrapolate incorrectly from a sample to the general population; in particular, to imagine that a sample is more representative of a population that it truly is (Hardin, 1999);
- **availability heuristic –** this heuristic lends more weight in decision-making to events and situations that the mind recalls easily. It also manifests itself in people couching a problem in terms of what is familiar to them, even though such familiarity may not in fact map onto the issue at hand (Ritter, 2003);
- **anchoring and adjustment heuristic –** this heuristic refers to the tendency of the mind to attach to a given fact for subsequent problem-solving. For example, you may be asked to guess whether a given property is worth more than $1 million or less than $1 million. If you are then asked to guess the actual value, you are likely to give an answer around $1 million, regardless of the true value of the property. Errors in judgement can occur from this initial attachment and subsequent adjustments (Hardin, 1999);
- **framing –** the phrasing of a choice can bias the selection of the course of action. Tversky and Kahneman, as recounted by Lewis (2017), as well as Ritter (2003), noted that physicians will make different choices depending on whether the term "survival probability" or "mortality rate" is used, even though the one is just the obverse of the other; and
- **confirmation bias –** this bias relates to the tendency of human beings to seek data and information that will reinforce existing beliefs. Evans (1989) also identifies this as "positivity bias".

Overconfidence is a staple of biases and heuristics. Though Tversky and Kahneman did not originally label it a bias, later scholarship (Croskerry, 2003; Beshears and Gino, 2015) include it as such. Regardless of whether it is included as a separate bias or not, it is so salient that it deserves a mention. Whether the experimental group contains doctors, lawyers, investors or entrepreneurs, the power of heuristics is such that people tend to be much more confident in their assessments than the available knowledge warrants (Lewis, 2017).

Another popular cognitive paradigm involves the difference between "fast" thinking and "slow" thinking (Kahneman, 2011). According to Kahneman, "fast" thinking is automatic and instinctive, relying on heuristics and biases to generate thoughts quickly. "Slow" thinking, on the other hand, is logical and methodical. Kahneman also terms these two cognitive modes as "System 1" and "System 2", respectively (ibid).

Behavioural real estate has followed the lead of economics and finance in focusing on cognitive psychology and the existence and impact of the aforementioned biases and heuristics on decision-making. As noted by Diaz (2002), real estate valuation has received the greatest amount of behavioural research attention, while other real estate areas have received less. Because of this development, behavioural real estate research has been relatively focused in terms of research scope, subject matter and methodologies. The common behavioural element is the desire to understand and manage the sources of departure from economic rationality attributable to human cognition (Wofford, Troilo and Dorchester, 2010, 2011a).

Following Wofford, Troilo and Dorchester (2010), behavioural real estate research in valuation has studied the extent to which cognitive factors influence valuation results to deviate from expected outcomes in the absence of such influences. In other real estate activity areas, such as development and equity investment, cognitive activity has the potential to create variability in assessing potential outcomes. Via problem-framing, problem-solving and decision-making, cognition also affects the variability of the actual outcomes themselves. This variability in perceiving potential outcomes and the influence of cognition on actual outcomes created by cognitive activity is defined as *cognitive risk* (Wofford, Troilo and Dorchester, 2010, 2011a). Cognitive risk includes a wide range of activities including assessment and evaluation, planning, problem-framing, problem-solving, decision-making and execution.

The cognitive sources of outcome variability include the familiar biases and heuristics of cognitive psychology, but there are others worth considering. Communication problems, organizational problems and dynamics, inattention, distractions, poor procedures, fatigue, systems issues, ergonomics, stress, information problems and time issues all find their locus in human cognition. When placed within the context of a real estate environment characterized by chaos or complexity, dynamism and ambiguity, the variability produced by these cognitive variations can be magnified in a nonlinear manner to produce significant risk exposure (Wofford, Troilo and Dorchester, 2010, 2011a).

6.3 Overcoming cognitive risk in decision-making

Recognizing the existence of cognitive risk is a necessary step, but not sufficient guidance, for managing REITs. In order to make decision-making effective, a holistic approach is essential for guarding against cognitive risk. Possible solutions may be found in a broad range of disciplines, from medicine to business management to nuclear engineering. A smooth translation of knowledge from theory to application is often the missing ingredient with the solution existing but either having not been communicated to practitioners or having been communicated in an incomplete way (Wofford, Troilo and Dorchester, 2011b).

An early, comprehensive attempt to catalogue cognitive errors and prevent them may be found in the work of Croskerry (2003) in the field of medicine. In addition to the biases listed previously, he elucidates other sources such as "premature closure", which is "the tendency to accept a diagnosis before it has been fully verified", and "search satisfying: the universal tendency to call off a search once something has been found" (Croskerry, 2003, p. 778). He follows this discussion by offering strategies such as considering alternatives, which he describes as a mandatory, routine exercise to imagine/examine other possibilities beyond the available diagnosis and metacognition, which is "training for a reflective approach to problem-solving; stepping back from the immediate problem to examine and reflect on the thinking process" (ibid). Such tactics would be valuable in almost any situation requiring judgement including, for example, in REIT management where a person should not be satisfied at the first explanation for a given situation but should consider alternatives and practice metacognition.

Additional work in business management, for example that of Zweig (2007), addresses aspects of cognitive risk and bias in decision-making. Soll, Milkman and Payne (2015) highlight the fact that even when managers engage System 2 thinking and set aside System 1 heuristics, their logic is often muddled. In order to clarify thinking, they suggest a "premortem", also known as "prospective hindsight" (ibid, p. 68). This technique involves "imagining a future failure, then explaining its cause" and they claim that premortems "help to identify potential problems that ordinary foresight won't bring to mind" (ibid, p. 68).

To counteract the natural tendency of managers to surround themselves with people who think as they do, Soyer and Hogarth (2015) proffer several suggestions. The first is to actively seek a group of people who will tend to disagree with one's opinions. The second, which is similar, is to retain a confidant who challenges one's thinking. The third is to establish anonymous feedback channels. In addition, they call for managers to engage in "disconfirming: asking yourself how you could know you were, in fact, wrong" (ibid, p. 77). These techniques and others enable managers to gain a wider perspective on problems and their possible solutions.

In the field of real estate, Wofford, Troilo and Dorchester (2010) specifically advise the use of continuous planning and strategic foresight to reduce both the likelihood of errors and failure induced by cognitive risk as well as the magnitude of these events. Wofford, Troilo and Dorchester (2011b) and Wofford and Troilo (2013) build on this work to advocate the use of evidence-based practices in both medicine and management to construct channels of translational research between academic theory and pragmatic application.

6.4 Cognitive risk and international REITs: toward a more holistic approach

In this section, the notion of managing cognitive risk in a more holistic manner is considered. A preliminary framework for framing and approaching cognitive risk management is developed in the spirit of stimulating the thinking necessary to develop a custom risk-management approach appropriate for a given enterprise, including REITs. It will be seen that a "one size fits all" approach simply will not work because contextual issues make every enterprise's cognitive risk exposure unique. Leaders and managers must assess their enterprise in order to identify the sources, importance and manageability of the cognitive risk inherent in their enterprise. Developing this framework begins with considering the REIT environment.

International REITs operate in difficult and varied property and financial markets, with varying legal, political, ethical and cultural environments. In short, they face challenging environments creating significant business risk and management challenges without even considering the possible risk introduced by cognitive issues. To further compound the management situation by introducing the notion of cognitive risk may turn a challenging situation into an overwhelming one, causing cognitive risk simply to be ignored or addressed in a cursory manner.

For start-up REITs, resources in terms of time, attention and expertise may be scarce. If the REIT has been around for some time and is doing well, the option of simply assuming that the enterprise has survived and continued to function adequately may be viewed as the best evidence for choosing not to consider cognitive risk explicitly. After all, "if it isn't broken, why fix it?" For other REITs, problems generated by poor decision-making may be apparent and compelling. For yet another group of REITs, poor performance may require a mindful assessment of decision-making processes in order to reveal any embedded elements of decision-making that require improvement. If successfully implemented, improvements in decision-making may reduce cognitive risk and contribute to the REIT's overall risk and return performance.

In discussing the development and operation of REITs, Parker (2011) notes that many REITs start as entrepreneurial enterprises and develop into larger, more complex organizations over time. In this regard he views REITs as "a major business enterprise which combines the skills common to all major business enterprises with roles and skills specific to the real estate sector" (Ibid, p. 3). In this chapter it is assumed that REITs share common characteristics with other business enterprises and that knowledge related to business in general has applicability to REITs. The remainder of this chapter builds on knowledge of cognition and decision-making in general business to develop a set of reasonable considerations for REIT leadership to apply in managing cognitive risk.

The goal is to outline the elements of a *customizable framework and associated process* for REIT problem-solving and decision-making that is effective in reducing cognitive risk. This framework is not perfect, but a framework that can increase the likelihood of consistently better decisions and decision processes is contended to be achievable, useful and valuable. Martin (2004, p. 17) notes that enterprise success may depend on superior processes, that is, processes that produce something better, cheaper and/or faster, as much as on a superior product itself. Given the variation between REITs in terms of size, organizational design and development, business models, scope of operations and overall complexity, there is no specific framework that is optimal for them all.

6.5 Fundamental elements of the decision-making landscape

Since REITs are business enterprises, it is helpful to develop a baseline understanding of critical elements of the problem-solving and decision-making landscape in which such enterprises must operate. These contextual elements provide a perspective for considering the nature of the decisions facing an enterprise and how the elements affect the presence, nature and extent of an enterprise's cognitive risk. As sources of cognitive risk, recognizing, assessing and managing these fundamental elements will provide an approach to managing cognitive risk. Fundamental elements related to the existence and management of cognitive risk include the following:

- problems and decisions vary in significance and urgency;
- enterprises operate within both internal and external contexts exhibiting varying degrees of risk and uncertainty;
- enterprises operate within contexts characterized by multiple interacting systems;
- time is an important consideration in almost all business matters;
- every enterprise has one or more implicit or explicit business model(s) that represents the logic underlying how it creates and captures value;
- every enterprise experiences identifiable stages of growth and development, with each stage having strategic, operational, leadership and decision-making implications;
- every enterprise has an implicit or explicit strategy that affects cognitive risk;
- formal organizational design, along with informal adjustments to it, affect the flow of information, the distribution of roles and responsibilities and decision-making expectations; and
- every organization has a culture embodying its norms and mores that also affects decision-making.

Collectively, the consideration of these fundamental elements creates a unique problem-solving and decision-making context for each REIT. While elements may be similar for two REITs, these elements will combine to create a unique problem-solving and decision-making context for each REIT. Thus, each contextually aware REIT will have a unique cognitive risk management context, making contextual awareness crucial. It is intended herein to offer a high level structure for understanding the opportunities and challenges of developing and incorporating these contextually sensitive elements. A brief discussion of each fundamental element follows.

6.5.1 Problems and decisions vary in significance and urgency

Every enterprise has limited resources in terms of time, energy, attention, passion and finances. Competing for these resources are an almost unlimited number of uses, including problem-solving and decision-making. Urgency often replaces significance as the decision criterion for the use of these limited resources. As a result, resources are not used in the most productive manner.

In the realm of problem-solving and decision-making, managers must exercise care to ensure that they address significant problems ahead of urgent or easy ones. Diverting limited resources from significant decisions may cause permanent harm to the enterprise. Consistent and mindful attention must be paid to keep enterprise efforts focused on the decisions that matter.

6.5.2 Internal and external problems and contexts

REITs encounter both internal and external situations. For example, developing and staffing a new GIS capability and integrating it into the REIT's investment analysis process is an internal problem. In contrast, the decision on whether to purchase a large Hong Kong office complex, including consideration of the possible financing options, is based primarily in an environment external to the REIT. Both internal and external contexts are important as they jointly determine the ability of an organization to solve problems and make decisions.

Imagine the cognitive risk produced by a complex or chaotic internal decision-making environment attempting to solve even a simple external problem. The pitfalls of such an environment are myriad: consistent processes would be minimal, information would be difficult to gather, its reliability and validity would be questionable, the enterprise vision, mission and goals likely would be unknown or constantly changing and decision-making authority would be ambiguous. Cognitive risk would be exceedingly high and good decision-making may be impossible. Generally, the goal is to develop and maintain ordered internal decision-making environments, especially when dealing with complex or chaotic external problem contexts.

Snowden and Boone (2007, p. 69) classify problems and circumstances into four categories: simple, complicated, complex or chaotic. They further assign these four categories into two types of systems, ordered and un-ordered, depending on whether causality is known or not. Simple and complicated categories are considered to be ordered in the sense that cause and effect are either known or can be established through analysis. Complex and chaotic environments are classified as un-ordered because causality is not known and cannot be determined by analysis (Snowden and Boone, 2007; Brougham, 2015).

Each of these classifications requires a different solution and decision approach, with the commitments of time and resources growing as one moves from simple to chaotic contexts. Similarly, the solutions and resulting decisions are progressively less certain as one moves from simple to chaotic contexts. Identifying the nature of the external and internal environments allows the REIT executive to estimate the resources required, including time and the nature of the expected solutions, if any. These expectations are important for everyone in the problem-solving and decision-making ecosystem.

6.5.3 Systems

REITs operate within numerous formal and informal systems. Meadows (2008, p. 2) defines a system as "A system is a set of things – people, cells, molecules, or whatever – interconnected in such a way that they produce their own pattern of behaviour over time". The interconnected nature of systems makes them very context sensitive. Depending on the nature of the interconnections and the "things" that are interconnected, there can be significant delays in effects. Thus, systems can make causality very difficult to discern. Systems exist in contexts ranging from simple to chaotic and those contexts influence the nature of problems and their solutions (Gharajedaghi, 2011).

Leaders and managers need not be systems experts, but they do need to understand the essentials of how systems operate, the importance of identifying relevant enterprise systems and the impact of systems on enterprise operation, including problem-solving and decision-making. They also need to be aware that, because of interconnectedness, a localized change may produce intended and unintended consequences throughout a given system and the systems connected to it. Time delays and the inability to know all of the connections may make causation difficult or impossible to estimate.

6.5.4 Time matters

Time is important in problem-solving and decision-making in several ways. First, cognitively speaking, humans find it challenging to evaluate long-term situations (Loewenstein, Read and Baumeister, 2003). Second, decisions often take time to implement, the implementation may be less than perfect and the consequences of the implementation may take significant time to materialize, making cause and effect difficult to discern in a complex environment. Third, dynamism can create pressure and stress affecting problem recognition and decision-making (Hammond, 2000; Svenson and Maule, 1993). When coupled with time pressure from unrealistic expectations in a complex decision environment, stress can significantly affect cognitive risk. Time is an important business element affecting cognitive risk.

6.5.5 Business model(s)

A business model summarizes how a firm creates and captures value, ultimately yielding operating cash flow. According to Osterwalder and Pigneur (2010), value creation and value capture result from a match between the value proposition and the customer segment. This drives pricing, revenues and profits after the deduction of expenses. Value creation costs are found by identifying the key activities necessary to deliver the value proposition, the key resources available within the enterprise and the relationships with key partners necessary to complete enterprise capabilities. Activities performed within the firm and the goods and services provided by key partners represent the cost structure of the firm (ibid). Given that revenue and expenditures are now established, the net cash flow of the firm can be estimated.

The business model determines the activities the firm performs and who will perform them. The activities performed within the enterprise generate the need for production, marketing, finance, accounting and management activities within the enterprise. The business model shapes the organizational design decisions a firm will make on an on-going basis and how they affect the ability of the firm to create and capture value. Of course, decision-making regarding the business model is dynamic. Strategy is a longer-term view of how the business model will change over time and the business model provides a context for considering the decisions that need to be made and how they interact with other elements of the enterprise.

6.5.6 Growth and development stages

As noted, REITs often begin as "small, localized entities" founded and led by an entrepreneur (Parker, 2011, p. 2). Over time, REITs can evolve into large and complex enterprises with management structures typical of many other international enterprises (ibid, pp. 5–7) but it is important to differentiate between enterprise growth and enterprise development. Growth generally refers to increases in revenue, which in the case of REITs primarily would be growth in the property portfolio and the related rental and other income. Enterprise development consists of the improvements in the supporting enterprise infrastructure, being the projects, processes and culture designed and implemented to support growth. Together, growth and development represent the "scaling" of the enterprise.

An enterprise generally follows the familiar "S" curve as it evolves from start-up to rapid growth to slower or no growth as the market is saturated. In order to develop a finer-grained understanding of the dynamics of the S curve, it has been divided into stages. The basis for defining stages and the number of stages differ, but this stage approach is a useful device for understanding the changing nature of problems and opportunities facing an enterprise (Lidow, 2014; Flamholtz and Randle, 2016). As an example, Lidow (2014, pp. 38–49) considers four stages: customer validation, operational validation, financial validation and self-sustainability.

The importance of these stages is that enterprise goals and activities are very different in each stage and leaders must adjust to the reality of each stage. This means that change is a critical element of any growing and developing enterprise and the ability to manage change differently in each stage is important. In this view, scaling a REIT is not simply a quantitative concept, in the sense of more of the same. It has a strong qualitative element in that different things need to be done in order to anticipate and respond to change. As enterprises near the transition point between stages, they may encounter signs of dysfunction, called growing pains by Flamholtz and Randle (2016). A sense of time pressure, inadequate numbers of competent managers, lack of clear goals and poor coordination and communications are among the growing pains they cite, with many of the growing pains adversely affecting problem-solving and decision-making abilities. The number and severity of growing pains present at any point in time are signals that the gap between growth and development is affecting the enterprise and the urgency of needed adaptation in order to maintain organizational health, including cognitive risk management (Ibid, p. 94).

Ideally, organizational infrastructure development should anticipate and facilitate growth, but this requires considerable foresight and a willingness to engage in the necessary investment of time, energy, attention and resources. Managing an enterprise to complement the implementation of enterprise strategy is an iterative operation in the sense that changes in organizational infrastructure will be effective only until growth once again produces new stresses and growing pains. Leadership, particularly entrepreneurial leadership, must manage present operations while simultaneously anticipating the future and adjusting projects, processes and culture to minimize the impact of organizational stresses that growth fosters.

Even in the final stage, that of becoming a sustainable enterprise, entrepreneurial leadership is required to keep the enterprise flexible, resilient and capable of pursuing the change necessary to survive in a challenging business environment (Lidow, 2014, pp. 46–49). Martin (2009, pp. 18–21) makes this point in terms of firms striving to move from a project to a process orientation and from heuristics to algorithms. Algorithms are more efficient because they require little human involvement and cognitive activity and are process-oriented. On the other hand, innovation is project-oriented and makes heavy use of human judgement and heuristics, making projects less efficient than processes based on repetition and algorithms. Further, the outcomes of project-based innovation are uncertain, as contrasted to the virtual certainty of algorithm-based processes. Rational managers may prefer a more efficient and certain, but lower, return from familiar process-based operations to a less efficient and uncertain return achieved through project-based innovation. This situation has contributed to the debate around the notion of "short-termism" in business decision-making.

Thus, it is not surprising that enterprises executing processes according to established internal procedures while simultaneously listening and responding to their best customers for existing products may miss major changes in the external environment representing both threats and opportunities. This situation makes it clear why the following discussions of organizational design and culture can be so important to the sustainability of an enterprise as executives must be aware of the need to reduce the cognitive risk created by the fixation on current operations. Coupled with the tendency to rely on habit to make decisions and to ignore disconfirming or negative information, innovation and the related decision-making is difficult in many organizations and can cause the enterprise to miss viable opportunities and become vulnerable to disruption (Cerulo, 2006; Duhigg, 2012; Christensen, 2003; Christensen and Raynor, 2003). Growth and development context matters in decision-making and the existence and management of cognitive risk.

6.5.7 Strategy

Every enterprise has a strategy, either implicit or explicit, that establishes the enterprise's vision, mission, goals and objectives and how they will be realized. In short, strategy answers questions concerning why the enterprise exists, how it will operate and what it will do. The development of crisp and clear notions of why, how, and what, in that order, provide clarity, guidance and motivation, enabling more effective leadership for the enterprise (Sinek, 2009).

Strategy is a dynamic activity, changing as internal and external environments warrant. This is not to say that strategy changes daily, but that mindful attention to strategy is warranted to assure that the organization strikes the proper balance between internal and external environments and consistency and change. The ultimate goal is self-sustainability but, as discussed, enterprises must grow and develop into self-sustainability (Lidow, 2014, pp. 46–49). This implies that until an enterprise achieves self-sustainability, strategy must focus on reaching the next stage of development (ibid., pp. 129–145). This contextual nature of strategy affects the determination of which enterprise decisions are more important, how those decisions are made and the potential for the existence of cognitive risk and its management.

6.5.8 Organizational design and decision-making

How a REIT is organized to solve problems and make decisions affects cognitive risk. Organizational structures can range from hierarchy-based models to matrix organizations with very different distributions of decision-making authority. Beyond the departments, functional areas, divisions, subsidiaries and other organizational units, it is important to understand the decisions that may be made in an organization and specify how they will be made. Especially important is a clear delineation of roles and who will play them in making decisions in various areas.

Rogers and Blenko (2006) develop the RAPID Decision Model in which they identify five key roles and the responsibilities of each in general decision-making processes to overcome the ambiguity as to responsibility and authority that often surrounds and adversely affects organizational decision-making. In an abbreviated form, the roles and responsibilities in the RAPID model are as follows (Rogers and Blenko, 2006, p. 52):

Role	*Responsibility*
Recommend	Proposing an analysis and recommendation on a decision.
Agree	Reviewing recommendations and agreeing, disagreeing or negotiating a proposed solution.
Perform	Executing a decision once it is made.
Input	Providing facts pertaining to feasibility and implications of alternatives to the recommender.
Decide	Considering available information and choosing a course of action.

An individual may play different roles in the RAPID model depending on the decision to be made.

Some very important activities precede the implementation of the RAPID framework to solve a given problem. Specifically, a problem must be identified, framed appropriately and selected to be solved. These are important activities. Considering the "right" problems is an important part of reducing cognitive risk for an enterprise as resource limitations make it infeasible to consider every possible problem. The IFS model, for the Identify, Frame and Select (IFS) roles, can be added to the RAPID model to provide a more complete picture of what problems get solved and how they get solved in an organization:

Role	*Responsibility*
Identify	Scanning the environment and finding important problems to be solved or decisions to be made.
Frame	Transforming a problematic situation into a problem by assessing the environment as context (Schon, 1983, p. 40) and the importance, relatively and absolutely, of the resultant problem.
Select	Deciding which problems merit the commitment of resources necessary to solve the problem or make the decision.

Clearly delineating organizational structure related to problem-solving and decision-making roles increases order in the internal environment. It should be noted that while the roles in the IFS and RAPID models may not strictly follow the formal organization structure, the formal organization design can have a profound impact on the flow of information and interaction. Care must be taken when dealing with project-oriented, cross-functional teams to ensure that the IFS and RAPID roles are clearly specified.

6.5.9 Culture

Enterprise culture can have strong impacts on how decisions are made. It does so by influencing how information is shared, the propensity to collaborate and cooperate and the method by which individuals and groups are evaluated and recognized, among many other ways. Culture can affect attitudes toward innovation and change, along with the orientation toward "process" (Flamholtz and Randle, 2016). If process is ignored or not valued, orderly decision-making processes may be impaired. Assessing and managing culture is an important element in managing cognitive risk.

Likewise, the culture surrounding decision-making is critical. Does the organization encourage decision-making by accepting decisions or does leadership often disregard organizational decisions? Are those in support roles recognized and celebrated or ignored? Are collaboration and cooperation rewarded? Misalignments between formal organizational design and culture produces organizational stresses, ambiguity with regard to decision-making roles and erratic and unpredictable problem-solving and decision-making behaviours, increasing cognitive risk.

6.6 The power of alignment

If a REIT executive were to consider a given problem or decision-making situation through the lens of the fundamental elements considered previously, managing cognitive risk may become more feasible. The likelihood of successfully managing cognitive risk would certainly be higher than if using an *ad hoc* approach. This finding derives from the realization that the fundamental elements considered previously collectively map a significant portion of the problem-solving/decision-making landscape within a REIT. These elements are all relevant to the creation and, therefore, the management of the cognitive risk facing a REIT.

Certainly, other fundamental elements may affect cognitive risk in any organization, but the elements considered previously represent enough of the landscape to support significant cognitive risk management efforts. The overarching principle in managing cognitive risk is alignment. Alignment simply means ensuring that the right people in the right roles are pursuing the right problem, supported by the right systems and the right culture and pursuing the right vision, mission, goals and objectives. If all of these pieces are in place, cognitive risk will be substantially reduced.

The most direct way to approach alignment is to use the fundamental elements as a checklist (Gawande, 2010; Gordon et al., 2013). The REIT executive(s) or leadership team can utilize the fundamental elements as a discussion guide and work on the list until it is felt that everything has been addressed as thoroughly as necessary. This will likely be an iterative approach and helps assure that all major considerations and the interactions between them have received the attention they deserve. Of course, the checklist can be edited or expanded as the executive(s) deem appropriate to customize it for their unique considerations. The fundamental elements are intended to be a good start to the cognitive risk management process.

The checklist is just the beginning. Some REIT executives may realize that a more thorough review of enterprise problem-solving and decision-making activities is warranted. Such a review may involve an assessment of the problems addressed, the quality of decisions, the extent to which solutions and decisions were implemented, a culture assessment and the clarity of decision roles among many other possibilities. Achieving alignment is the goal and enterprise cognitive risk management is the outcome. It should be noted that work to manage cognitive risk likely will produce positive externalities in other areas of REIT operations as improvements in communications, goal congruence, culture and strategic thinking permeate the enterprise. The transparency inherent in cognitive risk management promotes shared viewpoints and improved enterprise, team and individual relationships. In this sense, cognitive risk management can become part of a more comprehensive development of a virtuous cycle of improvement for a REIT.

Cognitive risk management requires a mindful approach. In this regard, the goal is to create what Schon (1983) refers to as reflective practitioners throughout the enterprise. A reflective practitioner recognizes that solving problems that are frequently described as messes or wicked requires more than just "technical rationality", it also requires understanding the "setting", or context, in which the problem-solving must occur and shaping thinking and actions accordingly (ibid, pp. 39–49). Mintzberg (2013, pp. 8–11) considers management as a practice combining varying combinations of science, art and craft to meld analysis, creative insights and experience capable of dealing with difficult problems. Clearly, problem-solving and decision-making and managing the cognitive risk associated therewith require a mindset beyond technical expertise (Wachs, 2016). REIT leaders must promote thinking outside the rigour box with an emphasis on relevance based on context awareness in problem-solving and decision-making. While not a panacea, cognitive risk management efforts can produce results ranging from simply good to transformative.

Cognitive risk management efforts require resources in the form of time, energy, attention and cash, resources that are scarce in every enterprise. Scarcity transforms the goal of cognitive risk management from eliminating all cognitive risk to reducing cognitive risk to acceptable levels in the areas that matter most. Thus, a mindful approach to balancing cognitive risk management efforts and the use of scarce resources is needed. Finding the appropriate balance of acceptable risk and expected return for addressing cognitive risk within the larger context of enterprise management is an important task for REIT executives.

6.7 Summary

Chapter 1 sought to identify critical contextual issues in international REITs, including the defining characteristics of REITs, the global evolution of REITs and the risk-return characteristics of real estate investment through REITs relative to non-REIT real estate investment with Chapter 2, the *Post-Modern REIT Era*, examining the evolution of the global REIT market, what is deemed to be best current market practice and how the market is expected to change going forward. Chapter 3, *Emerging Sector REITs*, investigated such sectors as timber and data centres with Chapter 4, *Sustainable REITs*, analysing REIT environmental performance and the cost of equity and Chapter 5 *Islamic REITs*, examined the evolution of global Islamic REITs through a study of the Malaysian market.

This chapter, *Behavioural Risk in REITs*, sought to develop a cognitive risk management framework for REIT entrepreneurs and executives. Beginning with a brief overview of the cognitive and behavioural landscape and the notion of cognitive risk, attention then turned to how to effectively manage cognitive risk in an enterprise setting. Building on the fundamental elements contributing to cognitive risk, the power of positive alignment between these fundamental elements was seen as critical to cognitive risk management. The use of checklists to aid in producing a unique context for each problem or decision was suggested as a mechanism for customizing and operationalizing meaningful cognitive risk management within a REIT. The power of mindfulness, the need for reflective practitioners and concept of management as a practice were identified as crucial elements of cognitive risk management implementation.

Part I now concludes with Chapter 7 which reviews recent research into *REIT Asset Allocation*.

Part II then includes six chapters each analysing REITs in a region of the world, using case studies from a developed, developing and emerging REIT sector in the region, concluding with Chapter 14 which considers *Directions for the Future of International REITs*.

References

Beshears, J. and Gino, F. 2015, 'Leaders as decision architects', *Harvard Business Review*, Vol. 93, pp. 51–62.

Black, R.T., Brown, M.G., Diaz, J., Gibler, K.M. and Grissom, T.V. 2003, 'Behavioural research in real estate: A search for the boundaries', *Journal of Real Estate Practice and Education*, Vol. 6, pp. 85–112.

Brougham, G. 2015, *The cynefin mini-book*, C4Media.
Cerulo, K.A. 2006, *Never saw it coming*, The University of Chicago Press, Chicago.
Christensen, C. 2003, *The innovator's dilemma*, HarperCollins Publishers, New York (a reprint of the original 1997 edition, Harvard Business School Press).
Christensen, C. and Raynor, M. 2003, *The innovator's solution: Creating and sustaining growth*, Harvard Business School Publishing Corporation, Boston.
Croskerry, P. 2003, 'The importance of cognitive errors in diagnosis and strategies to minimize them', *Academic Medicine*, Vol. 78, pp. 775–780.
Diaz, J. 1990, 'How appraisers do their work: A test of the appraisal process and the development of a descriptive model', *Journal of Real Estate Research*, Vol. 5, pp. 1–15.
Diaz, J. 2002, *Behavioural research in appraisal*, RICS Foundation, London.
Duhigg, C. 2012, *The power of habit*, Random House, New York.
Evans, J. 1989, *Biases in human reasoning: Causes and consequences*, Erlbaum, Hillsdale, NJ.
Fama, E. 1970, 'Efficient capital markets: A review of theory and empirical work', *Journal of Finance*, Vol. 25, pp. 383–417.
Flamholtz, E. and Randle, Y. 2016, *Growing pains: Building sustainably successful organisations*, Wiley, Hoboken, NJ.
Gawande, A. 2010, *The checklist manifesto*, Picador, New York.
Gharajedaghi, J. 2011, *Systems thinking: Managing chaos and complexity*, Third Edition, Morgan Kaufmann, Burlington, MA.
Gordon, S., Mendenhall, P. and O'Connor, B.B. 2013, *Beyond the checklist*, Cornell University Press, Ithaca, NY.
Hammond, K. 2000, *Judgments under stress*, Oxford University Press, Oxford.
Hardin, W. 1999, 'Behavioural research into heuristics and biases as an academic pursuit: Lessons from other disciplines and implications for real estate', *Journal of Property Investment & Finance*, Vol. 17, pp. 333–352.
Kahneman, D. 2011, *Thinking, fast and slow*, Penguin, London.
Lewis, M. 2017, *The undoing project: A friendship that changed our minds*, Norton, London.
Lidow, D. 2014, *Startup leadership: How savvy entrepreneurs turn their ideas into successful enterprises*, Josey-Bass (Wiley), San Francisco, CA.
Loewenstein, G., Read, D. and Baumeister, R.F. 2003, *Time and decision*, The Russell Sage Foundation, New York.
Martin, J. 2004, *Organisational culture*, Research Paper No. 1847, Stanford University, Stanford, CA.
Martin, R. 2009, *The design of business*, Harvard Business Press, Boston.
Meadows, D.H. 2008, *Thinking in systems*, Sustainability Institute, White River Junction, VT.
Miller, G.A. 1956, 'The magical number seven, plus or minus two: Some limits on our capacity for processing information', *Psychological Review*, Vol. 63, pp. 81–97.
Mintzberg, H. 2013, *Simply managing*, Barrett-Koehler Publishers, Inc., San Francisco.
Osterwalder, A. and Pigneur, Y. 2010, *Business model generation*, John Wiley & Sons, Hoboken, NJ.
Parker, D. 2011, *Global real estate investment trusts: People, process and management*, Wiley-Blackwell, West Sussex, UK.
Ratcliff, R.U. 1972, *Valuation for real estate decisions*, Democrat Press, Santa Cruz.
Ritter, J. 2003, 'Behavioural finance', *Pacific-Basin Finance Journal*, Vol. 11, pp. 429–37.
Rogers, P. and Blenko, M. 2006, 'Who has the "D"?', *Harvard Business Review*, January, pp. 52–61.
Schon, D.A. 1983, *The reflective practitioner*, Basic Books, Inc., New York.
Simon, H.A. 1955, 'A behavioural model of rational choice', *Quarterly Journal of Economics*, Vol. 69, pp. 99–118.
Simon, H.A. 1979, 'Rational decision-making in business organizations', *American Economic Review*, Vol. 69, pp. 493–513.
Simon, H.A. 1980, 'Bounded rationality', in *The New Palgrave: Utility and Probability*, ed. J. Eatwell, M. Milgate, and P. Newman, Norton, New York.
Sinek, S. 2009, *Start with why*, Penguin Group, New York.
Snowden, D.J. and Boone, M.E. 2007, 'A leader's framework for decision-making', *Harvard Business Review*, November, pp. 68–77.
Soll, J.B., Milkman, K.L. and Payne, J.W. 2015, 'Outsmart your own biases', *Harvard Business Review*, Vol. 93, pp. 64–71.
Soyer, E. and Hogarth, R.W. 2015, 'Fooled by experience', *Harvard Business Review*, Vol. 93, pp. 72–77.
Stigler, G. 1987, *The theory of price*, Macmillan, London.

Svenson, O. and Maule, A.J. 1993, *Time pressure and stress in human judgment and decision-making*, Plenum Press, New York.
Tversky, A. and Kahneman, D. 1971, 'Belief in the law of small numbers', *Psychological Bulletin*, Vol. 2, pp. 105–110.
Tversky, A. and Kahneman, D. 1972, 'Subjective probability: A judgment of representativeness', *Cognitive Psychology*, Vol. 3, pp. 430–454.
Tversky, A. and Kahneman, D. 1973, 'Availability: A heuristic for judging frequency and probability', *Cognitive Psychology*, Vol. 5, pp. 207–232.
Tversky, A. and Kahneman, D. 1974, 'Judgment under uncertainty: Heuristics and biases', *Science*, Vol. 185, pp. 1124–1131.
Tversky, A. and Kahneman, D. 1979, 'Prospect theory: An analysis of decision under risk', *Econometrica*, Vol. 47, pp. 263–291.
Wachs, M. 2016, 'Becoming a reflective planning educator', *Journal of the American Planning Association*, Vol. 82, pp. 363–370.
Wofford, L.E. 1985, 'Cognitive processes as determinants of real estate investment decisions', *Appraisal Journal*, Vol. 53, pp. 388–395.
Wofford, L.E. and Troilo, M.L. 2013, 'The academic-professional divide: Generating useful research and moving it to practice', *Journal of Property Investment & Finance*, Vol. 31, pp. 41–52.
Wofford, L.E., Troilo, M.L. and Dorchester, A.D. 2010, 'Managing cognitive risk in real estate', *Journal of Property Research*, Vol. 27, pp. 260–286.
Wofford, L.E., Troilo, M.L. and Dorchester, A.D. 2011a, 'Cognitive risk and real estate portfolio management', *Journal of Real Estate and Portfolio Management*, Vol. 17, pp. 69–73.
Wofford, L.E., Troilo, M.L. and Dorchester, A.D. 2011b, 'Real estate valuation, cognitive risk, and translational research', *Journal of Property Investment & Finance*, Vol. 29, pp. 385–408.
Zweig, J. 2007, *Your money & your brain*, Simon & Schuster, New York.

7

REIT ASSET ALLOCATION

Stephen Lee and Alex Moss

7.1 Introduction

This book aims to identify key areas for research in the REIT discipline for the next five to ten years by surveying the current state of the REIT discipline around the world and identifying emerging and cutting edge research areas through a thematic review of current contextual issues and a regional analysis based on case studies.

This book comprises two parts, the first part being a thematic review of emerging and cutting edge global research into current contextual issues in REITs internationally and the second part being a regional analysis of REITs around the world, each written by authoritative academic authors from the world's leading Universities and REIT industry experts.

Part I includes six chapters each reviewing a current theme of REIT evolution through the lens of contemporary research. Chapter 1 focused on critical contextual issues in international REITs while Chapter 2, the *Post-Modern REIT Era,* examined the evolution of the global REIT market, what is deemed to be best current market practice and how the market is expected to change going forward, with Chapter 3, *Emerging Sector REITs*, investigating such sectors as timber and data centres. Chapter 4, *Sustainable REITs*, analysed REIT environmental performance and the cost of equity with Chapter 5, *Islamic REITs*, examining the evolution of global Islamic REITs through a study of the Malaysian market.

Chapter 6, *Behavioural Risk in REITs*, addressed the management of behavioural risk in global REITs and Part I concludes with this chapter which reviews recent research into *REIT Asset Allocation* to consider whether REITs are real estate and the role of REITs in a mixed asset portfolio.

Part II includes six chapters each analysing REITs in a region of the world, using case studies from a developed, developing and emerging REIT sector in the region:

Chapter	Region	Developed **REIT Sector**	Developing **REIT Sector**	Emerging **REIT Sector**
8	North America	USA	Canada	Mexico
9	Latin America	Brazil	Argentina	Uruguay
10	Europe	UK	Spain	Poland

11	South East Asia	Singapore	Malaysia	Thailand
12	North Asia	Japan	Hong Kong	China
13	Oceania	Australia	South Africa	India

concluding with Chapter 14 which considers *Directions for the Future of International REITs*.

7.2 Are REITs real estate?

When investors look to diversify their stock-bond portfolio, the first alternative they usually consider is real estate, typically private commercial real estate. Investing in private real estate offers compelling portfolio advantages as it is a tangible asset with low volatility, an attractive income stream, long-term capital appreciation and particularly strong diversification benefits. Thus, there is extant literature showing that private real estate has a significant place in the mixed-asset portfolio.

Private real estate, however, has considerable drawbacks: it is illiquid, "lumpy", with low transparency and so needs expert management and significant capital to build a diversified portfolio. Therefore, private investment in real estate is only realistic for investors with hundreds of millions to invest. The disadvantages of private real estate investment have led investors to consider indirect real estate investment vehicles as an alternative.

Indirect real estate investment can be made through unlisted open- or closed-ended commingled funds or private partnerships but, like private real estate, these vehicles are still illiquid. Consequently, the focus of many investors has been investment in Real Estate Investment Trusts (REITs) for a number of reasons. First, REITs are income-oriented investment vehicles that derive most of their earnings from real estate activities. Second, REITs offer investors the ability to gain access to large property types such as regional malls, which would be prohibitively expensive for most investors. Last, investors can invest their money straight away, as REITs are listed on major stock markets and so are much more liquid than private investment.

In other words, investors have sought investment in REITs so as to gain the advantages of private real estate without its illiquidity disadvantages. Indeed, Ciochetti, Craft and Shilling (2002) found that institutional investors prefer to invest in REITs, as opposed to private real estate, due to liquidity arguments. However, since an investment in private real estate is quite different to an investment in a securitized vehicle "[T]he debate has long raged as to whether REITs are equity or real estate, and related to that judgement, where they fit in a portfolio" (Hudgins, 2012, p. 1).

This following seeks to address both of these questions by focusing on the performance of REITs in the US as it is, by far, the largest REIT market in the world and the model for REIT structures across the world with the performance of US REITs likely to provide a strong indication of how REITs might perform globally.

7.2.1 Short-run comparisons

Equity REITs (or simply REITs) will be considered in the following analysis as they dominate the market in terms of numbers and capital value. The standard approach to comparing the performance of REITs and private real estate is to use the National Association of Real Estate Investment Trust (NAREIT) index and the National Council for Real Estate Investment Fiduciaries (NCREIF) index, respectively. The NAREIT returns are available monthly from January 1972. NCREIF returns, however, are only available quarterly from 1978.Q1. In order

Table 7.1 Returns of REITs and Private Real Estate: 1994–2016: Quarterly Data

	REITs			*Private Real Estate*		
Panel A	*Total*	*Capital*	*Income*	*Total*	*Capital*	*Income*
Mean	3.13	1.72	1.41	2.35	0.61	1.74
Standard Deviation	9.78	9.71	0.43	2.18	2.17	0.36
Panel B	**Total**	**Capital**	**Income**	**Total**	**Capital**	**Income**
Return Ratio	100%	55%	45%	100%	26%	74%
Risk Ratio	100%	99%	1%	100%	98%	2%

Sources: NAREIT and NCREIF

to investigate the performance of REITs and private real estate investment in the short-run, quarterly returns over the period from 1994.Q1 to 2016.Q4 are used. While REITs have existed since the 1960s, the sample period selected covers the period of the Modern REIT Era and so is the most relevant to current REIT investors.

Panel A of Table 7.1 confirms the results of previous studies that REITs offer higher average returns than private real estate by 78 basis points per quarter, 3% per annum (see Tsai, 2007 and Ling and Naranjo, 2015, for reviews). The higher average return of REITs, however, is at the cost of a higher risk, which is almost five times that of private real estate.

Short-run return comparisons between REITs and private real estate, however, are complicated by a number of issues. First, the exceptionally low risk of the private real estate data is a common feature of international databases, due to the so-called appraisal-bias (see Fisher et al., 1994 and Corgel and deRoos, 1999, for comprehensive reviews). The downward bias in the standard deviation of appraisal-based private real estate market indices is usually attributed to the use of "stale" appraisals (i.e. dated appraisals), inaccurate appraisals due to lack of current market information and the temporal and cross sectional aggregation of individual property valuations into the market index (see, *inter alia*, Geltner, 1991 and Brown and Matysiak, 1998).

Second, for a long time the NAREIT Index consisted largely of retail and multifamily properties, whereas the NCREIF Index was mainly composed of office, retail and industrial properties. A number of studies also showed that REITs held properties in different geographical locations compared with the private real estate market. Since 1997, the composition of REIT portfolios has become more like that of the private market. Indeed, today REITs hold almost the same percentage of their property in the same six major cities as private real estate (Chicago, San Francisco, Boston, Los Angeles, Washington, D.C. and New York) though the NAREIT and NCREIF weights in these cities vary to some extent (Hudgins, 2012). Nonetheless, REITs are increasingly investing in non-traditional property-types, such as self-storage and health care, which display different risks and returns compared with the traditional sectors (Newell and Peng, 2006). In other words, the real estate indices behave differently as they are composed of different property types, in different locations, in different weights.

Third, the generally higher returns and risks of REITs are due, at least in part, to REITs tending to use 30% to 50% leverage. For instance, Pagliari, Scherer and Monopoli (2005) estimate that US REITs had an average debt-to-value ratio of around 40% in the period 1981–2001 (with the average rising to 50% towards the end of their analysis period). In other words, REITs typically offer investors the risk and return characteristics similar to those of a levered investment in the underlying real estate, whereas the private real estate data is unleveraged.

Since financial leverage increases the interest rate sensitivity of REIT returns, REITs are more affected by interest rate changes than the private real estate market. For instance, the mere fact that interest rates were potentially going to rise, after Ben Bernanke announced on the 22 May 2013 that the Fed would likely start tapering quantitative easing (QE) by slowing the pace of its bond purchases later in the year, led to the so-called Taper Tantrum resulting in the NAREIT index falling by 2.57% and continuing to decline for the rest of the year. In contrast, although stocks and bonds also fell on the day of the tapering announcement, they quickly recovered (Lee, 2016). In other words, the Taper Tantrum had a profound effect on the risk/return benefits of including REITs in the US mixed-asset portfolio due to REITs sensitivity to interest rate changes (see also Bohjalian, 2015).

Last, since the 1986 *Tax Reform Act* and the 1999 *REIT Modernization Act*, the higher risk of REITs also reflects the fact that REITs don't just passively hold a static real estate portfolio, but actively buy, sell and develop properties as well as land and engage in a variety of tenant services and other activities which increases volatility. In contrast, the data in the NCREIF database simply reflects the unlevered performance of standing investments in real estate portfolios.

To get a clearer picture of the similarity and differences between REIT and private real estate returns, the ratio of total return and risk attributed to capital and income returns was calculated. It is easy to calculate the contribution that income and capital returns make to total returns by dividing the total return by each return constituent. The calculation of the percentage contribution of total return risk from capital and income returns is more complicated due to the correlation between each return series but can be done using the approach of Menchero and Davis (2010). The results presented in Panel B of Table 7.1 show that the total return of REITs and private real estate mainly come from income rather than capital changes, especially for private real estate. That is, the total return of both indexes is driven primarily by a property's net operating income (NOI). This can be seen in Figure 7.1, which shows the correlation between private real estate data and REITs for the overall REIT index and for four property-types (office, retail, industrial and apartments).

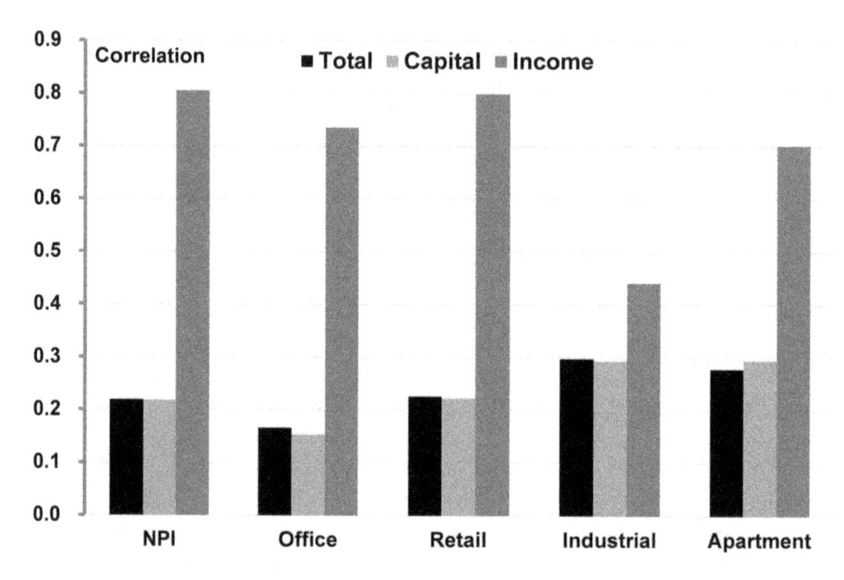

Figure 7.1 The Correlation between REIT Returns and Private Real Estate: Quarterly Data

Sources: NAREIT and NCREIF

Figure 7.1 shows that the income returns for all REIT sectors and for the market overall is significantly related to the same data for private real estate. In contrast, the total and capital returns of private real estate have an insignificant relationship with the corresponding REIT sectors. This supports the conclusion of Muhlhofer (2013) that in the short-run "REITs will only expose an investor to property income cash flows and not appreciation" (p. 855).

However, the most striking feature of Panel B in Table 7.1 is that capital returns contribute 99% to the risk of total returns for both REITs and private real estate, while income returns contribute only 1%. In other words, the lack of correlation between REITs and private real estate returns, in the short-run, can be largely attributed to the differences in capital appreciation. Indeed, Giliberto and Mengden (1996) believe that the different pricing mechanism used in each market disguises the "true" relationship between REIT and private real estate performance. This is due to the fact that, as financial securities, REITs follow a different return generating process than the underlying real estate market as a REIT is an ownership and operating construct wrapped around the properties that might otherwise be privately owned.

Consequently, as Pagliari, Scherer and Monopoli (2005) argue, "it is what's underneath the wrapper that matters in the long-run" (p. 181). The authors find that, after controlling for differences in property type mix and leverage, there is no statistical difference in the returns of REITs and private real estate investment. Ang et al. (2013) draw a similar conclusion and find that "[O]ver the full real estate cycle, the effects of these different innovation exposures largely disappear, and both public and private vehicles exhibit similar characteristics" (p. 130).

In particular, Pedersen et al. (2012) and Shepard, Liu and Dai (2014) find that, after adjustments are made for smoothing, leverage and portfolio composition, estimates of the correlation between REITs and private real estate returns increases as the holding period increases. For instance, Shepard, Liu and Dai (2014) find that, in the long run, the correlation between REITs and private real estate returns is about 80% for many countries with Pedersen et al. (2012) finding that, after correcting for smoothing, "the implied return correlations between REITs and private real estate investments fell in the range of 60% to 80% from January 1989 through June 2011" (p. 13). This suggests that, once measurement errors are controlled for, REITs and private real estate are closer substitutes than many may have believed. Indeed, since REITs are, fundamentally, primarily commercial property investment plays, over the long run it may be expected that the capital returns of REITs and private real estate will be substantially related.

Finally, investors in private real estate do not hold property for one month. Previous studies suggest holding periods are at least five years and in many cases longer, as real estate is a long-term investment by the very nature of the often long-term lease structures and the longevity of the physical asset. For instance, Fisher and Young (2000) suggest that US real estate analysts, owners, brokers and appraisers "habitually think of commercial property as having a tenure of 10 years" (p. 237). Any analysis as to whether REITs and private real estate display similar capital returns using monthly data therefore biases finding any link between REIT and private real estate capital appreciation.

7.2.2 Long-run comparisons

To investigate the long-run relationship between REITs and private real estate, the correlation between the capital returns of the two asset classes are examined over increasing holding periods from 1 month to 60 months. However, the same analysis is also repeated with stocks as a number of authors report that the REIT market is integrated with the general stock market and not just the private real estate market, although other authors find no significant evidence to support the cointegration of REITs and the stock market either domestically or internationally.

The REIT data is represented by the NAREIT price index, while the performance of the stock market is represented by two stock price series: the S&P 500 and the Russell 2000. The S&P 500 is one of the most commonly used benchmarks for the overall US stock market and is intended to reflect the risk/return characteristics of the large cap universe. The Russell 2000 is constructed to provide a comprehensive and unbiassed small-cap barometer and represents about 10% of the total market capitalization of the Russell 3000 Index.

However, NCREIF data is not used as, due to smoothing in the index, the variance and covariance of property returns are biassed towards zero which distorts the correlation coefficient calculations. To make the private real estate data more comparable with the market-based REIT and stock data, the appraisal-based data is often de-smoothed. While there are a number of methods that have been developed to de-smooth the appraisal-based data there is some doubt about the efficacy of such methods and whether the data should be de-smoothed at all (Lai and Wang, 1998).

Therefore, to compare the REIT and stock market data with private real estate returns, a real estate transaction based series is used, the CoStar Commercial Repeat-Sales Index (CCRSI), to avoid the problem of whether, or how, to de-smooth the appraisal-based data. The CCRSI is used as it is the longest running series, with data back to January 1996, and has the widest commercial property market coverage when compared with Moody's/RCA Commercial Property Price Index (MCPPI) and Green Street's CPPI (GSCPPI). The results are presented graphically in Figure 7.2.

Figure 7.2 shows a number of features of interest. First, the one-month correlation between REITs and private real estate, as represented by CCRSI, is -0.22. In other words, over the short run, capital value changes in REITs do to not equate to those in private real estate. Second, as the holding period increases up to 60 months (five years), the correlation between REITs and private real estate continues to increase as the holding period increases, peaking at 0.65, which is

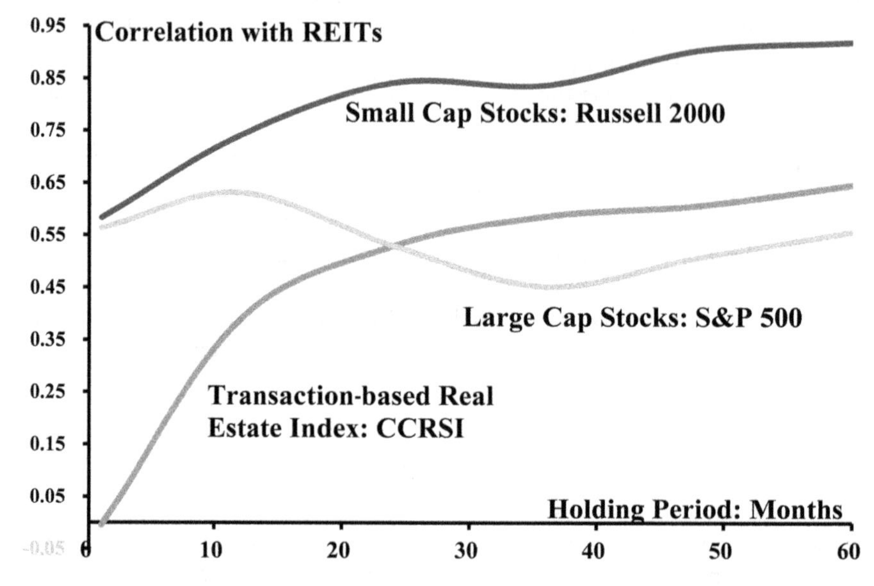

Figure 7.2 Holding Period and the Correlation between REITs, Private Real Estate and Stocks

Sources: NAREIT, CoStar, NCREIF and Yahoo Finance

supportive of previous research (see, *inter alia*, Giliberto, 1990 and Geltner and Kluger, 1998). This implies that, as the holding period increases, REITs tend to display more of the price appreciation of private real estate than the short-run data would suggest. A view shared by Oikarinen, Hoesli and Serrano (2011) along with many others who argue that "the longer the investment horizon is, the greater the degree of substitutability between REITs and direct real estate" (p. 96).

The one-month correlation of REITs with the S&P 500 is 0.56. Given that REITs have daily liquidity it is unsurprising to see that securitized real estate behaves more like stocks than private real estate in the short run. But, as the holding period increases, the correlation of REITs with the S&P 500 shows a slight downward trend, falling to its lowest value of 0.45 at the three-year holding period, after which the correlation increases back up to 0.56 at the five-year holding period. Thus, although REIT prices start to reflect the performance of private real estate in the long run, REITs are clearly still integrated with the general stock market.

However, the most notable feature of Figure 7.2 is that the correlation of REITs with a small cap stock index (R2000) starts at 0.58 and shows a continual increase as the holding period increases, peaking at 0.92 at the five-year holding period. This implies that REITs are a small cap stock in both the short run and long run, supportive of a large number of studies.

In summary, return comparisons between REITs and private real estate are complicated by data issues, property type mix and leverage. Previous research and the empirical results described herein, however, show that, once data issues are corrected for, the returns, risks and correlation between REITs and private real estate are similar, which has led many authors to conclude that REITs and private real estate are substitutes for one another, albeit in the long run.

However, even though in the long run REIT prices tend to behave more like those in the private real estate market, they still retain a significant stock market component, especially a small cap stock component. As Greg Whyte, Analyst, Morgan Stanley considers REITS smell like real estate, look like bonds and walk like equity. Thus, an investment in REITs requires the investor to have an adequate knowledge of both the stock market and the property market in order to get the most out of his or her investment.

7.3 REITs and the mixed-asset portfolio

Asset allocation is a strategic decision and has been shown to be the most important decision in portfolio management, determining about 80–90% of return variance (see, *inter alia*, Brinson, Singer and Beebower, 1991 and Ibbotson and Kaplan, 2000). Therefore, it is asset allocation that determines whether REITs have a significant place in the mixed-asset portfolio, so asset allocation is of prime importance.

What matters when evaluating the place of an asset in a portfolio is not the volatility of the asset but rather its correlation with the other assets in the portfolio. However, the correlation of REITs with stocks and bonds is time varying. For instance, the earliest study showed significant negative correlations between REITs and other assets, indicating large diversification benefits, while several studies conducted in the late 1990s reported weak but positive correlations between investments in REITs and stocks, indicating some potential for diversification.

In contrast, recent evidence finds that the declining trend observed in the 1990s has reversed since the start of the new millennium, suggesting a diminishing diversification potential for REITs. Importantly, the correlation of REITs with stocks strengthens in times of abnormal volatilities when diversification benefits are most needed. These changes in the correlation dynamics of REITs with other financial assets mean that it may be difficult to determine the actual diversification benefit of REITs in a mixed-asset portfolio, if relying solely on the long-run average.

For instance, Kuhle (1987) used monthly data to examine whether including REITs would have improved the performance of common stocks over the period 1980 to 1985, concluding that REITs do not add significant increases in Sharpe performance, with several studies coming to similar conclusions using data from the 1990s (see, *inter alia*, Gyourko and Nelling, 1996 and Paladino and Mayo, 1998). Mueller, Pauley and Morril (1994) show that REITs were only a valuable addition to the mixed-asset portfolio for the 1976–1980 and 1990–1993 time periods but not for the 1980–1990 sub-period, due to the high positive correlation that REITs show with small-cap stocks and the S&P 500 index but weak positive correlation with bonds.

Hudson-Wilson (2001) shows that REITs underperformed both bonds and stocks on a risk/return basis over the period from 1987 to 2000. In contrast, Ibbotson Associates, on behalf of NAREIT (NAREIT, 2002) found that the inclusion of REITs into a well-diversified stock and bond portfolio could have enhanced returns by up to 0.8% annually over the period from 1972 to 2001 and by 1.3% annually for the years 1992–2001. Nonetheless, Lee and Stevenson (2005) find that REITs consistently provide diversification benefits to the mixed-asset portfolio and that these benefits tend to increase as the investment horizon is extended.

Last, Lee (2010) finds that the return enhancement or diversification benefit of REITs to the mixed-asset portfolio has changed over time in line with the changes in the evolution of the REIT industry. Prior to 1999, REITs showed a strong diversification benefit to large cap growth and value stocks but a negative return benefit. However, since 1999, the benefit of REITs to large growth and value stocks came from their return enhancement benefits rather than any diversification benefit. The benefit of REITs to small cap growth stocks initially came from their diversification benefit but, after 1999, it came from both their return and diversification benefits, while the return benefit of REITs to small cap value stocks has always been negative. In contrast, REITs have always shown strong diversification and return benefits to bonds.

Thus, given the differences in time periods considered and the changes in the legal and regulatory structure of the REIT industry and the asset classes included in the original mixed-asset portfolio it is not surprising to see that the contradictory results have emerged as to whether REITs have a place in the mixed-asset portfolio (see, *inter alia*, Corgel, McIntosh and Ott, 1995 and Zietz, Sirmans and Friday, 2003). In other words, in attempting to answer the question as to whether REITs have a place in the mixed-asset portfolio it is necessary to have the appropriate composition of a benchmark portfolio and to separate the analysis into different capital market regimes to allow for the time-varying nature of the returns, otherwise erroneous conclusions are likely.

7.3.1 Empirical analysis – data

For the purposes of empirical analysis, the following asset classes are considered: large-cap stocks, small-cap stocks, long-term government bonds and REITs for the period January 1994 to December 2016. This sample period covers the Modern REIT Era and also encompasses several key capital market turmoil episodes including the "Dotcom Bubble", the "Global Financial Crisis" and the "Taper Tantrum".

REIT data is represented by the total returns of the US NAREIT index. To represent the performance of the stock market, both the S&P 500 and the Russell 2000 indexes are used. Bond data is represented by the Barclays US Aggregate Bond Index (AGG), formerly the Lehman Aggregate Bond Index, which is regarded as the benchmark index for long-term US bond investors and for many US index funds.

Summary statistics for the full sample period and for a number of capital market regimes is presented in Table 7.2. The first sub-period covers the period from the start of the Modern

REIT Era up to the start of the "Dotcom Bubble" in April 1997. This is followed by the period of the "Dotcom Bubble" from April 1997 to March 2000. The next period covers the time after the "Dotcom Bubble" until the start of the "Global Financial Crisis" in August 2007, being the period classified as the movement of REITs from "Main Street to Wall Street", as a result of the inclusion of REITs in the major US stock market indices, leading to the acceptance of REITs as a credible investment vehicle.

Next is the period of the "Global Financial Crisis" from August 2007 until May 2010. This is followed by the time after the "Global Financial Crisis" up to the start of the "Taper Tantrum" in May 2013. The final sub-period covers the period from the start of the "Taper Tantrum" up to the end of the sample period in December 2016. These different sample

Table 7.2 Summary Statistics: Monthly Data 1994.1 to 2016.12

Overall	*REITs*	*S&P*	*R2000*	*AGG*
Mean	0.85	0.72	0.86	0.59
SD	5.83	4.33	5.96	2.99
Sharpe	0.11	0.17	0.14	0.20
Correlation with REITs	-	0.56	0.59	0.01
New REIT Era	**REITs**	**S&P**	**R2000**	**AGG**
Mean	1.24	1.45	1.18	0.39
SD	2.78	2.89	3.78	2.63
Sharpe	0.30	0.50	0.31	0.15
Correlation with REITs	-	0.26	0.29	0.14
Dotcom Bubble	**REITs**	**S&P**	**R2000**	**AGG**
Mean	-0.11	2.02	1.70	0.82
SD	3.71	5.01	7.18	2.23
Sharpe	-0.14	0.40	0.24	0.37
Correlation with REITs	-	0.54	0.53	0.16
Main Street to Wall Street	**REITs**	**S&P**	**R2000**	**AGG**
Mean	1.47	0.12	0.77	0.59
SD	4.28	3.92	5.78	2.75
Sharpe	0.28	0.03	0.13	0.22
Correlation with REITs	-	0.30	0.37	-0.00
Global Financial Crisis	**REITs**	**S&P**	**R2000**	**AGG**
Mean	-0.80	-0.86	-0.62	0.66
SD	12.51	6.30	8.42	4.48
Sharpe	-0.07	-0.14	-0.07	0.15
Correlation with REITs	-	0.85	0.91	-0.03
Post-GFC	**REITs**	**S&P**	**R2000**	**AGG**
Mean	1.30	1.30	1.33	0.71
SD	4.80	3.95	5.49	3.30
Sharpe	0.27	0.33	0.24	0.21
Correlation with REITs	-	0.79	0.71	-0.35
Taper Tantrum	**REITs**	**S&P**	**R2000**	**AGG**
Mean	0.76	0.90	0.67	0.35
SD	4.11	3.10	4.16	2.56
Sharpe	0.19	0.29	0.16	0.14
Correlation with REITs	-	0.34	0.21	0.72

Source: Authors

periods are useful to gauge the stability of the contribution of REITs to the mixed-asset portfolio. Performance statistics include mean returns, standard deviations, Sharpe ratio and the correlation between REITs and the other asset classes, with the results presented in Table 7.2.

From the summary statistics, some stylized facts about the performance of REITs and the other capital market assets can be identified. First, over the sample period, small cap stocks offered the highest return, closely followed by REITs, while long-term government bonds showed the least return. Small cap stocks displayed the highest risk while bonds were the least volatile. The asset class with the highest Sharpe ratio was long-term government bonds, with REITs showing the worst risk-adjusted performance. Last, the correlation of REITs was the highest with small cap stocks and least with bonds, although the correlation of REITs with large cap stocks was a close second, supportive of previous studies.

However, the separation of the full time period into the six sub-periods provides interesting results. Using this 23-year history, these data show that REIT returns were highest when REITs moved from "Main Street to Wall Street" and least in the "Global Financial Crisis". In contrast, large-cap stocks and small-cap stocks produced their highest returns during the "Dotcom Bubble" and least in the "Global Financial Crisis". Bonds meanwhile showed their least return since the start of the "Taper Tantrum" and highest returns in the "Dotcom Bubble".

All asset classes showed their highest volatility in the "Global Financial Crisis" but, whereas REITs and stocks showed their least risk in the Modern REIT Era, long-term government bonds displayed their least risk in the "Dotcom Bubble".

The Sharpe ratios of REITs and stocks were highest in the Modern REIT Era and lowest in the "Global Financial Crisis". In contrast, long-term government bonds displayed their highest Sharpe ratio during the "Dotcom Bubble" and least since the start of the "Taper Tantrum".

Last, the correlation of REITs with the other asset classes shows a good deal of time-variation with its highest correlation with stocks in the "Global Financial Crisis", just when it was needed the most, and least in the Modern REIT Era. REITs displayed their highest correlation with long-term government bonds during the "Taper Tantrum", as REITs are often seen as an alternative to fixed interest savings (Downs, 1994), and least in the period following the "Global Financial Crisis".

7.3.2 Empirical analysis – mean-variance results

Because of its simplicity, analysis is limited to a simple mean-variance optimization approach. The mean-variance approach finds the combination of assets that for each level of risk offers the highest return or for each level of return offers the lowest risk. However, rather than simply analysing the benefits of REITs over the entire time period, the changing benefits of REITs to the mixed-asset portfolio are analysed over time by breaking the data down into the six sub-periods identified previously. Therefore, mean-variance analysis is also applied to the six sub-periods, allowing unambiguous comparisons of portfolio asset allocations and performance brought about by changes in market performance.

A basis of comparison is required to examine the place of REITs in the mixed-asset portfolio. Accordingly, two stock-bond portfolios are considered based on the overall sample period:

- a standard 60/40 stock-bond portfolio (40% large cap stocks and 20% small cap stocks) and 40% long-term government bonds; and
- the optimal Sharpe ratio portfolio of the stock-bond portfolio.

Table 7.3 Including REITs in the US Mixed-asset Portfolio: Monthly Data 1994.1 to 2016.12

Base Portfolios	*REITs*	*S&P*	*R2000*	*AGG*	*Mean*	*SD*	*Sharpe*
60/40	-	40%	20%	40%	0.69	2.78	0.173
Sharpe	-	22%	17%	62%	0.66	2.32	0.193
Adding REITs	**REITs**	**S&P**	**R2000**	**AGG**	**Mean**	**SD**	**Sharpe**
Overall	5%	20%	14%	60%	0.67	2.34	0.194
"New REIT Era"	43%	57%	-	-	1.36	2.27	0.421
"Dotcom Bubble"	-	37%	8%	55%	1.34	2.51	0.371
"Main Street to Wall Street"	57%	-	2%	41%	1.10	2.72	0.308
"Global Financial Crisis"	-	-	-	100%	0.66	4.48	0.128
"Post-GFC"	-	41%	5%	54%	0.98	1.35	0.724
"Taper Tantrum"	-	60%	-	40%	0.68	2.00	0.339

Source: Authors

The 60/40 stock-bond portfolio is consistent with the previous studies examining the inclusion of REITs in the mixed-asset portfolio. The optimal Sharpe ratio stock-bond portfolio was chosen as it represents the portfolio with the greatest risk-return trade-off over the sample period and so is the portfolio that all investors would have liked to own in the long-run (i.e. it is independent of the investors' preference structure).

Table 7.3 compares the enhanced mixed-asset portfolio, including REITs, against the risk-return performance of the base portfolios and presents a number of features of interest. First, the optimal allocation to REITs (5%) over the entire sample period would have resulted in a higher Sharpe ratio than the 60/40 or Sharpe stock-bond portfolio, but for different reasons. Compared to the 60/40 stock-bond portfolio, the optimal portfolio, including REITs, would have shown higher returns and lower risk, leading to a substantial increase in risk-adjusted performance.

In contrast, when the performance of the portfolio including REITs is compared with the optimal Sharpe stock-bond portfolio, although REITs increased returns this was at the expense of higher risk, leading to a reduction in risk-adjusted performance, which implies that REITs are a return enhancer rather than a risk reducer.

The results in Table 7.3 indicate that the benefits of REITs to the mixed-asset portfolio are likely to be time-varying, a hypothesis confirmed when the data is broken down into the six sub-periods. A casual examination of Table 7.3 shows that REITs would have only entered the capital market mixed-asset portfolio in two sub-periods: the Modern REIT Era and when REITs moved from "Main Street to Wall Street", albeit with very large allocations. In all other sub-periods REITs would have played no place in the mixed-asset portfolio.

7.4 Summary

Chapter 1 sought to identify critical contextual issues in international REITs, including the defining characteristics of REITs, the global evolution of REITs and the risk-return characteristics of real estate investment through REITs relative to non-REIT real estate investment with Chapter 2, the *Post-Modern REIT Era*, examining the evolution of the global REIT market, what is deemed to be best current market practice and how the market is expected to change going forward. Chapter 3, *Emerging Sector REITs*, investigated such sectors as timber and data centres with Chapter 4, *Sustainable REITs*, analysing REIT environmental performance and the cost of equity. Chapter 5 *Islamic REITs*, examined the evolution of global Islamic REITs through a study of the Malaysian market and Chapter 6, *Behavioural Risk in REITs*, addressed the management of behavioural risk in global REITs.

This chapter, *REIT Asset Allocation*, reviewed recent research into asset allocation and sought to answer two questions which have been the subject of academic and investor research for four decades:

- are REITs real estate; and
- what place, if any, do REITs have in the mixed-asset portfolio?

The answer to the first question seems to be that REITs are real estate, albeit in the long run. So long-run investors should expect total returns approximating those achieved from the private real estate market. Indeed, since the performance of REITs is ultimately dependent upon the performance of the underlying property, a close link between markets is to be expected. It should be noted that "long-term" in this context may be at least five years, which would seem to negate the liquidity advantage of REITs. Nonetheless, REITs can provide some of the diversification opportunities of private real estate not normally available to smaller investors, plus new investment opportunities with the added bonus of liquidity. However, these advantages come at increased volatility and loss of control.

The answer to the second question, what place, if any, do REITs have in the mixed-asset portfolio, is more problematical as previous research and the results of the empirical test herein are mixed such that no clear recommendation for adding REITs to a mixed-asset portfolio can be given. On the one hand, the results suggest that the inclusion of REITs into a portfolio of stocks and bonds over the long-run increases risk-adjusted performance, due to the return enhancement of REITs. So for those investors with an appetite for more return REITs may be an attractive addition to their mixed-asset portfolio, if they are willing to accept greater volatility. The enhanced performance from including REITs in the mixed-asset portfolio however can be traced to just two time periods when REITs performed exceptionally well, being the Modern REIT Era and when REITs moved from "Main Street to Wall Street". Therefore, deciding if REITs will offer such exceptional performance in the future is a real challenge for investors.

Nonetheless, there are a number of characteristics of REITs that provide clues to investors when considering including REITs in their mixed-asset portfolio. First, since in the long-run REITs tend to behave more like private real estate, even allowing for the presence of a stock market component, REITs will perform well when private real estate performs well and vice versa. Second, due to their use of leverage, REITs they will perform badly in times of financial distress and vice versa. Third, as REIT returns are more correlated with small cap stocks, especially small cap value stocks, rather than the stock market as a whole because of the typically small market capitalization of REIT issues, they will perform relatively well when value stocks perform well and vice versa. These pointers as to when REITs are likely to perform well can be incorporated into an asset allocation framework and provide the potential to generate increased returns. Therefore, how good investors are in identifying these capital market regimes will determine the performance of their portfolios.

This chapter concludes Part I of this book with Part II including six chapters each analysing REITs in a region of the world, using case studies from a developed, developing and emerging REIT sector in the region, concluding with Chapter 14 which considers *Directions for the Future of International REITs*.

References

Ang, A., Nabar, N. and Wald, S.J. 2013, 'Searching for a common factor in public and private real estate returns', *Journal of Portfolio Management*, Vol. 39, pp. 120–133.

Bohjalian, T. 2015, *REIT opportunities in a rising-rate market*, Viewpoint, Cohen and Steers, New York.

Brinson, G.P., Singer, B.D. and Beebower, G.P. 1991, 'Determinants of portfolio performance II', *Financial Analysts Journal*, Vol. 47, No. 3, pp. 40–48.

Brown, G.R. and Matysiak, G.A. 1998, 'Valuation smoothing without temporal aggregation', *Journal of Property Research*, Vol. 15, pp. 89–103.

Ciochetti, B.A., Craft, T.M. and Shilling, J.D. 2002, 'Institutional investors' preferences for REIT stocks', *Real Estate Economics*, Vol. 30, No. 4, pp. 567–594.

Corgel, J.B. and deRoos, J.A. 1999, 'Recovery of real estate returns for portfolio allocation', *The Journal of Real Estate Finance and Economics*, Vol. 18, No. 3, pp. 279–296.

Corgel, J.B., McIntosh, W. and Ott, S.H. 1995, 'Real estate investment trusts: A review of the financial economic literature', *Journal of Real Estate Literature*, Vol. 3, pp. 13–43.

Downs, A. 1994, *The REIT 'explosion'. What does it mean?*, United States Real Estate Research, Market Perspective, Salomon Brothers, New York.

Fisher, J. D., D. M. Geltner, and R. B. Webb. (1994). Value Indices of Commercial Real Estate: A Comparison of Index Construction Methods, *Journal of Real Estate Finance and Economics*, Vol. 9, No. 2, pp. 137–164.

Fisher, J.D. and Young, M.S. 2000, 'Institutional property tenure: Evidence from the NCREIF database', *Journal of Real Estate Portfolio Management*, Vol. 6, No. 4, pp. 327–338.

Geltner, D. 1991, 'Smoothing in appraisal-based returns', *Journal of Real Estate Finance and Economics*, Vol. 4, pp. 327–345.

Geltner, D. and Kluger, B. 1998, 'REIT-based pure-play portfolios: The case of property types', *Real Estate Economics*, Vol. 26, pp. 581–612.

Giliberto, S.M. 1990, 'Equity real estate investment trusts and real estate returns', *Journal of Real Estate Research*, Vol. 5, pp. 259–263.

Giliberto, S.M. and Mengden, A. 1996, 'REITs and real estate: Two markets re-examined', *Real Estate Finance*, Vol. 13. pp. 56–60.

Gyourko, J. and Nelling, E. 1996, 'Systematic risk and diversification in the equity REIT market', *Real Estate Economics*, Vol. 24, pp. 493–515.

Hudgins, M.C. 2012, *The role of REITs in a portfolio: A 'core plus plus' real estate allocation*, JP Morgan, New York.

Hudson-Wilson, S. 2001, 'Why real estate?', *The Journal of Portfolio Management*, Vol. 28, No. 1, pp. 20–32.

Ibbotson, R.G. and Kaplan, P.D. 2000, 'Does asset allocation policy explain 40, 90, or 100 percent of performance?' *Financial Analysts Journal*, Vol. 56, No. 1, pp. 26–33.

Kuhle, J. 1987, 'Portfolio diversification and return benefit – common stocks vs. real estate investment trusts (REITs)', *The Journal of Real Estate Research*, Vol. 2, pp. 1–9.

Lai, T.Y. and Wang, K. 1998, 'Appraisal smoothing: The other side of story', *Real Estate Economics*, Vol. 26, No. 3. pp. 511–535.

Lee, S. 2010, 'The changing benefit of REITs to the mixed-asset portfolio', *Journal of Real Estate Portfolio Management*, Vol. 16, No. 3, pp. 201–215.

Lee, S. 2016, 'REITs and the taper tantrum', *Journal of Property Investment and Finance*, Vol. 34, No. 5, pp. 457–464.

Lee, S. and Stevenson, S. 2005, 'The case for REITs in the mixed-asset portfolio in the short and long run', *Journal of Real Estate Portfolio Management*, Vol. 11, No. 1, pp. 55–80.

Ling, D. and Naranjo, A. 2015, 'Returns and information transmission dynamics in public and private real estate markets', *Real Estate Economics*, Vol. 43, pp. 163–208.

Menchero, J. and Davis, B. 2010, 'Risk contribution is exposure times volatility times correlation: Decomposing risk using the *x-sigma-rho* formula', *Journal of Portfolio Management*, Vol. 37, No. 2, pp. 97–106.

Mueller, G.R., Pauley, K.R. and Morrill Jr., W.K. 1994, 'Should REITs be included in a mixed-asset portfolio?', *Real Estate Finance*, Vol. 11. pp. 23–28.

Muhlhofer, T. 2013, 'Why do REIT returns poorly reflect property returns? Unrealizable appreciation gains due to trading constraints as the solution to the short-term disparity', *Real Estate Economics*, Vol. 41, No. 4, pp. 814–857.

NAREIT 2002, *Diversification benefits of REITs – An analysis by Ibbotson Associates*, html//www.nareit.com

Newell, G. and Peng, H. 2006, 'The role of non-traditional real estate sectors in REIT portfolios', *Journal of Real Estate Portfolio Management*, Vol. 12, No. 2, pp. 156–166.

Oikarinen, E., Hoesli, M. and Serrano, C. 2011, 'The long-run dynamics between direct and securitized real estate', *Journal of Real Estate Research*, Vol. 33, pp. 73–104.

Pagliari Jr., J.L., Scherer, K.S. and Monopoli, R.T. 2005, 'Public versus private real estate equities: A more refined, long-term comparison', *Real Estate Economics*, Vol. 33, No. 1, pp. 147–187.

Paladino, M. and Mayo, H. 1998, 'REIT stocks do not diversify stock portfolios: An update', *Real Estate Review*, Vol. 27, pp. 39–40.

Pedersen, N., He, F., Tiwari, A. and Hofmann, A. 2012, *Modelling the risk characteristics of real estate investments*, Technical Report, PIMCO.

Shepard, P., Liu, Y. and Dai, Y. 2014, *The Barra private real estate model* (PRE2), Research Notes, MSCI.

Tsai, J.P. 2007, *A successive effort on performance comparison between public and private real estate equity investment*, Master's Thesis, MIT Center for Real Estate.

Zietz, E.M., Sirmans, G.S. and Friday, H.S. 2003, 'The environment and performance of real estate investment trusts', *Journal of Real Estate Portfolio Management*, Vol. 9, No. 2, pp. 127–165.

PART II

International REITs

Regional review of issues

8
NORTH AMERICA

Brad Case

8.1 Introduction

This book aims to identify key areas for research in the REIT discipline for the next five to ten years by surveying the current state of the REIT discipline around the world and identifying emerging and cutting edge research areas through a thematic review of current contextual issues and a regional analysis based on case studies.

This book comprises two parts, the first part being a thematic review of emerging and cutting-edge global research into current contextual issues in REITs internationally and the second part being a regional analysis of REITs around the world, each written by authoritative academic authors from the world's leading Universities and REIT industry experts.

Part I included six chapters each reviewing a current theme of REIT evolution through the lens of contemporary research. Chapter 1 focused on critical contextual issues in international REITs while Chapter 2, the *Post-Modern REIT Era*, examined the evolution of the global REIT market, what is deemed to be best current market practice and how the market is expected to change going forward and Chapter 3, *Emerging Sector REITs*, investigated such sectors as residential, health care, self-storage, timber, infrastructure and data centres generally and then specifically through a case study of US REITs.

Chapter 4, *Sustainable REITs*, analysed REIT environmental performance and the cost of equity and Chapter 5, *Islamic REITs*, examined the evolution of global Islamic REITs through a study of the Malaysian market. Chapter 6, *Behavioural Risk in REITs*, addressed the management of behavioural risk in global REITs and Part I concluded with Chapter 7 which reviewed recent research into *REIT Asset Allocation*.

Part II includes six chapters each analysing REITs in a region of the world, using case studies from a developed, developing and emerging REIT sector in the region. This chapter analyses REITs in the North American region, using case studies from the USA (being a developed REIT sector in the region), Canada (being a developing REIT sector in the region) and Mexico (being an emerging REIT sector in the region). Chapter 9 will then analyse REITs in the Latin American region, using case studies from Brazil, Argentina and Uruguay for the developed, developing and emerging REIT sectors in the region, respectively.

Chapter 10 analyses REITS in the European region, using case studies from the UK, Spain and Poland with Chapter 11 analysing REITs in the South East Asian region, using case studies from Singapore, Malaysia and Thailand. Chapter 12 analyses REITS in the North Asian region, using case studies from Japan, Hong Kong and China with Chapter 13 analysing REITs in the Oceania region, using case studies from Australia, South Africa and India. Chapter 14 concludes Part II by considering *Directions for the Future of International REITs*.

8.2 REITs in the US

This chapter analyses REITs in the North American region, using case studies from the US (being a developed REIT sector in the region), Canada (being a developing REIT sector in the region) and Mexico (being an emerging REIT sector in the region).

8.2.1 Status of the US REIT industry

The United States is clearly the most highly developed REIT market in the world, with only Australia laying claim to some comparability. The REIT approach to real estate investing was developed in the United States, with the key legislation passing in 1960 as one of the last actions of the Presidential administration of Dwight D. Eisenhower. Six REITs formed immediately, three of which – Pennsylvania REIT, Washington REIT and First Union Real Estate (now called Winthrop Realty Trust) – are still in operation more than 56 years later. In June 1965 another of the inaugural REITs, Continental Mortgage Investors, became the first REIT listed on the New York Stock Exchange.

The *Wall Street Journal* published a research article focused on REITs for the first time in 1969, and 1970 saw the first publication focused specifically on REIT investing, Realty Trust Review. At the beginning of 1972 – a time when only four other nations (the Netherlands, New Zealand, Taiwan and Australia) had yet passed REIT legislation – the National Association of Real Estate Investment Trusts (NAREIT) started computing the first index of REIT returns for investment decision-making purposes. The beginning of 1985 saw the formation of the first mutual fund focused exclusively on REIT investments, while 1986 saw the formation of what became the largest REIT-dedicated asset management company, Cohen & Steers.

By the end of 2016, REIT structures substantially similar to the US model had been established in 36 countries, but the US still dominated the global REIT market with 224 REITs listed on US stock exchanges valued at an aggregate market capitalization in excess of US$1 trillion. The 132 exchange-traded US equity REITs that were constituents of the FTSE EPRA/NAREIT Global Real Estate Index Series (which also includes non-REITs) accounted for half of its total market cap of US$1.5 trillion. The growth of the US REIT industry has been impressive with the aggregate exchange-traded US REIT industry now nearly 700 times as large as it was when data first became available at the end of 1971, with market cap having grown by almost 16%pa over that 45-year period.

The long-term performance of the US REIT market has been comparable to that of the broad US stock market: total returns of the entire industry as measured by the FTSE NAREIT All REIT Index have averaged 9.74% with volatility (based on monthly returns) of 17.6%, compared to average total returns and annualized volatility of 10.36% and 15.2% for the Standard & Poor's 500 Index of large-cap stocks and 10.42% and 15.6% for the Dow Jones US Total Stock Market Index (DJUSTM), over the 45-year period from the end of 1971 to the end of 2016.

Exchange-traded equity REITs have provided stronger returns than mortgage REITs both absolutely and on a risk-adjusted basis: over the same period the FTSE NAREIT All Equity

REIT Index measured total returns averaging 11.93% with annualized volatility of 17.0%, while the FTSE NAREIT Mortgage REIT Index measured returns of 5.26% with 19.9% volatility.

As noted later, important developments in the early 1990s marked the beginning of what is often referred to as the Modern REIT Era, commonly dated from the initial public offering (IPO) of Kimco Realty near the end of 1991. Over the 25-year period from the end of 1991 through to the end of 2016 the US REIT industry outperformed the broad US stock market with the FTSE NAREIT All REIT Index showing average annual total returns of 10.58% compared to 9.15% for the S&P 500 and 9.28% for the DJUSTM. Volatility was also slightly higher for the US REIT market (18.1%) than for the much larger US stock market (14.2% for the S&P 500 and 14.5% for the DJUSTM) and, as a result, risk-adjusted returns were essentially identical with Sharpe ratios of 0.52 for all three indices. As was true over the longest available historical period, equity REITs outperformed mortgage REITs over the 25-year Modern REIT Era, with equity REITs showing average total returns of 11.13% and volatility of 18.9% compared to 6.21% and 19.9% for mortgage REITs.

The long-term average correlation between US REITs (FTSE NAREIT All REIT Index) and the broad stock market (based on monthly returns) was just 56% during the 25-year Modern REIT Era and REIT returns showed a beta of 0.73 with annualized alpha of +4.55% relative to the Dow Jones US Total Stock Market Index, indicating that exchange-traded REITs in the US have provided very strong diversification benefits to investors in a broadly diversified portfolio of US stocks.

8.2.2 Composition of the US REIT industry

The US REIT market has included not only equity REITs and mortgage REITs but also companies that hold a combination of real estate equity assets and real estate debt assets, called hybrid REITs. At the end of 1971, the exchange-traded US REIT market consisted of 12 equity REITs whose market cap of US$332 million was 22% of the total, another 12 mortgage REITs whose market cap of US$571 million was 38% of the total and 10 hybrid REITs whose market cap of US$592 million made them the largest segment of the market at 40% of the total.

The growth of the industry since then, however, has been almost exclusively in the equity REIT segment: by the end of 2016, the equity REIT segment had grown by more than 19%pa on average and had come to dominate the industry with 95% of the aggregate market cap, while the mortgage REIT segment had grown by an average of less than 11%pa and totalled just 5% of the industry aggregate. Hybrid REITs, meanwhile, dwindled until by the end of 2010 just three companies remained with an aggregate market cap that was just 0.1% of the industry, prompting the separate recognition of that part of the market to cease during December of that year.

The composition of the equity REIT segment of the US REIT industry has been uneven, with much of it coming from an expanding appetite among investors for different property types. For example, during the early part of the industry there were several companies with holdings concentrated in both the office and industrial property types, so a "Mixed Office/Industrial" segment was recognized; by late 2015, however, the eight such companies accounted for just 2% of the total market cap of equity REITs, so the "Mixed Office/Industrial" segment was discontinued and the remaining companies re-identified as Industrial, Office or Diversified REITs. In fact, more generally, the propensity of US REITs to hold diversified property portfolios seems to have declined as the industry has matured: the "Diversified REITs" segment, for example, has declined in relative terms from 8% of the market as of the end of 1999 to 6% at the end of 2016 following growth in market cap of 11.6% pa, less than the overall equity REIT average of 13.1% during the same period.

Conversely, new segments of the US equity REIT industry have been created to recognize the increasing importance of what used to be considered "specialty" property types. The infrastructure REIT segment was added at the beginning of 2012 and comprised 8% of the equity REIT total as of the end of 2016; data centre REITs were added near the end of 2015 and now account for 6% of the total; timber REITs were recognized at the end of 2010 and accounted for 3% at the end of 2016.

A separate category called "Specialty REITs", which included 4% of the equity REIT total as of the end of 2016, encompasses companies whose holdings focus specifically on another property type such as document storage (Iron Mountain), casinos (Gaming & Leisure Properties), outdoor advertising (Lamar Advertising, Outfront Media), entertainment and recreation (EPR Properties), detention facilities (CoreCivic, Geo Group) or agricultural land (Farmland Partners, American Farmland, and Gladstone Land). Any of these property types, or another not currently part of the exchange-traded US REIT industry, may become a large enough segment of the total to merit the creation of a new category in the future.

REITs whose holdings were primarily office properties accounted for 20% of the equity REIT industry as of the end of 1999; 17 years later, by the end of 2016, office REITs accounted for only 10% of the industry with market cap having grown by 8.7%pa on average. Residential REITs – including those whose holdings consist primarily of apartment buildings, student housing, manufactured home communities, and, later, single-family residential properties, accounted for 21% of the equity REIT industry at the end of 1999 but just 14% at the end of 2016, having grown by 10.2%pa. Industrial REITs grew by 12.3%pa but declined as a share of the overall equity REIT industry from 7% at the end of 1999 to 6% at the end of 2016, while Retail REITs – including those whose holdings consisted primarily of regional malls, neighbourhood shopping centres or free-standing retail properties – grew by 12.8% but declined from 22% of the total to 21%; hotel REITs similarly grew by 12.8%pa but declined slightly from 5.5% of the total to 5.2%.

Aside from the newly created infrastructure, data centre and timber categories, the most rapid growth in the US equity REIT industry from 1999 to 2016 came in the self-storage and health care segments. The market cap of self-storage REITs grew by 15.9%pa and expanded from 4% of the equity REIT market at the end of 1999 to 6% at the end of 2016, while health care REITs grew by 21.5% and expanded from 3% to 10% of the total.

As noted briefly, the relative share of REITs with portfolios diversified across multiple property types has declined at least over the 17 years since the end of 1999. In fact, observers had noted repeatedly before 1999 that the US REIT market seemed to be transforming gradually from a norm in which individual companies tended to have portfolios that were concentrated geographically but diversified by property type into one with the opposite norm in which portfolios tended to be diversified geographically but concentrated by property type and even sub-type.

The difference between these two approaches has given rise to several empirical studies of performance at both property and company levels as well as questions regarding the effect of focus or diversification on cost of capital, general and administrative costs and other impacts. Capozza and Seguin (1999), for example, found that higher costs tended to make diversified REITs less valuable than focused REITs, while increased opacity tended to make them less liquid as well. Ro and Ziobrowski (2011, 2012) published two analyses using different methodologies, one of which supported the Capozza and Seguin result while the other opposed it. Similar analyses have been conducted by Danielsen and Harrison (2007), Benefield, Anderson and Zumpano (2009), Hartzell, Sun and Titman (2014) and Anderson, Benefield and Hurst (2015), among others, but the questions have certainly not yet yielded to a consensus.

8.2.3 Research focused on the evolution of the REIT structure

The US has played a leading role in the development of key aspects of the REIT approach to real estate investment and the long available history of data concerning the US REIT industry – including accurate monthly returns (and daily returns beginning in January 1990) as well as data on REIT portfolio holdings, operating performance, capital structures and many other items of interest – has encouraged researchers to investigate a rich variety of aspects of the real estate market by focusing on the impacts of changes in REIT rules on the behaviour and performance of US REITs.

The original 1960 legislation required REITs to be organized as business trusts, but amendments passed as part of the Tax Reform Act of 1986 permitted REITs to be organized as corporations. While there are still some REITs organized as business trusts, the corporate form came to dominate the US REIT industry so quickly that there seems to have been no academic research comparing the two.

The original legislation also required that REIT-owned assets be externally advised and managed, a requirement that still characterizes many non-US REIT regimes, but that was quickly recognized as giving rise to potential conflicts of interest, as noted by Sagalyn (1996). Amendments passed as part of the Tax Reform Act of 1986 permitted REITs to internalize these functions and – as with the corporate forms – internally advised and managed REITs quickly came to dominate the US industry. The continued existence after 1986 of both internally and externally advised REITs has, however, provided a particularly fruitful set of research topics with numerous published studies comparing internal and external management approaches including those by Cannon and Vogt (1995), Capozza and Seguin (2000), Ambrose and Linneman (2001), Chen and Lu (2006), Hardin and Hill (2008), Benefield and Pyles (2009), Hardin et al. (2009), Sun (2010), Brockman, French and Tamm (2014) and Deng, Hu and Srinivasan (2016a and 2016b).

In 1991, in response to recognition that "earnings" as defined under Generally Accepted Accounting Principles (GAAP) are less informative for evaluating real estate investments, NAREIT developed a definition of Funds From Operations (FFO) which rapidly became the industry-standard REIT performance metric, especially after modifications announced in 1999 to reinforce comparability across REITs. Several studies have examined the information content of FFO, including those published by Graham and Knight (2000), Stunda and Typpo (2004), Higgins, Ott and Van Ness (2006), Baik, Billings and Morton (2008) and Ben-Shahar, Sulganik and Tsang (2011).

In 1992 Taubman Centers developed the Umbrella Partnership REIT (UPREIT) structure to form an "operating partnership" between property owners and a REIT that would enable property owners to benefit from high-quality professional management of their assets. This important advance helped trigger the growth of the industry during what is often called the Modern REIT Era. Although many studies control for differences between UPREITs and REITs that did not adopt the UPREIT form in empirical studies focusing on other questions, very few appear to have focused specifically on the effect of the UPREIT structure for REIT performance comparison. Wu and Yavas (2005) concluded that concern regarding a "tax timing" conflict of interest among UPREITs was likely overstated, while Ebrahim and Mathur (2013) credited UPREITs with enhancing investor welfare during the 2008–2009 liquidity crisis.

The original legislation required that the top five shareholders of any REIT could own no more than 50% of its shares, but left ambiguous whether institutional investors – especially pension funds and mutual funds – should be considered "shareholders" themselves or entities acting on behalf of shareholders. Amendments passed in 1993 clarified that the so-called five and fifty rule should not apply to such investors. This change provided another major catalyst for growth of the US REIT industry which, by opening investment in the industry fully to institutional as well as individual sources of capital, seems to have prompted a sharp downward movement in the correlation between REITs and non-REIT stocks as Case, Yang and Yildiray (2012) documented.

A series of developments – including a 1996 letter ruling from the Internal Revenue Service, the REIT Simplification Act of 1997 and the REIT Modernization Act of 1999 – expanded the services that could be regarded as contributing to generating "qualifying real estate rental income" and permitted REITs to engage in non-qualifying activities through "taxable REIT subsidiaries".

These changes were especially important to the development of the hotel REIT sector, which grew to 5% of the total industry by the end of 2016. The 2008 passage of the REIT Investment and Diversification Act provided a similar stimulus to the development of the health care REIT sector, which by the end of 2016 constituted 10% of the overall industry. While Howe and Jain (2004) documented both a general gain in wealth and a general decline in systematic risk associated with the passage of the REIT Modernization Act, most other studies of the impact of this series of developments have focused more specifically on the hotel or health care segments, such as Kim, Gu and Mattila (2002a, 2002b) and Jackson (2008).

Amendments passed as part of the REIT Improvement Act of 2004 eliminated a provision of the 1980 Foreign Investment in Real Property Tax Act (FIRPTA) that had presented a barrier to foreign investors seeking to buy stock in exchange-traded US REITs with another amendment, passed in 2015, going further to eliminate the problem. The only empirical study to date of the effect of the 2004 amendment to FIRPTA is Howard, Pancak and Shackelford (2016), who documented responses among both foreign investors and REIT managers, but additional research is surely warranted in response to the 2015 amendment.

8.2.4 REITs in the US – summary

This case study of REITs in the US, being a developed REIT sector in the North American region, provides a clear example of how a REIT market grows and evolves with the passage of time. Further, the US REIT sector illustrates how evolution in the REIT sector can drive legislative change and also how legislative change can drive evolution in the REIT sector. Significantly, given the maturity of the US REIT sector, a deep body of academic research into REITs now exists, with each legislative change subject to further academic research into both its impact on the REIT sector and the response of the REIT sector.

8.3 REITs in Canada

This chapter analyses REITs in the North American region, using case studies from the US (being a developed REIT sector in the region), Canada (being a developing REIT sector in the region) and Mexico (being an emerging REIT sector in the region).

8.3.1 Status of the Canadian REIT industry

Canada, unlike Mexico but like the US, has a long history of a prosperous economy with highly developed economic institutions, including a highly developed real estate market structure. Like Mexico and unlike the US, however, the REIT industry in Canada developed much more recently; as a result, although the Canadian REIT market is already much larger than the Mexican REIT market, many research questions have not yet been addressed using data specific to Canadian REITs.

Canada adopted a REIT structure in 1993 as a response to the sharp early 1990s downturn in the Canadian real estate market. At that time, many Canadian real estate investors held positions in open-end real estate mutual funds which were required to redeem shares at values determined by the most recently completed property appraisals. As market values of their properties collapsed, however, many investors took advantage of the redemption rules to "sell" their property ownership shares at what had become obviously inflated valuations and the funds found themselves unable to meet redemption requests as liquidity suddenly dried up in the property market. As soon as the Canadian REIT rules were finalized, three of the existing real estate mutual funds converted to become the first Canadian REITs: RealFund, Canadian REIT and RioCan REIT. RealFund was later acquired by RioCan, but the other two inaugural REITs are still operating independently.

As of the end of 2016, there were 45 REITs listed on stock exchanges in Canada with an aggregate market cap of CA$66 billion. The largest share of the market – 18 companies worth one-third of the aggregate market cap – consisted of REITs holding diversified property portfolios. The next largest segment was retail REITs, with seven companies valued at CA$17 billion (one-fourth of the industry total) including the largest, RioCan. Eight residential REITs worth CA$12 billion accounted for 18% of the total, while four office REITs worth CA$8 billion comprised another 12%. The smallest segments of the Canadian market were health care REITs (two companies worth CA$4 billion or 6% of the total), industrial REITs (five companies worth CA$5 billion or 4% of the total) and hotel REITs (just one company worth CA$600 million).

8.3.2 Performance of the Canadian REIT industry

The performance of the Canadian REIT industry has been significantly better than that of the broader Canadian stock market over the 19-year available historical period from the end of 1997 through the end of 2016.

According to the S&P/TSX Capped REIT Index, total returns of Canadian REITs averaged 11.61%pa over that period with annualized volatility of 14.3% compared to returns of just 7.86%pa with very slightly higher volatility of 15.0% for the broad stock market (Canada S&P/TSX Composite Index), indicating that Canadian REITs performed about 70% better than the broad stock market on a risk-adjusted basis.

The correlation in monthly returns between the REIT series and the broad market series was just 55% and the REIT beta relative to the broad stock market was just 0.53 with an annualized alpha of +7.47%, indicating that Canadian REITs had provided very strong diversification benefits to investors in a broadly diversified Canadian stock portfolio.

8.3.3 REITs in Canada – summary

This case study of REITs in Canada, being a developing REIT sector in the North American region, shows how quickly a REIT industry may emerge out of a property market downturn that suddenly generates a supply of suitable investment stock. In only 25 years, the Canadian REIT industry has grown to CA$66 billion by late 2016 and generated performance about 70% better than the broad stock market on a risk-adjusted basis.

8.4 REITs in Mexico

This chapter analyses REITs in the North American region, using case studies from the US (being a developed REIT sector in the region), Canada (being a developing REIT sector in the region) and Mexico (being an emerging REIT sector in the region).

8.4.1 Status of the Mexican REIT industry

Mexico presents a good case study of an emerging-market economy with an emerging REIT sector. Mexico adopted the precursor to REIT-type legislation in 2004, but important changes in 2010 marked the actual birth of the industry with Fibra Uno becoming the first Fideicomiso de Inversión en Bienes Raices (FIBRA) listed on the Mexican stock exchange in March 2011.

By the end of 2016, five Fibras were constituents of the FTSE EPRA/NAREIT Global Real Estate Index Series with an aggregate market cap of US$8.3 billion and, collectively, a weight of 0.5% in the Global REIT Index (on a free float cap-weighted basis) and 0.7% in the Americas REIT Index.

Although Fibra Uno remains the largest participant in the Mexican REIT industry, at the end of 2016 there were 10 exchange-traded Fibras with an aggregate market cap of about 230 billion Mexican Pesos (US$11 billion):

- Fibra Uno, with a market cap of US$4.7 billion (41% of the industry), owns 17 retail, industrial, office and mixed-use properties in seven states;
- Fibra Danhos, with a market cap of US$2.3 billion (20%), owns 14 retail, office and mixed-use properties in nine states;
- Fibra Prologis, valued at US$930 million (8%), owns a portfolio of nearly 200 industrial properties in six major markets;
- Terrafina, valued at US$900 million (8%), holds more than 200 industrial properties in 16 states;
- Fibra Macquarie, valued at US$850 (7%), holds nearly 300 mostly industrial properties in 19 states;
- Fideicomiso Hipotecaria (FHipo), valued at US$530 million (5%), is the only Mexican mortgage REIT with a portfolio of more than 65,000 loans;
- Fibra Hotel, valued at US$370 million (3%), holds more than 80 hotels operated under several Mexican and American brands in 26 states;
- Fibra Shop, valued at US$310 million (3%), owns 17 retail properties in 13 states;
- Fibra Monterrey (Fibra MTY), valued at US$270 million (2%), owns 35 office, industrial, and retail properties; and
- Fibra Inn, valued at US$230 million (2%), owns 19 hotels operated under several brands throughout the country.

8.4.2 Performance of the Mexican REIT industry

The investment performance of Fibras was strong during their first two years through to early 2013 and solidly better than the broad Mexican stock market, perhaps benefiting from a novelty effect. Since then, however, Fibra investment performance has been uneven: according to the BMV Fibra Index, Fibras lost some 25% from mid-May 2013 through August 2013, regained their value over the ensuing year and then continued to post modest gains through early November 2016, but lost more than 12% during the last two months of 2016.

With only five years of performance data and just 10 exchange-traded companies to date, it is challenging to conduct research on the Mexican REIT market comparable to the analyses that have been done using data from more mature markets. The variations among the existing Fibras, however – including several with portfolios concentrated by property type, others with property type-diversified portfolios, variations in geographic concentration and one mortgage REIT – suggests several interesting research questions to be addressed over the next several years.

Perhaps an overriding question, however, is the long-term relative influence of developments in the Mexican economy versus those in the much larger American economy next door in driving Fibra investment performance: the decline of Fibra stock prices during the last two months of 2016, for example, may very well have reflected a response to political developments in the US.

8.4.3 REITs in Mexico – summary

This case study of REITs in Mexico, being an emerging REIT sector in the North American region, shows how much progress may be achieved in less than a decade with 10 exchange-traded REITs by late 2016 having an aggregate market cap of around US$11 billion. Significantly, in the emerging REIT stage, 62% of REITs by capitalization are diversified providing investors with a broad portfolio spread across several property sectors though, interestingly, 24% of REITs by capitalization are industrial sector specific with other sector-specific REITs nascent.

It will, however, require the passage of time and the creation of further Mexican REITs before an adequate sample exists with which to undertake comparative research relative to the developing and developed REIT sectors in the North American region.

8.5 Summary

Part II includes six chapters each analysing REITs in a region of the world, using case studies from a developed, developing and emerging REIT sector in the region.

This chapter analysed REITs in the North American region, using case studies from the US (being a developed REIT sector in the region), Canada (being a developing REIT sector in the region) and Mexico (being an emerging REIT sector in the region).

In the North American region, it is the difference in size between developed, developing and emerging REIT sectors in the world's regions that is the most stark. With the developed US REIT sector at US$1 trillion, the developing Canadian REIT sector at CA$66 billion and the emerging Mexican REIT sector at US$11 billion in late 2016, the gap between the size of each is startlingly evident.

There is, however, a commonality in evolution with each REIT sector in the region starting its life cycle with a greater proportion of diversified REITs plus a sector specific REIT, then broadening to other large sectors, then further broadening to other smaller sectors. The US REIT sector started with diversified REITs, then broadened into industrial and office REITs before further broadening into infrastructure, data centres and timber REITs. Similarly, the Canadian REIT sector started with diversified REITs and retail REITs, then broadened to residential and office REITs before further broadening into health care, industrial and hotel REITs. The Mexican REIT sector, at a much earlier stage if its evolution, started with diversified and industrial REITs, then broadened to retail and hotel REITs but with the further broadening stage yet to come as the sector grows and evolves.

Chapter 9 will analyse REITs in the Latin American region, using case studies from Brazil, Argentina and Uruguay for the developed, developing and emerging REIT sectors in the region, respectively.

Chapter 10 analyses REITs in the European region, using case studies from the UK, Spain and Poland with Chapter 11 analysing REITs in the South East Asian region, using case studies from Singapore, Malaysia and Thailand. Chapter 12 analyses REITS in the North Asian region, using case studies from Japan, Hong Kong and China with Chapter 13 analysing REITs in the Oceania region, using case studies from Australia, South Africa and India and Chapter 14 concluding Part II by considering *Directions for the Future of International REITs*.

References

Ambrose, B. and Linneman, P. 2001, 'REIT organizational structure and operating characteristics', *Journal of Real Estate Research*, Vol. 21, No. 3, pp. 141–162.

Anderson, R.I., Benefield, J.D. and Hurst, M.E. 2015, 'Property-type diversification and REIT performance: An analysis of operating performance and abnormal returns', *Journal of Economics and Finance*, Vol. 39, No. 1, pp. 48–74.

Baik, B., Billings, B.K. and Morton, R.M. 2008, *Manipulation, increased transparency, and value relevance of non-GAAP disclosures for real estate investment trusts (REITs)*, September 20, 2005, AAA 2006 Financial Accounting and Reporting Section (FARS) Meeting Paper, available at SSRN: https://ssrn.com/abstract=811427

Benefield, J.D., Anderson, R.I. and Zumpano, L.V. 2009, 'Performance differences in property-type diversified versus specialized real estate investment trusts (REITs)', *Review of Financial Economics*, Vol. 18, No. 2, pp. 70–79.

Benefield, J. and Pyles, M.K. 2009, 'Internally versus externally advised non-brokerage real estate firms', *Journal of Alternative Investments*, Vol. 12, No. 1, pp. 39–49.

Ben-Shahar, D., Sulganik, E. and Tsang, D. 2011, 'Funds from operations versus net income: Examining the dividend relevance of REIT performance measures', *Journal of Real Estate Research*, Vol. 33, No. 3, pp. 415–441.

Brockman, P., French, D. and Tamm, C. 2014, 'REIT organizational structure, institutional ownership, and stock performance', *Journal of Real Estate Portfolio Management*, Vol. 20, No. 1, pp. 21–36.

Cannon, S. and Vogt, S. 1995, 'REITs and their management: An analysis of organizational structure, performance and management compensation', *Journal of Real Estate Research*, Vol. 10, No. 3, pp. 297–317.

Capozza, D.R. and Seguin, P.J. 1999, 'Focus, transparency and value: The REIT evidence', *Real Estate Economics*, Vol. 27, No. 4, pp. 587–619.

Capozza, D.R. and Seguin, P.J. 2000, 'Debt, agency, and management contracts in REITs: The external advisor puzzle', *Journal of Real Estate Finance and Economics*, Vol. 20, No. 91.

Case, B., Yang, Y. and Yildiray, Y. 2012, 'Dynamic correlations among asset classes: REIT and stock returns', *Journal of Real Estate Finance and Economics, Vol.* 44, No. 3.

Chen, H.C. and Lu, C. 2006, 'How much do REITs pay for their IPOs?', *Journal of Real Estate Finance and Economics*, Vol. 33, No. 105.

Danielsen, B. and Harrison, D. 2007, 'The impact of property type diversification on REIT liquidity', *Journal of Real Estate Portfolio Management*, Vol. 13, No. 4, pp. 329–344.

Deng, Y.H., Hu, M.R. and Srinivasan, A. 2016a, 'Information asymmetry and organizational structure: Evidence from REITs', *Journal of Real Estate Finance and Economics*, Vol. 38, pp. 1–29.

Deng, Y.H., Hu, M.R. and Srinivasan, A. 2016b, 'Hold-up versus benefits in relationship banking: A natural experiment using REIT organizational form', unpublished manuscript.

Ebrahim, M.S. and Mathur, I. 2013, 'On the efficiency of the UPREIT organizational form: Implications for the subprime crisis and CDO's', *Journal of Economic Behaviour & Organization*, Vol. 85, pp. 286–305.

Graham, C. and Knight, J. 2000, 'Cash flows vs. earnings in the valuation of equity REITs', *Journal of Real Estate Portfolio Management*, Vol. 6, No. 1, pp. 17–25.

Hardin, W.G., Highfield, M.J., Hill, M.D. and Kelly, G.W. 2009, 'The determinants of REIT cash holdings', *Journal of Real Estate Finance and Economics*, Vol. 39, No. 1, pp. 39–57.

Hardin, W.G. and Hill, M.D. 2008, 'REIT dividend determinants: Excess dividends and capital markets', *Real Estate Economics*, Vol. 36, No. 2, pp. 349–369.
Hartzell, J.C., Sun, L. and Titman, S. 2014, 'Institutional investors as monitors of corporate diversification decisions: Evidence from real estate investment trusts', *Journal of Corporate Finance*, Vol. 25, pp. 61–72.
Higgins, E., Ott, R., and Van Ness, R. 2006, 'The information content of the 1999 announcement of funds from operations changes for real estate investment trusts', *Journal of Real Estate Research*, Vol. 28, No. 3, pp. 241–256.
Howard, M., Pancak, K.A. and Shackelford, D.A. 2016, 'Taxes, investors, and managers: Exploring the taxation of foreign investors in US REITs', *The Journal of the American Taxation Association*, Vol. 38, No. 2, pp. 1–19.
Howe, J.S. and Jain, R. 2004, 'The REIT Modernization Act of 1999', *Journal of Real Estate Finance and Economics*, Vol. 28, p. 369.
Jackson, L.A. 2008, 'The structure and performance of US hotel real estate investment trusts', *Journal of Retail & Leisure Property*, Vol. 7, No. 4, pp. 275–290.
Kim, H., Gu, Z. and Mattila, A.S. 2002a, 'Hotel real estate investment trusts' risk features and beta determinants', *Journal of Hospitality & Tourism Research*, Vol. 26, No. 2.
Kim, H., Gu, Z. and Mattila, A.S. 2002b, 'Performance of hotel real estate investment trusts: A comparative analysis of Jensen indexes', *International Journal of Hospitality Management*, Vol. 21, No. 1.
Ro, S.H. and Ziobrowski, A.J. 2011, 'Does focus really matter? Specialized vs. diversified REITs', *Journal of Real Estate Finance and Economics*, Vol. 42, No. 1, pp. 68–83.
Ro, S.H. and Ziobrowski, A.J. 2012, 'Wealth effects of REIT property-type focus changes: Evidence from property transactions and joint ventures', *Journal of Property Research*, Vol. 29, No. 3.
Sagalyn, L.B. 1996, 'Conflicts of interest in the structure of REITs', *Real Estate Finance*, Vol. 13, No. 2, pp. 34–51.
Stunda, R. and Typpo, E. 2004, 'The relevance of earnings and funds flow from operations in the presence of transitory earnings', *Journal of Real Estate Portfolio Management*, Vol. 10, No. 1, pp. 37–45.
Sun, H. 2010, 'A theory on REIT's advisor choice and the optimal compensation mechanism', *Journal of Real Estate Finance and Economics*, Vol. 40, p. 387.
Wu, F. and Yavas, A. 2005, *Do UPREITs suffer tax-timing conflict of interest?* Working paper.

9
LATIN AMERICA

Paloma Taltavull de La Paz, Joao da Rocha Lima Jr and Claudio Tavares Alencar

9.1 Introduction

This book aims to identify key areas for research in the REIT discipline for the next five to ten years by surveying the current state of the REIT discipline around the world and identifying emerging and cutting edge research areas through a thematic review of current contextual issues and a regional analysis based on case studies.

This book comprises two parts, the first part being a thematic review of emerging and cutting edge global research into current contextual issues in REITs internationally and the second part being a regional analysis of REITs around the world, each written by authoritative academic authors from the world's leading Universities and REIT industry experts.

Part I included six chapters each reviewing a current theme of REIT evolution through the lens of contemporary research. Chapter 1 focused on critical contextual issues in international REITs while Chapter 2, the *Post-Modern REIT Era,* examined the evolution of the global REIT market, what is deemed to be best current market practice and how the market is expected to change going forward and Chapter 3, *Emerging Sector REITs,* investigated such sectors as residential, health care, self-storage, timber, infrastructure and data centres generally and then specifically through a case study of US REITs.

Chapter 4, *Sustainable REITs,* analysed REIT environmental performance and the cost of equity and Chapter 5, *Islamic REITs,* examined the evolution of global Islamic REITs through a study of the Malaysian market. Chapter 6, *Behavioural Risk in REITs,* addressed the management of behavioural risk in global REITs and Part I concluded with Chapter 7 which reviewed recent research into *REIT Asset Allocation.*

Part II includes six chapters each analysing REITs in a region of the world, using case studies from a developed, developing and emerging REIT sector in the region. Chapter 8 analysed REITs in the North American region, using case studies from the USA, Canada and Mexico for the developed, developing and emerging REIT sectors in the region, respectively. This chapter analyses REITs in the Latin American region, using case studies from Brazil (being a developed REIT sector in the region), Argentina (being a developing REIT sector in the region) and Uruguay (being an emerging REIT sector in the region), respectively.

Chapter 10 then analyses REITS in the European region, using case studies from the UK, Spain and Poland with Chapter 11 analysing REITs in the South East Asian region, using case studies from Singapore, Malaysia and Thailand. Chapter 12 analyses REITS in the North Asian region, using case studies from Japan, Hong Kong and China with Chapter 13 analysing REITs in the Oceania region, using case studies from Australia, South Africa and India and Chapter 14 concludes Part II by considering *Directions for the Future of International REITs*.

9.1.1 Property investment funds in Latin America and the fideicomiso

There is relatively limited research into REITs in the Latin American region with EPRA (2016a) noting the existence of REITs in Brazil, Costa Rica and Mexico (EPRA, 2016b) and Martinez (2011) noting those REITs existing in Argentina, Chile and Colombia.

In Latin America, REITs are often confused with similar real estate investment funds existing in various countries, which have evolved over time with financial structures suiting the legal and customary environment and the needs of the investment markets in such countries.

Rapid change in the world's capital markets, coinciding with periods of economic weakness and financial and monetary distress, have prevented countries from adopting more modern structures for real estate investment during the last few decades (McGreal and Sotelo, 2008; Sotelo and McGreal, 2013). However, capital markets and legal environments have changed substantially in recent years (IMF, 2016), allowing most Latin American countries to create specific regulation for their capital markets and permitting modernization of stock exchanges and related operations. This process has led to a generalization of the security regulations across Latin America, facilitated capital movement and fostered the start-up or modernization of investment funds.

Today, all Latin American countries have regulated their stock exchanges, including provision of online information concerning listed assets. Further, all investment funds have been defined and regulated, including provision of prior administrative authorization when required, to facilitate fund raising and asset acquisition. The limitations on these investment funds are clearly defined in the regulations, which are also accessible through the regulator's websites. From this viewpoint, transparency and the provision of information have improved substantially in Latin America.

These changes have facilitated a clearer regulatory definition of investment funds across Latin America as a form of indirect investment vehicle based on the popular, long-standing and well-recognized legal entity, the *fideicomiso*, which is common across Latin America and combines attractive characteristics in risk management and investment capability. With modernization, the *fideicomiso* may coexist with advanced investment fund structures and facilitate potential conversions into REITs.

This chapter evaluates these issues prior to providing country case studies from Brazil (being a developed REIT sector in the region), Argentina (being a developing REIT sector in the region) and Uruguay (being an emerging REIT sector in the region), respectively. The following section provides an overview of regulatory changes in the Latin American capital markets, the structures adopted by investment funds and the specific definition of real estate investment funds in Latin American countries. The next section then evaluates the types of real estate investment funds in Latin America, followed by a discussion of the characteristics of REITs, their similarities with investment funds existing in some Latin American countries and the benefits of REITs for Latin American investors, with the country case studies of Brazil, Argentina and Uruguay following thereafter.

9.2 Real estate investment funds and regulatory changes in Latin America

In most Latin American countries, real estate investment funds can be legally constituted as mutual funds and are generally known as Collective Investment Institutions (IICs), including

Investment Funds, Mutual Funds, Trusts, Fideicomisos, Fiducia, Collective Funds and Investment Societies, amongst others, having the ability to offer financial participation and issue certificates to raise funds for their activities which are specified in the constitution document. These activities basically reflect the nature and specialization of the funds, such that where the fund activities include real estate investment, the assets can only be real estate if it is defined as such in the respective national regulations.

Investment funds have existed for several decades (sometimes longer) in Latin American regulations and most of them were born along with the regulations that developed the operating rules for the capital markets, often mirroring some within their foundation regulations, as shown in Table 9.1. More recently, a number of stock exchange regulations came into force, including the creation (or modification) of rules for Investment Funds (IFs) so that investment funds could be adapted to meet the requirements of modern international capital markets.

Table 9.1 Basic IIC+ regulation (general and real estate)

	1st IIC+	*General regulation*	*Regulator/Supervisor*
			Institution
Argentina		– Ley 24,083/1992 de Fondos Comunes de Inversión++, Decreto Reglamentario 174/93 y – Resolución general 368/01, cap XI, XII y XIII. Ley 17811 de creación de La Bolsa de Valores+++ – Ley de Fideicomisos financieros++++, num 24441 /95 modified by the new Civil Code, Ley 26994 of 1–08–2015 in art 1666–1700	CAFCI/BCBA
Bolivia		– Stock Exchange Act ★★★ Law No. 1834 of 31–03–1998, Titulo IX★★, Decreto Supremo num 25022, includes regulation of IIC++ in Resolución Admva 421 del 13–08–2004	ASFI/BBV
Brazil	1957 – Fundos de investimento, 1984 Club de investimento	– Ley Federal 638/76, LF 10303/2001 de creación del CVM+++ – Instrucc 409/04 del 18–08–2004 para FI++. – Instrucción 205 para fondos del mercado inmobiliario (for Real Estate Funds) (2004) ++	ANBID,BOVESPA
Chile	1961 – Fondos de Inversión abiertos (open Funds) 1976 – Commercial Act (Ley 1328, arts 433–434)	– Decreto Supremo 249/1982 for Open Funds++ – Ley 18815/1989 y Decreto Supremo 864/1989 for Closed funds++ – Ley 18045 de 1981 del Mercado de Valores+++ – Ley 18815 de Fondos de Inversión++ – Rg 1327/76 and DS 249 of Hacienda/1982 de Fondos Mutuos (Mutual Funds)++	AAFM, ACAFI

	1st IIC[+]	General regulation	Regulator/Supervisor Institution
Colombia		– Resolución 400 of 22–05–1995 de Regulación de fondos de Valores e Inversión[+++], – Decree 663/1993, arts. 29 y 146 y ss, que regula los fondos comunes (Common Funds regulation)[++] – Decree 1242/2013 of new regime of IIC (nuevo regimen de Fondos de Inversión Colectiva), Circular Externa 026/2014[++]	SB/BVC
Costa Rica		– Law 7732 del 27–03–1998 del Mecado de Valores[+++], Título V para FI[++]. – Fideicomisos regulation is in the Commerce Act, Chapter XII and Ley 8634, Banking Act, chapter III and DIR-TN-01–2016 of 16–02–2016[++++]	BNV/ SUGEVAL
Ecuador		– Ley Orgánica 006–2014 del 13 de Mayo de creación del Mercado de Capitales[+++] Título XIV for FI[++]. – Fideicomisos Act, 21–11–2001 (Reglamento de Administradoras de Fondos y Fideicomisos)[++++]	AAFF
El Salvador	1912 Commercial Act (Código de Comercio) with fideicomiso regulation	– Decree 776 del 21,08,2017, Ley de fondos de inversión[++] – Decree 809 del 11,09,1996 de creación de la Bolsa de Valores[+++]	BVES – SSF
Guatemala	1970, Commercial Act, Decree 2–70	– Ley de Mercado de Valores y Mercancias, Decreto 34–96[+++] and its reform in Decree 49–2008, – Ley de Fondos de Inversión del 10,02,2014[++] – Fideicomisos: Commercial Act and, Acuerdo Ministerial 25–2010 of 27–04–2010[++++]	BVNSA/SB
Honduras		– Decree 8 of 20,02,2001, de Mercado de Capitales[+++], Titulo III, arts 71–138, for fondos Mutuos y de Inversión[++]. – Regulation and resolution num 590/06–05–2003 for investment Funds commercialization[++]	CNBS (SUPERVISION DE 2016)
México	1955 – Equity Funds (Fondos de renta variable) 1961 – Mutual Open Funds	– Ley de Sociedades de Inversión del 4,06,2001 y disposiciones del marco normativo de las sociedades de Inversión[++], – Ley del mercado de Valores[+++] of 10–01–2014-	CNBV/COSSIF

(Continued)

Table 9.1 (Continued)

	1st IIC[+]	*General regulation*	*Regulator/Supervisor Institution*
Nicaragua		– Ley de fideicomisos -Resolucion CB-SIBOITF-677–2 de 16,05,2011[++], – Ley 587 del Mercado de Capitales de 26.10.2006[+++], Titulo V, fondos de inversión[++]	SMV
Panamá		– Decree -Act 1 del 8, 07,1999 of Stock Exchange[+++], Titulo IX regula las IIC[++]. – Acuerdo num 5–2004 del 23,07,2004, y Acuerdo 2–2005 del 9,03,2005 both develop the FI[++]. – Posterior changes are in the Ley 23 de 27 de abril de 2015 Reforma, ley 66 de 9 de3 13 dic de 2016, for Investment Funds regulation[++]	SB
Paraguay		-Ley 921/96 de negocios fiduciarios[++++] de 9,04,1996, – Reglamento 921/96, modif en 15,02,2011. (reg limitada a bancos y agencias o a sociedades creadas por ellos) – Ley 1284/98 del mercado de Valores[+++]	CNV/SB
Perú	1979- LEY 772 of Capital Market	– Open Funds[++], Decreto supremo num 093–2002-EF y Rgl de Fondos Mutuos de Inversión, and Resolución CONASEV num 026–2000-EF/94,10, modified by Resolución CONASEV num 071–2005-EF/94,10. – Closed Funds[++] -Ley de Fondos de Inversión, Decreto Legislativo num 862 and reglamento CONASEV num 042–2003-EF/94,10. – Regulación de fideicomisos del 11–11–1999.[++++] (Fideicomisos Act) – Ley 94/91 de Modificación del Mercado de Capitales y Creación de la CN de Valores[+++]	CONASEV
Uruguay	Ley Orgánica 5343/1915 del Bco Hipotecario Uruguay	– Ley 16749 del 31–05–1993 and Ley 18627 de 2–12–2009 del Mercado de Valores[+++] – Ley 16774 de Fondos de Inversión del 27–09–1996[++] – Ley 16906 de la promoción y protección de inversiones. Decreto 002/012 de Régimen General de Promoción de Inversiones[++]	

	1st IIC[+]	General regulation	Regulator/Supervisor Institution
		– Decree 175/003 and Decreto 404/010 promoción a la explotación inmobiliaria turisticos y hoteles (promoting tourist real estate and hotel management)[++] – Ley 18795 inversión en viviendas de interés social of 17–08–2011 (social Housing Act) – Decree 355/011 of 6–10–2011- Ley 16779 del 27,09,96 sobre Fondos de Inversión, securitización y factoraje[++], modified by Ley 17202 del 24,09,99. – Fideicomiso Act: Ley 17703 de fideicomiso del 4–11–2003, modif por Ley 18127 del 22–05–2007[++++]	SB/BC (BVM)
Venezuela	1988 – Mixed Open-Ended Funds in Bolivars	– Ley de Entidades de Inversión Colectiva, Gaceta of num 36027 del 28,08,1996[++], Normas relativas a las Entidades de Inversión Colectiva, 30,05,1996[++] and updated in Reforma del 23,07,1998 -. – Ley 18627 del Mercado de Valores, of 18,12,2009 [+++]. – Ley de Fideicomiso, Gaceta of Extraord num 496 del 17,08,1976 [++++] and Normas reguladoras de la operación de fideicomiso, resol 083–12, Gaceta of 39941, del 11,06,2012 [++++]	CNV/AVAF

[+] in Spanish *Instituciones de Inversión Colectiva*, which refers to Investment Funds, Mutual Funds and other collective investment figures
[++] Investment Funds Act
[+++] Stock Exchange creation Act (or modification)
[++++] Financial "*fideicomisos*" Act or regulation
Codes:

IIC
FI Collective Investment Institutions/investment trusts
Investment Funds
LEY
LM_MV Act, main Law
Stock Exchange Main Regulation (Ley Marco reguladora del Mercado de Valores)

SMV
SB Superintendent of Stock Exchange
Superintendent of Bank System

Sources: Instituto Iberoamericano del Mercado de Valores – IIMV (2006, 2010), www.iimv.org, For Costa Rica, EPRA (2016b), Felaban (www.felaban.net), regulations existing in each countries' Stock Exchange websites

Some Latin American countries, such as Argentina (IF appears in 1992), Peru, and Uruguay, were very progressive, regulating IFs far earlier than their capital markets. However, the most common approach has tended to be the creation of modern funds subsequent to the creation of capital markets, as happened in Brazil, Chile, Colombia, El Salvador, Guatemala, Mexico, Panama (which also presented reforms in other funds), Paraguay and Venezuela.

Countries where Collective Investment Institutions and investment funds have been regulated along with the creation of capital market legislation include Bolivia, Costa Rica, Ecuador, Honduras, Nicaragua and Panama. Those maintaining active regulation of fideicomisos are Argentina, Brazil, Costa Rica, Ecuador, Guatemala, Mexico, Nicaragua, Paraguay, Peru, Uruguay and Venezuela (see Table 9.1).

Capital markets, investment funds and operations with financial assets are supervised by the relevant institutions of the banking or financial systems in each country. Such institutions are listed in the last column in Table 9.1, as well as the fundamental regulations governing the country's capital markets, though the list is by no means exhaustive and should be considered indicative only. This regulatory framework was then adapted to develop the regulations governing REITs since the beginning of the millennium.

9.3 Types of real estate investment funds in Latin America

IFs created in Latin American regulation are commonly authorized to issue shares or participations that will be approved and registered by the financial authorities for their listing on the stock exchange, with the typology of assets then depending on the type of fund constituted. Regulation in Latin American countries permits the existence of open and closed funds, with open funds being the most commonly accepted form in most countries (except for Colombia, Ecuador and Guatemala). In addition, IFs may include *fideicomisos* having been regulated by bringing their structure closer to that of an open fund, with the nature of modern *fideicomisos* discussed in the following section.

Investment fund creation generally allows the pooling of capital to provide larger amounts to invest. However, governments have focused on different capital sources in the regulatory process, with some regulations directed at individual investors in certain countries (with data for 2005 suggesting 76% of total fund assets belong to private investors in Bolivia, with 71% in Costa Rica and 70% in Peru) and some regulations mainly directed at institutions and societies (88% investment in Argentina and 61% investment in Colombia) (IIMV, 2010, p. 163).

This could be one reason why regulation favouring the emergence of REITs has not advanced further and may explain why Latin American regulations have concentrated on developing general real estate investment structures which can adapt to every type of fund. Another reason for the lower prevalence of REITs may be the strength of the banking sector in the financial system. Mirroring the continental European model, which encourages the banking system to take an active part in the management of investment funds, the regulation has started to change (IIMV, 2006, p. 9). This has resulted in an efficient extension of the intermediation activity for capital markets which has resulted in the delay of the development of independent investment structures such as REITs, similar to the experience in Spain.

Table 9.2 shows the types of generic Investment Funds created within each country's national regulation, as well as those specializing in the real estate sector. The first group of countries that may be defined (Argentina, Bolivia, Chile, El Salvador and Nicaragua) are where Open Funds (OFs) exist. In an OF, the investor makes the contribution directly to the IF, which reinvests the capital on their behalf and compensates, when required, within the same fund. The normal

Table 9.2 Typology of Investment Funds (General and Real Estate)

	Types of Funds		*Investment vehicles (allowed to include RE) and Real Estate investment specific figures*	
	Open	*Closed*	*From IIC*	*Others*
Argentina	X	X	Open FI. General Fondos Comunes de Inversión (Common Investment Funds), abiertos y cerrados Fondos comunes con objeto específicos (cerrados) Fondo común de inversión inmobiliaria (art 2) (abiertos) Fondos de Infraestructura Fondos Raíces (real estate funds)	Fideicomisos financieros (1995, Ley 24441) Fideicomiso de Garantía (cap 30, art 1666–1700 del Código de Comercio, Ley 26994)
Bolivia	X	X	Open FI, General (1992) Closed FI, General	
Brazil	X	X	Fundos de Investimento Inmobiliario (1994) (fideicomisos)	
Chile	X	X	Open FI Fondos Mutuos – Mutual Funds (2002) Closed FI (1990), all general	Housing Investment Funds (Fondos para la vivienda, 1995)
Colombia	no	X	Collective Real Estate Portfolio- Cartera Colectiva Inmobiliaria (2008) Fondos de Capital Privado- Fondos Inmobiliarios	
Costa Rica	X	X	Closed Real Estate FI (1998), Real Estate Development Funds (Fondos de Desarrollo Inmobiliario)	Fideicomisos – Investment Funds (2000)
Ecuador	no	X	Managed Investment Funds (Fondos Administrados de Inversión, 1994), General Collective Funds (Fondos Colectivos,1995), General	
El Salvador	X	X-RE	Open Investment Funds, General Closed Real Estate Investment Funds	
Guatemala	X	X	Investment Fund Contracts (Contratos de Fondos de Inversión) Investment Corporations (Sociedades de inversión)	Fideicomiso Contracts (Contrato de fideicomiso de inversión**)
Honduras	no (set in the Act)	X	Closed Funds. General but allowed to invest in real estate and RE rights located in Honduras (art 128, 6) Mix Closed Funds, General (with legal constrains)	
México	X	X	Closed Investment Corporation (of limited objectives), General	Fideicomiso Certificados bursátiles fiduciarios inmobilairios (art 63 bis1 Stock Exchange Act)

(*Continued*)

Table 9.2 (Continued)

	Types of Funds		*Investment vehicles (allowed to include RE) and Real Estate investment specific figures*	
	Open	*Closed*	*From IIC*	*Others*
Nicaragua	X	X	Investment Funds (Ley 587 art 89) Open, Closed Financial IF Open and closed, Non-Financial IF, mega-funds Real Estate Investment Funds- (Fondos de inversión inmobiliaria, art 104), defined as Non-Financial IF***	Participation Certificates (Certificados de Participación) Fideicomiso,
Panamá	X	X	Real Estate Investment Corporations (Sociedades de inversión inmobiliarias,2000)	Stock Houses (Casas de valores)
Paraguay	X	X	Real Estate Common Funds (type 2-Fiducia Inmobiliaria, by authorization)* Management of Real Estate Common Investment Funds (Subtype 3 of Fiducia management: Administración de fondos comunes de inversión inmobiliarios)	Real Estate Fideicomisos (Fideicomisos inmobiliarios)* Investment Funds/ Fideicomisos to develop infrastructures
Perú	X	X	Closed Real Estate Funds (Fondos Cerrados Inmobiliarios, 1997)	Real Estate Capitalization Companies (Empresas de capitalización inmobiliaria, 272 Act, cap IV, art 295 y ss,)
Uruguay	X	no		- Fondo de garantía para Desarrollos Inmobiliarios (FOGADI) 2016 – (to promote non-residential rent properties) – Fideicomiso Financiero de Inversiones y Rentas Inmobiliarias – Financial Fideicomiso (art 21)
Venezuela	X	X[1]	Mutual Funds, General (Fondos mutuales) [1]Only closed Real Estate Funds	Fideicomisos

* Paraguay: Common Funds are financial mechanism of vehicles to manage funds from several investors whom are considered collective managers. They are NOT a legal person (art 13, Titulo II). They are thought mainly to develop Project (construction). The Real Estate Common Funds are those with 60% of their assets being properties in Paraguay or abroad. They can invest in other funds, sell and build up to the limits defined in the contract.

** Guatemala: All certificates issued by a Fideicomiso should be registered as a public offer.

*** By law requirements, the Fund managers act on behalf of investors. They perform like *fideicomisos*. The duration could not be limited, that is, regulation let those funds to be either open or closed.

Sources: Instituto Iberoamericano del Mercado de Valores – IIMV (2006, 2010), www.iimv.org, for Costa Rica, EPRA (2016b), Felaban (www.felaban.net) regulations existing on each countries' stock exchange websites

design for an OF is to be "general", although contributions in property assets or real estate rights are normally accepted.

Some countries (like Ecuador) may define general Open Funds as "Managed Funds", or as Real Estate Investment Funds (as in Nicaragua) and also as specialized OFs for real estate funding. In Paraguay, real estate OFs are called Fiducia Inmobiliaria (its creation must be authorized) or Fiducia where management includes real estate investment common funds administration. Honduras is the only country that has defined a Joint Investment Fund with a general character. Mutual Funds exist as specific OFs in Venezuela and Chile.

Other countries have prioritized Closed Funds (CFs), being funds with their own legal status and able to issue shares that can be sold through the stock exchange to raise capital. Many countries have defined property/real estate funds specifically as a Closed Fund, but under different names. There are generic definitions in Chile, Costa Rica, Ecuador (Collective Funds), in Guatemala (with Investment Fund contracts and investment companies), in Honduras (where they can invest in real estate but only in properties located in that country) and in Mexico (Closed Investment Corporations).

Closed Real Estate Funds (FICs) exist in Colombia (Real Estate Collective Portfolio), Costa Rica (FIC and Real Estate Development Funds), El Salvador (FIC), Panama (Real Estate Investment Companies), Peru (FIC) and Venezuela (FIC). In El Salvador and Venezuela, only Closed Funds are allowed if they are for real estate investment. Some segment-specific real estate funds exist, such as in Chile (Housing Investment Funds), Peru (Real Estate Capitalization Companies) or the Uruguayan FOGADI (Real Estate Development Guarantee Funds), which promote non-residential housing construction (Uruguayan Stock Exchange, 2017).

Most countries have the financial *fideicomiso* structure (see Table 9.2, right column), based on customary law (the common tradition having been the *fideicomiso* as a fundamental basis for generating capital for investment). These are usually defined without specialization (as generic) in Argentina, Costa Rica, Guatemala (investment trust agreements), Nicaragua and Uruguay. However, the regulations allow for the creation of real estate *fideicomisos* issuing stock real estate certificates ("Certificados bursátiles Fiduciarios inmobiliarios") in Mexico (real estate *fideicomisos* and Investment Funds – *fideicomisos* for the development of infrastructure) and in Paraguay. The Brazilian *fideicomiso* (Real Estate Investment Funds) is one of the most advanced structures with a profile closer to that of an investment fund, thus marking a transition in financial structuring.

In general, the possibility exists to include real estate assets in the form of their rights or their charges (mortgages) to comprise part of the contributions to the fund, demonstrably when the fund is defined as a real estate fund and with restrictions in other cases.

As Table 9.2 shows, most countries have tended to define Latin American investment funds as Open and Closed IFs or as Investment Companies and not as REITs. However, as in Spain, their formative structure may permit their conversion to REITs when suitable conditions exist (Taltavull and Peña, 2013). In some cases, the characteristics of funds closely resemble those of REITs, such as in Brazil (in Instruction 205, on the investment of funds in the real estate market and later in funds to guarantee leases).

It is generally evident that those IFs created to be real estate vehicles are often adapted structures which combine the characteristics of both REITs and traditional fideicomisos. Accordingly, such combination suggests that it may be helpful to identify the precise characteristics of REITs that will be beneficial for the expansion of real estate investment across Latin America.

9.4 Characteristics of REITs and trusts – *fideicomisos*

The main difference between those funds defined in Latin America regulations and REITs lies in their creation structure. Following Taltavull and Peña (2013, p. 178) (Figure 9.1), the Open

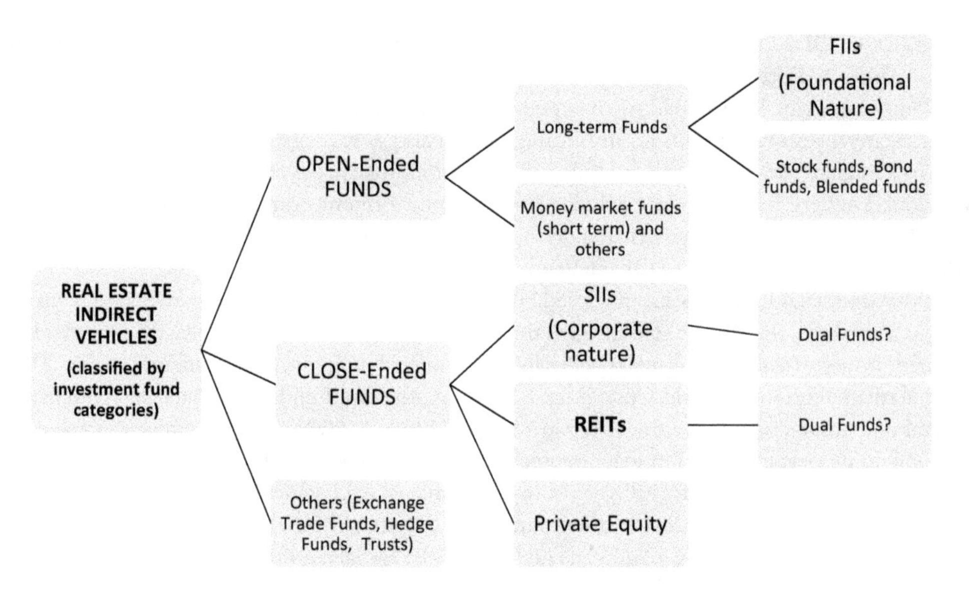

Figure 9.1 Alternative Fund Structures

Source: Author, Based on Taltavull and Peña (2013, p. 178)

Fund (OF) has a foundation basis which usually includes limitations when it comes to carrying out investments (for example, regarding the number of participants). These funds have an obligation to distribute dividends, but the transmission of their assets happens within the fund itself and it may result in liquidity problems during negative economic periods.

REITs follow the Closed Fund design with a corporate nature (known as Real Estate Investment Companies – SII-REITs trust) and a fixed capital in the constitution, managed by a board of trustees, with the REIT having a series of obligations and fiscal benefits. REITs issue units that are placed on the market at the prevailing price and do not intervene in the unit transaction. The existence of FIIs and SIIs is a tradition in the Latin American world, being of little difference to REITS in this case but with the distinction lying in fiscal advantages.

REITs are investment funds that issue units or shares and allocate resources to finance and develop real estate assets with the obligation to obtain a significant part of their revenue through rental income. They invest in owned property assets which generate income, buy and rent properties, buy debt and finance and construct buildings for those purposes. In some cases, they can sell the buildings with a deadline. Characteristics such as the obligation to distribute income to investors and high tax benefits distinguish them from other funds.

Furthermore, IIC-FIs are investment funds that issue unlisted shares whose trading price does not depend on the price in the stock market but on the activity of the fund itself and its liquidity. They are not properly risk-supported during periods when valuation standards lead to a gap between asset price and book value, generating arbitrage opportunities for holders to require redemption and reduce the liquidity of the fund. Nevertheless, all regulations envisage the possibility of suspending redemptions for a certain period of time if necessary. As shown in Table 9.2, only a few Latin American countries have regulated closed real estate investment funds, even though it seems that the most successful structure to drive real estate activity has been the financial one – fideicomiso.

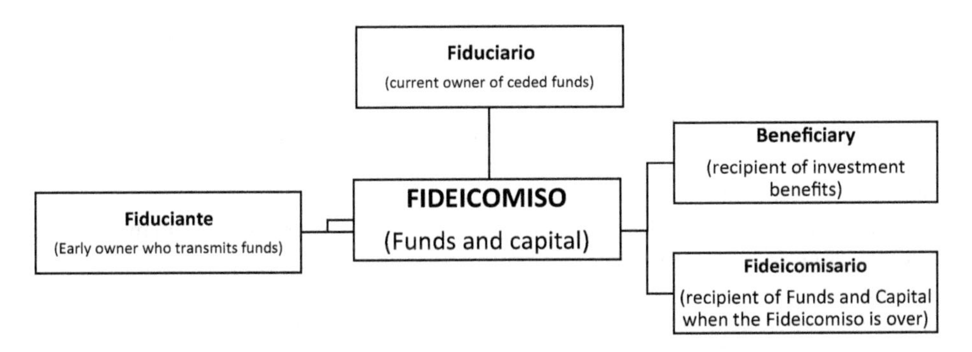

Figure 9.2 Fideicomiso Structure

Source: Author, Based on Porcaro and Malumian (2003) and Garcia and Porcaro (2003)

The "ordinary *fideicomiso*" is illustrated in Figure 9.2. Most of the funds match what is known as the fideicomiso fund, being a fideicomiso which meets two conditions:

- that beneficiaries have their rights represented in securities known as "titulus *fiduciarios*"; and
- that the trustee is a financial institution or a specially authorized company. Thus, financial trusts have the rights of their beneficiaries represented in trust certificates, their fiduciary being a professional entity authorized for such purposes.

The financial *fideicomiso* is a legal business:

> under which a person (fideicomitente o fiduciante) transfers the fiduciary property of certain goods to another person (fiduciario), thus allowing the latter to exercise it in accordance with the purpose determined by the fidecomitente and for the benefit of the person designated by him at the time of constitution (beneficiary or fideicomisario) and . . . shall transfer them to the fideicomisario, fideicomitente or the beneficiary upon completion of a term or condition.
>
> (Porcaro and Malumian, 2003)

The *fideicomiso* does not have a legal person, but it implies a transfer of ownership which makes it possible "to create an equity whose assets do not respond to the obligations of the fiduciario, the fideicomitente or the beneficiary, but only to the debts and liabilities of the project that is carried out with these goods" (Porcaro and Malumian, 2003). It is a very versatile structure applied under a variety of legal systems.

Figure 9.2 shows the transfer of fiduciary property and its management on behalf of third parties:

The *fideicomiso* is extinguished when it meets the intended purpose or the term established in the Constitution, or by means of an agreement between the *fiduciante* and the beneficiary (or when it meets the revocation of the shareholder who was left with a "golden share"). It is also extinguished when either a cessation of obligation payments or another established cause in the document occurs.

While the *fideicomiso* structure does relate directly to that of REITs, there are numerous similarities:

- the way in which capital contributions are made is similar, insofar as REITs allow the initial participants to add properties that they own to the entity;
- management tasks should be undertaken by a (registered) entity;
- the main aim of a *fideicomiso* is to manage funds and capital, whereas a REITs mainly focuses on managing real estate through rent, construction and the search for yields; and
- the beneficiary should be defined in the *fideicomiso* and the possibility exists to determine the participation holders, like the unitholders in REITs. Such similarities, of course, need to be defined in legal documents though trusts are widely considered as Open Funds in Latin American regulation.

9.5 Benefits of REITs for Latin American investors

In general terms, the economies of Latin American countries are largely dependent on the banking system. Accordingly, finance for major projects is supplied through bank credit with small investors having few opportunities to invest in profitable projects. In this respect, regulatory changes to create more flexible structures should assist in solving this problem by creating mechanisms that multiply the available sources of cash flow for project financing, while also providing an opportunity for small investors.

This is particularly relevant in the real estate market since it would allow companies owning real estate assets, being buildings not related directly to their core business activity that generate no returns by themselves for the company, to obtain liquidity for reinvestment in their business.

Both REITs and Real Estate *Fideicomisos* can play the role required to channel such specialized investment in real estate to provide liquidity through the "conversion" of property into funds. Furthermore, both of them are long-term investors, despite REITs being time limited, as well as being guaranteed and transparent due to the control exerted by security market regulators or the financial system.

To the extent permitted by specific regulations, these two structures could give small investors a chance to participate in real estate opportunities of a larger magnitude than they could ordinarily achieve, with such investment potentially diversified to include all real estate sectors in both cases.

REIT benefits generally include higher-than-average yields, the stability of long-term lease contracts, simple tax treatment and liquidity, with the ability for investors to easily sell their units, making the structure attractive to investors. A further benefit of REITs for small investors arises from the requirements to distribute most or all net income annually whereas *fideicomisos* do not have such an obligation, unless otherwise specified in the regulations. From this perspective, the small investor is more likely to be interested in REITs.

Two additional differences that would also make REITs preferable would be liquidity and the facility and ability to exchange REITs in the capital market, compared to the impossibility of recovering the shares of a fideicomiso. Tax treatment also makes a difference, since it is much more favourable in REITs than in *fideicomisos*, being more significant for corporate investors rather than individual investors.

However, the lower level of development of REITs in Latin America and the preference to adopt a *fideicomiso* structure in real estate investment vehicles continues to be a reality, essentially due to perceptions of investment risk, guarantee and security. It is widely believed that *fideicomisos* are very safe and that their structure enhances investment value, insofar as it eliminates the principle of "unique patrimony" (Porcaro and Malumian, 2003, p. 14) that safeguards investors' assets in the event that a project should have losses.

According to Payne and Waters (2005), the greater reluctance to adopt REITs in Latin America has to do with the greater ease of REITs for the transmission of bubbles in prices:

> Given a REIT is a company that owns and sometimes finances and operates income generating real estate, price bubbles associated with the underlying real estate asset may be transmitted into price bubbles for the securitized real estate assets known as REITs. Moreover, a relatively close connection exists between the income generated from the real estate asset and the dividends distributed to REIT shareholders. More precisely, at least 90 percent of a REIT taxable income must be distributed in the form of dividends to shareholders on an annual basis (www.nareit.com). Thus, a positive bubble in real estate prices would . . . lead to a sharp increase in dividends.
>
> Payne and Waters (2005, p. 66)

Martinez (2011) supports this view, stating that in countries where the risk associated with the financial market as well as with volatility is high, traditional real estate funds can act as collateral and a security. REITs, on the other hand, would be dependent both on market values and on changes in the stock market. This would generate assets characterized by their high volatility (in Latin American markets) because they depend on local markets.

The greater volatility of stock markets in South American countries, real estate regulations and valuations along with fiscal requirements, make it easier to understand the reluctance to implement REITs with all their characteristics (including fiscal ones). Such facts, together with a lower degree of transparency, have contributed to the lack of development of REITs in Latin America.

Further, is should be stressed that each country has regulated funds differently in terms of the details authorized by the investment structures and the particular conditions associated with the assets. The predominance of structures lying between REITs and *fideicomisos* in the definition of investment funds therefore becomes apparent.

9.6 REITs in Brazil

9.6.1 Structure and legislation

REITs in Brazil are joint-owned investment funds whose purpose is to apply funds to real estate undertakings and their derivatives. The majority of these funds, called Fundo de Investimento Imobiliário (FII), follow international styles and focus on long-term income generated from commercial real estate. The funds strive to offer the investor an investment environment based on assets which produce a stable, long-term, monthly income.

An FII is the typical structure available in the Brazilian market for investment sharing in real estate, which offers fiscal advantages not found in other forms of securitization in Brazil but whose structure differs from REITs in the US. All the financial operations are regulated by the Rules for Fund Operation, being registered and submitted for approval by the CMV (Comissão de Valores Mobiliários or Brazilian Securities Commission), a government agency that regulates capital market operations, being in some ways similar to the US SEC.

The legal concept of FIIs creates some restrictions with respect to real estate maintenance, registration and transactions, as well as to certain business-development activities which the FIIs are free to carry out. These restrictions were relaxed by Federal Law 8.668 (1993), which governs the creation and operation of the FIIs, and by CVM Instruction 205 (1994), which outlined the FII's constitution, operations and management, providing detailed management and operation benchmarks.

Law 8.668 (1993) specified that the operations (buying/selling of assets) and profit sharing of FIIs is tax-free, which was not the case earlier. Under current legislation, private investors are exempt as long as they comply with certain rules of distribution such as not owning more than 10% of the total shares in an FII.

In Brazil, where high taxation can discourage investing, the tax exemptions offered to the FIIs and their investors are so many and so significant that other real estate-based portfolio securitization solutions end up being ignored, even if they present a more advanced structural design than that of the FIIs. In general, FII shareholders are only taxed after shares have appreciated and been sold at a profit.

However, under Brazilian legislation, structures for investment sharing in real estate may be established under concepts equivalent to those of the REIT using an SPE (Sole Purpose Company) which may shelter the portfolio and attract investors through shares with a mandatory dividend or perennial result-sharing debentures. However, an SPE must pay a tax on income and earnings which leaves the SPE less attractive than a real estate portfolio sheltered in an FII.

Originally, FIIs' portfolios included only real estate properties. Recently, however, portfolios may include a range of by products such as participation in SPC real estate developers and even participation in receivables embedded in CRIs (Certificados de Recebíveis Imobiliários), which are Brazilian, mortgage-backed securities. This broadening of fund investments has changed the character of Brazilian funds compared to other real estate investment funds in first world economies. Although the broadening of portfolio options may be interpreted as a means of bolstering the Brazilian real estate market, which it desperately needs, it must be understood that, under this diffused configuration, real estate investment has lost its identity in relation to applications in commercial real estate, which is the mainstay of equivalent real estate investment structures in countries like the US, Japan and parts of Europe.

One other important difference between FIIs and US REITs lies in the legislation covering management whereby FIIs must necessarily be under the administration of financial institutions rather than controlled by an entrepreneurial type of executive while the management of SPEs and REITs is in the hands of such individuals.

9.6.2 Management issues

The manager of an FII administers a portfolio having little real estate management specialization which, in real estate deals, should be central rather than financial management skills. FII administrators typically manage portfolios having little or no flexibility in that the FII is comprised of a single building. The administration of such a portfolio does not involve the dynamism of a continual change in portfolio positions, but merely requires topical actions and the administration of rental contracts and leases.

For example, at BOVESPA (The São Paulo Stock Exchange), units may be transacted in an FII whose portfolio comprises a single office building leased to Caixa Econômica Federal (a state-owned bank) for ten years and with the tenant having the option to renew for another ten years. In such cases, the managerial skills required by the administrator comprise collecting and distributing the monthly rental among the unitholders and renegotiating the rental, perhaps every five years. Other FIIs listed on BOVESPA are similar, comprising office buildings (one per FII), shopping centres (one per FII) or a hotel.

When setting up a portfolio of commercial real estate, the REIT can be classified as either "noble" or "poor". A "noble" portfolio would consist of a group of properties belonging to a particular market segment (hotels, offices, shopping centres) so that any financial oscillation in a particular market segment would be absorbed by the overall group and so maintain a steady, balanced monthly income for the investor(s). A "noble" portfolio allows continual renewal of assets through buying and selling, whereas a "poor" portfolio only allows managers to sell a property when its market value has appreciated more than its capacity to generate income, which will, in turn, appreciate the portfolio's stock when such a sale results in the acquisition of asset(s) that are more profitable than those averaged by the group as a whole.

A portfolio designated as "poor" is that where the sole function of the REIT is to distribute the investment into one particular property, or in a small group of properties, therefore, leaving the investor(s) vulnerable to oscillating market behaviour, with no chance of hedging investments. Brazilian REITs can mostly be classified as "poor", keeping in mind that the term "poor" does not in any way refer to the implicit quality of underlying property but to the investment strategy used by the REIT.

Those responsible for starting the REIT have chosen a primary method of creating opportunities where small and medium investors have a chance to invest in commercial real estate of significant financial size, whose market performance (stable value and income) surpasses investing in small office properties which are all that their small and medium savings may access. A global review of real estate investment strategies reveals that a share of a REIT with a non-diverse portfolio often has a greater value than an investment in small properties, although investing in a REIT having a portfolio with a diverse range of properties is, from the viewpoint of any conservative investor, much more advantageous.

9.6.3 FIIs and REITs size issues

While US REITs may be very large and hold multiple properties in different market sectors, Brazilian REITs, in contrast, are used as a vehicle for sharing the investment in a single property or a small group of properties. There has been no effort to make the Brazilian REIT an instrument for the concentration of savings or for the development of commercial real estate, as is the case with REITs in the US. In other words, the Brazilian REIT, which could be used in a nobler manner, is used in the poorest form as a vehicle for the gathering of funds for profitable exits or for a specific project. For example, a REIT was specifically created to participate in a project, Shopping Center 13 in São Paulo, simply in order to raise funds for completion of the project itself.

A typical US REIT has a market value of US$4.0 billion, whereas the FIIs on the Brazilian stock exchange average R$196 million Brazilian Reais (the current exchange rate of US$1 = 3.1 Reais (R$) – source: Brazil Central Bank [www.bcb.gov.br]). In relation to the Brazilian stock market capitalization of R$9.7 billion for 129 FIIs, the 175 listed US REITs represent a market capitalization 62 times greater.

Further, Simon Properties (shopping centres) in the US alone has a portfolio 2.4 times greater than all the Gross Leasable Area (GLA) recorded by Abrasce (Brazilian Association of Shopping Centers). Among the largest office holding REITs, Boston Properties holds a portfolio which is valued at more than the entire area of Faria Lima Avenue in São Paulo.

The lack of attention by developers to attracting funds by way of FIIs may be, in part, responsible for the slow evolution of the Brazilian market. Considering that the REIT equity market capitalization in the US is equivalent to 2.8% of the US GDP, this suggests a market potential in

Brazil of R$60 billion, being nearly 20 times greater than the market's current size. To illustrate the magnitude of this potential, consider the region of Luiz Berrini Avenue where the total value of new triple A office buildings does not exceed R$6 billion.

9.6.4 The future of the FIIs

The question for Brazil is who is responsible for reversing the tendency to create small entities for the sharing of investments that are sometimes only a part of a building or property and to take the necessary steps to establish a modern market where REITs represent authentic channels of investment in commercial properties?

Nobly structured REITs could promote the inclusion of a range of commercial properties, from office buildings to hospitals, by taking advantage of the not too abundant, but available, resources which are currently invested in mediocre, high-risk investments like small offices, such as those which regularly sell in São Paulo.

Is channelling available savings into modern, safer investment instruments the role of the financial market or of the real estate market? Modest efforts in this direction have been taken by participants within the financial market, real estate agents and investment banks to promote REITs with the proposition of creating investment portfolios of commercial buildings, but these have been relatively small portfolios. At the same time, however, the sale of stock in specialized property companies (BR-Properties, Iguatemi, BR-Malls, etc.), having the same objectives as a REIT but without the tax advantages and under greater management costs than a REIT, achieved higher volumes of trading than those funds that propose creating a portfolio of commercial buildings.

As long as promoters remain focused on raising money for funds comprising small commercial office space using traditional methods and do not understand that they could unite all the raising under one REIT and offer the market an authentic product for investment rather than just an everyday real estate investment, it seems unlikely that REITs will be able to escape the category of poor funds and migrate to a position in the Brazilian market that is equivalent to that of first world economies.

The Brazilian market functions as if the action of real estate brokers is to deliver the final product, which leads real estate development companies to not see themselves nor take a position as suppliers of real estate investment channels. On the other hand, agents in the financial markets take it upon themselves to offer any type of investment, even those secured in real estate, as evidenced by legislation granting the administration of REITs to financial entities.

There is no specialization within the financial markets indicating that there are many financial entities focused on real estate, but none with a standing that could lend a brand to a REIT with a large portfolio. However, investment products could be presented with a guarantee to seek partnerships or the acquisition of brand undertakings chosen at a later date. This may not give the investor the confidence to buy, but he or she may be more comfortable when buying the image of the product with a defined local brand quality identified (the developer).

It would benefit developers to understand that the sale of investments in commercial real estate is more than just the delivery of the product on time, within the stipulated specifications and for the contracted price. There is an implicit opportunity to offer a product capable of protecting the investor and producing a stable income, which will facilitate the formation of partnerships between developers and agents from the financial market and subsequently lead to the formation of REITs, with portfolios representing modern investments such as those of REITs in more developed markets.

Sadly, Brazilian hesitation will not allow the REIT product to fulfill the noble role for which it was conceived, being limited to presenting itself solely as a means for holding one or a few buildings.

9.6.5 Summary – Brazilian REITs

As a developed REIT sector, Brazil offers the Fundo de Investimento Imobiliário (FII), in either "noble" (multiple properties) or "poor" (single property) form, which provides most of the internationally familiar REIT characteristics, including tax transparency, as well as opportunities to broaden the fund asset base to include mortgage backed securities.

Indeed, in Brazil, where high taxation can discourage investing, the tax exemptions offered to the FIIs and their investors are so many and so significant that other real estate–based portfolio securitization solutions end up being ignored, even if they present a more advanced structural design than that of the FIIs.

There would appear to be considerable scope for growth in REITs/FIIs in Brazil with opportunities, in particular, for developers to create a portfolio of investment properties for the REIT market as commonly occurs in the US, Europe, Asia and Oceania.

9.7 REITs in Argentina

Recent decades have witnessed high levels of volatility in the Argentinian economy, which underwent a serious financial crisis in 2000–2001 as a result of which debt had to be defaulted and restructured. The Republic of Argentina imposed tight controls on its foreign exchange market as of 2011, limiting foreign currency outflow through formal and informal regulations. The strict currency controls, trade protectionism and heavy taxes introduced from November 2011 to protect the country's low foreign reserves have been gradually changed since December 2015 with the introduction of a single exchange rate applied to all cross-border transactions, reductions in international capital transfer constraints for capital movements, principal loans payments and increasing foreign currency deposits.

9.7.1 Changes to legislation and regulation

As a result of the imbalances existing in the financial system and the need for its economy to receive funds, Argentina modified and modernized its capital market regulations and securities, adapting them to modern financing structures in recent years with substantial updates in the Argentinian Civil Code (Law 26994 of 1 August 2015) and in the regulation of *fideicomisos* and investment funds. The previous regulatory framework, to which changes have been made, stems from Law No. 24.083 of 1992 on Common Investment Funds, Regulatory Decree No. 174/93 and the regulations of the National Securities Commission (General Resolution No. 368 [Chapter XI], Closed Funds [Chapter XII] and other funds), as well as issues relating to liquidation, cancellation and marketing and to the Financial Information Expressway (AIF, being its initials in Spanish) for the electronic referral of securities and agent information. Further legal information can be found on the CNV's website (National Securities Commission, http://cnv.org.ar/web/secciones/marco/leyes.aspx).

The creation of the Common Real Estate Investment Fund (RE-IF), within the regulation of the Common Investment Fund, appears in Art. 2 of the Ordinance of Law 24.083 of Common Investment Funds. The RE-IF is an open-ended fund with no legal personality created with the capital contribution of partners with a maximum number of shares (*cuotapartes*) and administered by a management company. In this case, the contributions may include property rights, mortgage loans and antichresis rights in limited proportions. (Antichresis rights comprise a contract for security between a debtor and a creditor whereby there is a transfer of possession of the pledged real property from the debtor to the creditor, including the fruits or rental income therefrom, in lieu of payments on the loan, including interest, for any such time period as is provided for in the contract.) The manager of these funds has the mandate to act as a *fiduciario* in the contributors' interest, to manage the real estate (including to manage the leases, sell, mortgage or constitute real rights to preserve the assets), retain custody of the remaining assets that make up the common fund and also to assume fund organization and party registration.

The Argentinian regulation also allows for the existence of (open) real estate funds in trusts and mixed real estate funds could also exist pursuant to the previous regulation. A *fideicomiso* is a univocal business consisting of two estates: the contract that determines the legal relationship between the *fiduciante* and the fiduciary; and the transfer of fiduciary property or goods assigned to the business in favour of a third-party beneficiary. The regulation has expanded the fiduciary object, which can now be any trade good including goodwill, the risks, the liability to which the trust asset corresponds and the liquidation mechanisms. A guarantee trust is created.

However, Argentina has not developed a REIT regulatory regime nor specific regulation for real estate funds which can be described as more detailed than that defined in the aforementioned Article 2.

9.7.2 Common funds

In the Argentinian capital market, foreign funds have been operating for many years and amongst them are REITs from other countries that take positions in Common Funds. There is one structure, the Exchange Trade Fund (LARE – Tierra XP Latin America Real Estate) where REITs account for 55% of the portfolio. The similarity between the existing funds in Argentina and international REITS can be seen in Table 9.3.

The Common Funds of Real Estate Investment are similar to REITs in essence, although they present a number of fundamental differences that distinguish them and make them require a regulatory change in the event that an attempt should be made to develop the structure. These differences include the following:

- the different corporate entities, since the absence of a typical legal personality in trusts and investment funds goes against the essence of a REIT, which does have legal personality;
- the liquidity derived from the intermediation in the capital market of a REIT since it can participate in the market, unlike funds and trusts since they are changed in the own fund and their shares must be directly rescued;
- the duration of the fund is indefinite, while that of REITs is not;
- dividend share is required for REITs but is not compulsory for investment funds; and
- the tax treatment is fundamental to REITs, with fiscal advantages, but non-existent for funds and *fideicomisos*.

Another feature that distinguishes REITs and *fideicomisos*'s assets is the existence of restrictions on investment, since limitations may affect the building or transferring of property (Table 9.3).

As for the other characteristics, both funds and REITs have similarities in the case of Argentina. For example, they offer the possibility of diversification in real estate investment assets, as

Table 9.3 Comparison Between REITs and Argentinian Real Estate Funds

Characteristic	*International REITs (main features)*	*Common Real Estate Investment Fund*	*Financial Fideicomiso*	*Comments*
Type	Closed Fund	Open Fund	Open Fund	
Legal person	Yes	No	No	
Diversification	YES	YES	YES	All types of Real Estate allowed
	Rent, leasing, sell and loan management	Rent, leasing, sell and loan management	Rent, leasing, sell and loan management	Activity
Leverage Restrictions	NO	NO	NO	Indebtedness is allowed or not
Assets/participations	Non-nominative	Nominative	Nominative	
Transparency	Quoted RE Society	NO Quoted	NO Quoted	Yes
	Regulated market (financial supervisor)	Regulated market (financial Superintendent)	Regulated market (Civil Law)	Stock Exchange Market or capital market
	Yes	yes in listed	Yes	Information requirement, published by the regulating Authority, website and Official Register
	Yes	Yes	Yes	Public selling (market)
	Limited	Un-limited	Un-limited	Time of life
Investment restrictions	Few (country dependent)	yes (art 7)	Yes	
Operational restrictions	Yes	Yes	Yes	Financial management
	Development sometimes	No	No	Limits to operational restrictions
	Own management	listed manager ("sociedad gerente-fiduciaria")	Listed manager	Listed Management approved by regulator
	Market	Fund	regulated	redemptions/ transactions
Liquidity	Yes	depending on the Fund	No	
Dividends	90%	75%	–	
Fiscal exceptions	No	Yes	Yes	Corporate tax
	No	Yes	Yes	Personal Income tax
	Yes	Redemption	No (if public supply)	Transaction cost (real estate)
	Normally yes	Yes	Yes	Withholding Tax

Source: Author

well as the capacity for rent management and the commercialization of goods. Similarly, all of them have to face control procedures through the securities market, both in the issue and in the transfer of shares, being subject to public transparency and information requirements.

9.7.3 Summary – Argentinian REITs

While a developing REIT sector, Argentina has not yet established a REIT regulatory regime nor specific regulations, with the existing Common Real Estate Investment Fund (RE-IF) providing an attractive alternative structure. It may, therefore, be some considerable time before a fully operational REIT market develops in Argentina.

9.8 REITs in Uruguay

Uruguay's economy is heavily based on its banking sector, which provides most sources of finance. The financial crisis at the beginning of the millennium severely affected the economy of Uruguay, whose banking system became disrupted by the payment chain and credit was reduced to its lowest known levels (Porcaro and Malumian, 2003, p. 2), which affected the real economy. The economic crisis generated as a result of this collapse caused a reaction against the traditional credit shortage and lack of financing, leading the government to redesign the regulatory system and allowing the mobilization of investment capital with new legal instruments as well as guarantee systems and information.

9.8.1 Changes to legislation and regulation

The updating of the regulation on investment funds in Uruguay had already begun with the standard published in 1996 and has been very dynamic ever since, with numerous amendments to the Laws on Investment Funds and *Fideicomisos* making the capital market more flexible and modernized. This has allowed the country to maintain the guarantees for investors by subjecting them to the supervision of the financial regulator (the Central Bank) and the information requirements of capital markets. As noted in Table 9.1, regulations evolved in the 2000s leading towards the current legal environment reached in 2007 and towards facilitating the issue of real estate trusts, REITs, in 2012.

The 16774/1996 Law on Investment Funds Act opened the possibility of investment through the creation of common securities in Uruguay. This law creates open-ended funds, without legal personality, with limited liability and mandatorily managed by specialized fund managers authorized by the central bank, which cannot be the owner.

These funds issue shares known as "*cuotapartes*" which are placed directly and registered, notably with the authorization of the Central Bank (both in their issue and in their transfer), in the Securities Registry. A feature of funds is that they cannot sell their capital contributions – considered undivided (Art. 3) – and that the investment must comply with a rule of maximum percentages according to the object of investment.

Despite not including the identification of specific real estate funds, the general regulation on Investment Funds did make the contribution of assets other than securities, deposits or shares dependent on the decision of the Central Bank (Art. 21), even though it was incorporated in Law 17202/1999, Art. 30, which created the form of mortgage-backed investment funds, able to issue securities that grant credit rights, joint ownership or both. Article 12 explicitly indicates that the management company will not be able to buy or lease that real estate which makes up the assets of the fund, which means that its activity is not conceived as that of a real estate fund.

Following the 2001 financial crisis, numerous amendments were made to investment legislation with the aim not only of making it more flexible and specialized but also of creating ways to facilitate safe investment by private agents and individuals towards profitable projects by channelling off-balance sheet financing. Significant amongst those amendments are those concerning trust regulation (2003–2007), promotion and protection of investments (2012), promotion and real estate exploitation (2003–2010) and housing protection (2011).

The regulation on trusts was approved in 2003 (Act 17703/2003) and includes the Private Investment Real Estate Fideicomiso (FPII), which has as its purpose to build investor funds integrated by real estate that is commercially exploited by means of sale and rent. Management is one of the objectives, in addition to the diversification and construction of quality real estate assets.

This regulation also allows for the creation of several types of trusts adapted to different aspects of the real estate sector – guarantee, administration, construction at cost and real estate (financial and ordinary) investment – which can be variously used in investment projects, such that a financial trust scheme can be backed up in a guarantee trust (Porcaro, 2006).

9.8.2 Financial Fideicomiso of Investments and Real Estate Income

The regulation on trusts opens the possibility of being applied to any property, with the first Financial Fideicomiso of Investments and Real Estate Income constituted in May 2012 and having US$60 million of capital from individual and institutional investors. It was created as a pool of funds that included the contribution of existing properties in offices (Class A offices in Montevideo), warehouses, distribution centres and logistic depots. The comparative characteristics of an international REIT and a Financial Real Estate Trust are shown in Table 9.4.

Table 9.4 Comparison between REITs and Uruguay Financial Real Estate Trust

Characteristic	*International REITs (main features)*	*Financial Real Estate Fideicomiso*	*Comments*
Type	Closed Fund	Contract – Open Fund	
Legal-person form	Yes	No	
Diversification	YES	Yes	All types of Real Estate allowed
	Rent, leasing, sell and loan management	Yes	Activity
Leverage Restrictions	No	No	Indebtedness is allowed or not
Assets/ participations	Non-nominative	Nominative	
Transparency	Quoted RE Society participations	Not quoted	
	Regulated market (financial supervisor)	Regulated market (financial supervisor)	Stock Exchange Market or capital market
	Yes	Yes	Information requirement, published by the regulating Authority, website and Official Register

(*Continued*)

Table 9.4 (Continued)

Characteristic	*International REITs (main features)*	*Financial Real Estate Fideicomiso*	*Comments*
	Yes	Yes	Public selling (market)
	Limited – long-term	Limited – long-term	Time of life
Investment restrictions	Few (country dependent)	Yes (forbidden to sell capital)	
Operational restrictions	Yes	No	Financial management
	Development sometimes	Defined in the contract	Limits to operational restrictions
	Own management	Yes	Listed Management approved by regulator
	Market	NO Market transaction-redemption in the fideicomiso	redemptions/transactions
Liquidity	Yes	Depends on the fideicomiso	
Dividends	90%	Few (defined in the contract)	
Fiscal exceptions	No	Yes	Corporate tax
	NO	Yes	Personal Income tax
	Yes	Non-exonerated (mainly for "Guarantee *fideicomiso*" and if the titles issued are publicly offered.	Transaction cost (real estate)
	Normally yes	Non-exonerated	Withholding Tax

Source; Author

By using the structure of the *fideicomiso*, the Uruguayan system has developed specific regulations to promote the financing of social housing, infrastructure and real estate projects, such as the guarantee fund for Desarrollos Inmobiliarios – FOGADI (in 2016). The current development of such funds is scarce, with the creation of funds to build social housing (e.g. CasasURU), funds for roads and infrastructures in some regions (e.g. CANALONES II) and city areas (e.g. FF Expansion Boulevard Batlle or Ordonez). In general, FDII activity is very limited in the financial market.

9.8.3 Summary – Uruguayan REITs

While an emerging sector, REITs as such do not yet exist in Uruguay with the Real Estate Finance Fideicomiso structure providing most REIT roles to facilitate investment in real estate. Given the long-standing and widely accepted nature of the *fideicomiso* structure, it may some considerable time before REITs become established in Uruguay.

9.9 Summary

Part II includes six chapters each analysing REITs in a region of the world, using case studies from a developed, developing and emerging REIT sector in the region.

Chapter 8 analysed REITs in the North American region, using case studies from the USA, Canada and Mexico for the developed, developing and emerging REIT sectors in the region, respectively. This chapter analysed REITs in the Latin American region, using case studies from Brazil (being a developed REIT sector in the region), Argentina (being a developing REIT sector in the region) and Uruguay (being an emerging REIT sector in the region), respectively.

Recent legislative and regulatory changes in various Latin American countries have facilitated a clearer regulatory definition of investment funds across Latin America as a form of indirect vehicle based on the popular, long standing and well recognized legal entity, the *fideicomiso*, which is common across Latin America and offers attractive characteristics in risk management and investment capability. With modernization, the *fideicomiso* may coexist with advanced investment fund structures and facilitate potential conversions into REITs, though the *fideicomiso* appears likely to predominate as the preferred structure for the foreseeable future.

Since the main difference between a *fideicomiso* (closer to an open fund) and a REIT is the legal status, changes in regulation (or specifically a new regulation) will be required to allow (closed) funds to have legal personality so that they can manage real estate funds. Closed funds also appear in the regulation of the main relevant countries and, in those cases where closed real estate funds are defined by the legislation, they are subject to large constraints.

While there is no logical reason why REITs cannot develop in the current Latin American environment, the financial market instability associated with more risk in long-term investment tarnishes the attractiveness of REITs while the stronger guarantee and lower uncertainty associated with fideicomisos is possibly the key to their ongoing success (at least in the legal environment).

Chapter 10 will analyse REITs in the European region, using case studies from the UK, Spain and Poland with Chapter 11 analysing REITs in the South East Asian region, using case studies from Singapore, Malaysia and Thailand. Chapter 12 analyses REITS in the North Asian region, using case studies from Japan, Hong Kong and China with Chapter 13 analysing REITs in the Oceania region, using case studies from Australia, South Africa and India with Chapter 14 concluding Part II by considering *Directions for the Future of International REITs*.

References

EPRA 2016a, *REITs reports*, EPRA, Brussels, available at www.epra.com

EPRA 2016b, *REITs en Brazil, Costa Rica y México*, EPRA, Brussels, available at www.epra.com

Garcia, D. and Porcaro, D. 2003, *Fideicomiso en Uruguay*, Price Waterhouse Coopers – available at http://wwwpwc.uy/es/servicios-fiduciarios/assets/leyfiduciaria.pdf, 21 January 2017.

IIMV 2006, *Estudio comparativo de la industria de la inversión colectiva en Iberoamérica*, Instituto Iberoamericano de Mercados de Valores, caps 1–7 más anexo, available at www.iimv.org/estudios/estudio-comparativo-de-la-industria-de-la-inversion-colectiva-en-Iberoamerica/, 02 January 2017.

IIMV 2010, *Estudio comparativo de la industria de la inversión colectiva en Iberoamérica*, Instituto Iberoamericano de Mercados de Valores, caps 1–7, actualización, available at http://eee.iimv.org/estudios/estudio-comparativo-de-la-industria-de-la-inversion-colectiva-en-Iberoamerica/, 02 January 2017.

International Monetary Fund 2016, *Financial integration in Latin America*, March 4, available at www.imf.org/external/pp/ppindex.aspx, 02 February 2017.

Martinez, A. 2011, *Liquid real estate investment fund in Latin America: Analysis of worldwide best practices and portfolio proposal*, program in RE Development and Center for Real Estate of Massachusetts Institute of Technology (MIT), September 2011.

McGreal, S. and Sotelo, R. (eds) 2008, *The introduction of REITs in Europe: A global perspective*, Competence Circle Service, Austria.

Payne, J.E. and Waters, G.A. 2005, 'REIT markets: Periodically collapsing negative bubbles?', *Applied Financial Economic Letters*, Vol. 1, No. 2, pp. 65–69.

Porcaro, D. 2006, *Fideicomisos aplicados en el sector inmobiliario*, available at http://fa.ort.edu.uy/innovaportal/file/13006/1/fideicomisos_aplicados_en_el_sector_inmobili.pdf, 06 January 2017.
Porcaro, D. and Malumian, N. 2003, *Fideicomiso en Uruguay. Análisis del Proyecto de Ley*, Price Waterhouse Coopers, available at http://wwwpwc.uy/es/servicios-fiduciarios/assets/pub-libroproyectoleyfideicomiso.pdf, 21 January 2017.
Sotelo, R. and McGreal, S. (eds) 2013, *Real estate investment trusts in Europe*, Springer-Verlag, Berlin Heidelberg.
Taltavull de La Paz, P. and Peña Cuenca, I. 2013, 'Spanish REITs: The new regulated SOCIMIs', in *Real estate investment trusts in Europe*, ed. R. Sotelo and S. McGreal, Chapter 13, Springer-Verlag Berlin Heidelberg.
Uruguayan Stock Exchange, 2017, available at www.bvm.com.uy, 10 February 2017.

10
EUROPE

Alex Moss

10.1 Introduction

This book aims to identify key areas for research in the REIT discipline for the next five to ten years by surveying the current state of the REIT discipline around the world and identifying emerging and cutting edge research areas through a thematic review of current contextual issues and a regional analysis based on case studies.

This book comprises two parts, the first part being a thematic review of emerging and cutting edge global research into current contextual issues in REITs internationally and the second part being a regional analysis of REITs around the world, each written by authoritative academic authors from the world's leading Universities and REIT industry experts.

Part I included six chapters each reviewing a current theme of REIT evolution through the lens of contemporary research. Chapter 1 focused on critical contextual issues in international REITs while Chapter 2, the *Post-Modern REIT Era*, examined the evolution of the global REIT market, what is deemed to be best current market practice and how the market is expected to change going forward and Chapter 3, *Emerging Sector REITs*, investigated such sectors as residential, health care, self-storage, timber, infrastructure and data centres generally and then specifically through a case study of US REITs.

Chapter 4, *Sustainable REITs*, analysed REIT environmental performance and the cost of equity and Chapter 5, *Islamic REITs*, examined the evolution of global Islamic REITs through a study of the Malaysian market. Chapter 6, *Behavioural Risk in REITs*, addressed the management of behavioural risk in global REITs and Part I concluded with Chapter 7 which reviewed recent research into *REIT Asset Allocation*.

Part II includes six chapters each analysing REITs in a region of the world, using case studies from a developed, developing and emerging REIT sector in the region. Chapter 8 analysed REITs in the North American region, using case studies from the USA, Canada and Mexico, with Chapter 9 analysing REITs in the Latin American region, using case studies from Brazil, Argentina and Uruguay for the developed, developing and emerging REIT sectors in the region, respectively.

This chapter analyses REITs in the European region, using case studies from the UK (being a developed REIT sector in the region), Spain (being a developing REIT sector in the region) and Poland (being an emerging REIT sector in the region).

Chapter 11 then analyses REITs in the South East Asian region, using case studies from Singapore, Malaysia and Thailand with Chapter 12 analysing REITs in the North Asian region, using case studies from Japan, Hong Kong and China and Chapter 13 analysing REITs in the Oceania region, using case studies from Australia, South Africa and India with Chapter 14 concluding Part II by considering *Directions for the Future of International REITs*.

10.2 Overview of the European REIT sector

This chapter will cover the background, current position and outlook for the European REIT sector, both in absolute terms and relative to the global market, being divided into three principal sections:

10.3 background and context;
10.4 determinants of the growth of REIT markets in Europe; and
10.5 REITs in Europe.

REITs in Europe comprise case studies of three specific countries, which are at various stages of development and represent different aspects of the future growth of the European market, including:

10.5.1 the UK – being a developed market, which converted to REIT status in 2007;
10.5.2 Spain – being a developing market that has re-emerged post-GFC, with a REIT regime that started in 2013; and
10.5.3 Poland – being an emerging market that is looking to establish a REIT regime for the first time.

10.3 Background and context

To understand the background and context of the European REIT sector, the following questions will be considered:

- what is the impact on and relevance of the European real estate market to the economy?;
- what is the current structure of the European REIT market?;
- how does the European sector fit into the global framework of REITs?;
- do European REITs act as real estate or equities?;
- how are European REITs classified and used by fund managers?;
- can European REITs act as a hedge against inflation?; and
- how time-variant are the equity market valuations of European REITs?

10.3.1 Impact on and relevance to the European economy

The key to understanding a REIT market is to understand the size and relevance of the domestic commercial real estate market within the economy. In their joint report with INREV (EPRA and INREV, 2016), EPRA identified the following factors which are important to consider when looking at the European REIT market:

10.3.1.1 *Contribution to the economy*

The commercial property industry directly contributed €329 billion to the European economy in 2015, representing about 2.5% of the total economy and comparable to the combined size of the European automotive industry and telecommunications sector. It employs 3.7 million people, which is not only more than the auto manufacturing industry and the telecommunications sectors combined, but also greater than banking.

10.3.1.2 *Contribution of property management*

Investment, fund and portfolio management are small but disproportionately high value-added activities, contributing 6.5 times more per worker than the overall European average value-added per worker.

10.3.1.3 *Size*

The market value of commercial real estate in Europe in 2015 was approximately €6.2 trillion. This is greater than the value of the plant, machinery and equipment used by Europe's businesses and manufacturers. Offices are the largest property type, although retail is also substantial. The total value of residential, at €25.5 trillion, however far exceeds other property sectors.

10.3.1.4 *Relative holdings of REITs*

Around 40% of all commercial property – with a total market value of over €2.5 trillion – is held as an investment. Of this, around €312 billion (12.5%) is held by listed REITs or property companies.

10.3.1.5 *Increasingly important source of global investment*

The amount of commercial property held by non-EU institutions, including sovereign wealth funds, is estimated to be €156 billion, being twice the value in 2011.

10.3.1.6 *Importance of role in pension funds*

The long-term cash flows generated from property investment provide an important source of diversified income in the portfolios of European savers and pensioners. Property, in its various forms, accounts for €824 billion of European pension funds and insurance companies' investments, being an allocation of nearly 5.5%.

10.3.1.7 *Summary – impact on and relevance to the European economy*

It can, therefore, be seen that the underlying European commercial real estate market plays a significant role in the wider European economy, being an important component of the investment universe of the European savings institutions and pension fund market. As will be observed later in this chapter, a sophisticated, high quality, commercial real estate market is one of the key precursors to a successful REIT market.

10.3.2 Current structure of the European REIT market

Having established the significance of the underlying real estate market, it is now important to understand the dynamics of the European REIT sector. Europe has a long-established listed real estate sector and more recently developed REIT sector, with 12 out of the 28 EU member states now having a REIT regime. These 12 countries represent 83% of the EU GDP so it can be seen that REIT regimes have a high penetration amongst the largest countries in economic terms. However, not all of these regimes have been successful in achieving critical mass as far as investors are concerned. Indeed, some countries have a REIT structure in place but little in the way of a listed REIT market.

Table 10.1 shows the countries which have a REIT market in place and their relative size. As can be seen, the UK and France dominate. The date given in the second column is when REIT legislation was enacted, although, as can be seen from the entries for Hungary, Israel, Lithuania and Luxembourg, this does not necessarily mean that a REIT sector has actually been established. Legislation is only one step in the process and needs to be accompanied by market demand. It should be noted that these figures relate only to REITs (not Prop Cos) and that the market capitalization figure given is gross, whereas figures provided by EPRA for Index purposes are free float weighted (i.e. related party and insider holdings are taken out of the figure to provide a sense of true investability and liquidity in the shares).

The different types of REIT markets in Europe and their development may be classified as follows:

10.3.2.1 Conversion

For example, the UK and France both had an existing listed real estate sector (non-REIT, PropCo sector), which converted to REIT status. Subsequently, all the largest companies are REITs. On conversion the companies paid a charge to the government (a fixed percentage of gross assets at the time of conversion) in exchange for the tax transparency of REIT status.

Table 10.1 European REIT Markets

	Mkt. Started	*No. of REITs*	*Mkt cap (€bn)*	*In EPRA Index*
UK	2007	36	23	56.59
France	2003	32	8	49.36
Netherlands	1969	5	5	21.20
Belgium	1995	17	7	11.00
Spain	2009	5	4	7.80
Turkey	1995	30	4	6.79
Germany	2007	4	2	2.42
Ireland	2013	3	3	2.17
Italy	2007	3	2	1.82
Greece	1999	4	1	0.08
Bulgaria	2004	53	0	0.00
Finland	2009	1	0	0.00
Hungary	2011	0	0	0.00
Israel	2006	0	0	0.00
Lithuania	2013	0	0	0.00
Luxembourg	2007	0	0	0.00

Source: EPRA Global REIT Survey September 2016

10.3.2.2 Long Established

For example, the Netherlands and Belgium, where a structure had been utilized before the Modern REIT Era structure was established in the US (1992–1993) and remains the dominant structure. Typically, these REITs have a long and successful track record of dividend payments with less focus on capital growth.

10.3.2.3 Post-GFC Stimulant

For example, Ireland and Spain both had a structure that was used to transfer assets from an over-extended banking sector that was an unwilling holder of real estate (via foreclosure) to long-term equity market investors.

10.3.2.4 Parallel markets

For example, Germany and Italy, where a listed sector already exists and REIT legislation has subsequently been introduced. Interestingly, although the German-listed sector has seen dramatic growth in absolute and relative terms since 2013, this has not occurred through the REIT market, but rather through the energetic expansion of the PropCos specializing in the residential sector. Although Italy has yet to take off, a number of conditions already exist for it to be the next REIT growth market in Europe.

10.3.2.5 Emerging markets

For example, Greece, Bulgaria, Finland and Turkey where REIT legislation has been introduced in an attempt to stimulate inward investment into the commercial real estate market.

10.3.2.6 Stalled markets

For example, Hungary, Israel, Lithuania and Luxembourg are all places where legislation has been introduced, but the market has yet to have any listed vehicles.

Table 10.2 shows the relative size of the developed markets (REITs and non-REITs) as measured by free float market capitalization. It should be noted that there is often a wide

Table 10.2 Relative Size of European-Listed Markets

Market	*Size (€bn)*	*% of Europe*
UK	61.7	31.32%
Germany	41.9	21.27%
Netherlands	25.5	12.94%
France	18.2	9.24%
Sweden	15.6	7.92%
Switzerland	11.8	5.99%
Spain	7.6	3.86%
Belgium	6.3	3.20%
Austria	4	2.03%
Finland	2.3	1.17%
Norway	1.2	0.61%
Italy	0.9	0.46%

Source: EPRA (2017)

divergence between free float market cap and gross market capitalization. It should also be noted that, for Index purposes, the largest stock in Europe, Unibail-Rodamco, is classified under the Netherlands (where its main listing is) whereas, in practice, it is a French REIT.

10.3.3 European REITs in a global context

Having observed the country-specific composition of the European REIT market, how does the region compare globally? In a global context it should be noted that, as at end February 2017, Europe accounted for only around 13.85% of the global market, as measured by the EPRA Developed Global Market Index (EPRA, 2017).

Relative to the global market where 75% by market capitalization are REITs and 25% non-REITs, Europe has a lower proportion of REITs (57%) vs non-REITs (43%) which reflects the growth of the (particularly residential-based) non-REIT market in the second largest country, Germany.

As Figure 10.1 shows, the European market has grown from US$50bn in 1989, when it represented over 30% of the global market (as measured by the EPRA Global Developed Index) to over US$200bn at the end of 2016, although its share of the global market has fallen to under 15%. This trend of absolute growth and relative decline reflects improving direct property market values (ex GFC) and some equity issuance, which in relative terms has been overshadowed by the growth in the US market.

10.3.4 Do European REITs act as real estate or equities?

This is a topic that is, obviously, also relevant globally and the following briefly summarizes evidence from three papers which have been sponsored by the EPRA Research Committee with each seeking to answer a slightly different variation of this question.

10.3.4.1 Are public and private real estate returns and risks the same?

In their seminal paper, Hoesli and Oikarinen (2012) examined the evidence for the UK (and US) markets. Their analysis shows that while, in the short run, the observed REIT and direct real estate returns can substantially deviate from each other due to factors such as data complications, market frictions and slow adjustment to changes in the fundamentals in the private market, in the long term (c. three years), public and private real estate returns are similar after allowing for the effects of property type, leverage and management costs.

10.3.4.2 What is the breakdown of REIT returns?

Kroencke, Schindler and Steinenger (2015) examined the question in terms of the composition of REIT returns. Investors who are interested in obtaining real estate exposure in their stock- and bond-dominated portfolios often try to achieve this by investing in publicly traded REITs. But it is questionable as to the extent to which they really invest in the underlying real estate market by using this vehicle. In other words, are REITs real estate or stocks?

With their asset pricing model, the authors quantitatively show to what extent REIT returns can be explained by a combination of the pure stock market risk, pure real estate market risk and business cycle risk. This result helps investors to reallocate their multi-asset portfolios to their actual desired exposure to the different risk factors. According to the author's calibration,

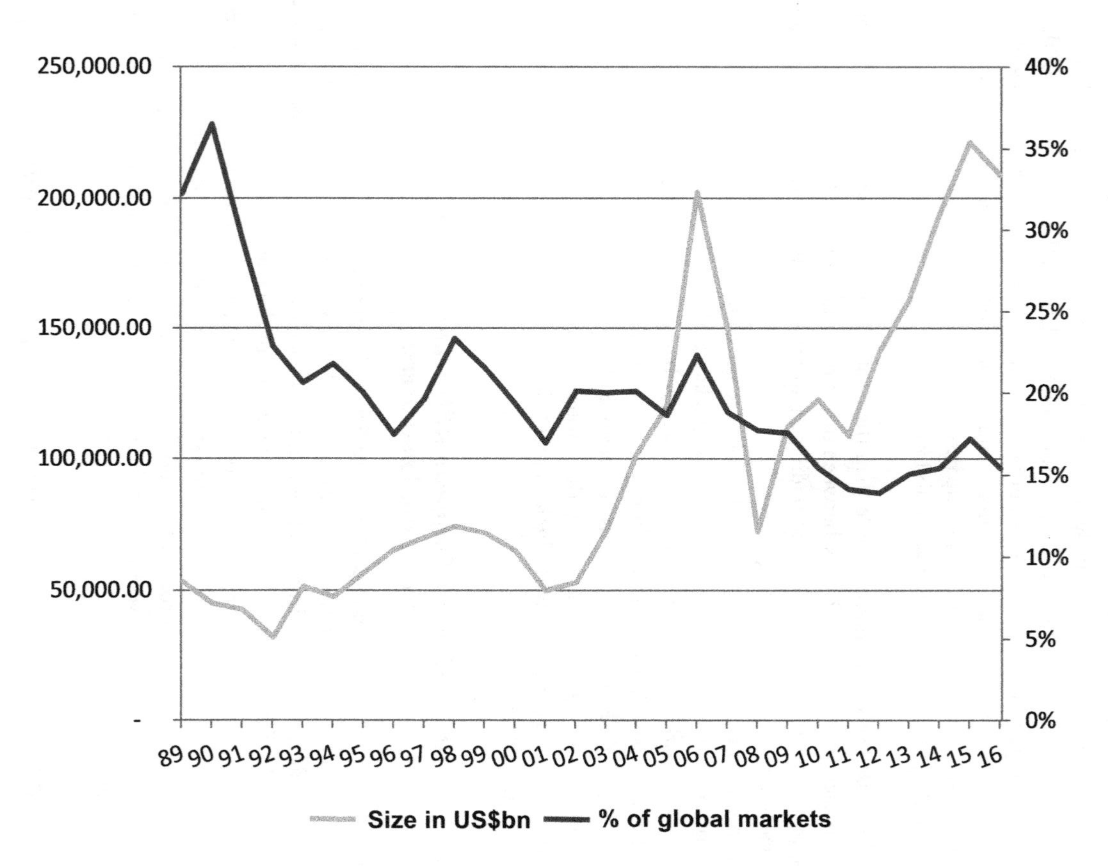

Figure 10.1 Growth of the European Market – Absolute and Relative Terms

Source: EPRA/Author

the expected listed real estate premium consists of 36% stock market risk, 40% real estate risk and 24% business cycle risk.

10.3.4.3 How are listed returns related to the direct market?

A study by Devaney, Xiao and Clacy-Jones (2013) investigated the extent to which returns from the listed real estate sector are related to returns in the direct real estate market for the US and for six European countries: France, Germany, the Netherlands, Sweden, Switzerland and the UK.

Past research has often used valuation-based indices for the direct real estate market, but these are criticized with regard to their perceived smoothing and lagging of market performance. In contrast, this study uses transaction based/linked indices of direct real estate prices, as well as valuation-based indices for the purpose of comparison.

The main findings from the research are as follows:

- transaction-based series are more volatile than their valuation-based counterparts and thus imply a smaller difference in volatility between direct real estate and the listed sector. Other factors, such as gearing and the different trading environment, continue to distinguish indices for each form of real estate;
- movements in listed sector returns lead those in the direct market regardless of whether a transaction or valuation-based series is used for the latter. Hence, the lead is not solely due to valuation smoothing. The extent of the lead varies, being two to four quarters in France, the Netherlands, Sweden and the US, but zero to two quarters in the UK; and
- in terms of the recent major cycle, listed sector indices exhibited both an earlier peak (either Q4.2006 or Q1.2007) and an earlier trough (Q1.2009) than their direct market counterparts. This is consistent with the idea that price discovery occurs first in one market and then in the other.

10.3.5 How are European REITs classified and used by fund managers?

Moss and Baum (2013a, 2013b) undertook two studies to determine how listed real estate was used by fund managers. For some European investors and managers, listed real estate is clearly part of the equity allocation. For others, there is some evidence that pension funds and consultants regard (or would like to regard) listed real estate as part of the real estate allocation.

However, there is strong evidence to suggest that asset managers (with their greater experience of execution as well as a propensity for business unit separation) may not have developed a satisfactory integrated investment process. The authors found that only 8 of 56 interviewees, or 14% of their sample, claimed to have an internally integrated approach to the management of listed and direct/unlisted real estate.

10.3.6 Do European real estate stocks hedge inflation?

Lee and Lee (2014) examined the long-run and short-run inflation-hedging properties of real estate stocks for five European markets (the United Kingdom, France, Germany, Poland and the Czech Republic) over the period January 1990 to July 2011.

The authors proposed two key findings:

- it appears that, for real estate stock investors, it is very difficult to hedge the short-run inflation risk. The empirical results show little inflation-hedging ability of European real estate stocks over the short run. The results are consistent with the findings of previous studies of US REITs. This also implies that real estate stocks are probably a better hedge against longer-term inflation rather than short-term inflation risk; and

- that real estate stocks in developed markets do provide a positive inflation hedge against expected inflation over the long run – strong, long-run inflation-hedging results for real estate stocks were evident in the UK, France and Germany, suggesting that real estate stocks in these developed markets do serve as a good hedge against expected inflation in the long-run.

10.3.7 How time-variant are the equity market valuations of European REITs?

The final area for examination is how time-variant the valuations of listed European real estate stocks are? Why is this important? Predominantly because it will determine, at least in part, the level of capital market activity in the region. Put simply, when shares are trading at a premium to their underlying net asset value (NAV) it indicates that the stock market is ascribing more value to the corporate entity than to the underlying real estate.

This may be for a number of reasons, including expectation of future capital value growth, management expertise, etc. However, the impact is the same, i.e. when shares trade at a premium the level of equity issuance increases as does the amount of IPOs. Similarly, when shares are trading at a deep discount, this implies falling property values with limited possibilities for equity raising. Figure 10.2 shows the variation in equity market valuations of European real estate companies, indicating a range of 30% premium to 50% discount.

10.4 Determinants of the growth of REIT markets in Europe

This section investigates what are the common factors behind growth in REIT markets in Europe? This is examined both in terms of factors relating to the domestic market as well as within a global context.

10.4.1 Is there a "normalized" relative weighting for Europe?

To provide some historical context, consider the historic trading range of Europe's listed sector as a fraction of the global total shown in Figure 10.1. Prior to the turbo-charged US REIT growth of the early 1990s, the UK and Eurozone listed sectors together accounted for a peak share of over 30% of the global universe, as measured by the EPRA Global Index. Subsequent to that date, there was an initial US REIT IPO explosion of 1993–1994, which became known as the birth of the Modern REIT Era, being followed by a series of later offerings which continued to boost the relative size of the US sector up to 1998.

Accordingly, the share of the UK and Eurozone fell steadily, reaching a trough of 12.1% in August 1997. From the beginning of 1998 to January 2010 the UK and Eurozone share never fell below 15% nor rose above 20%. It may be contended, therefore, that 15–20% comprises the "normal" weighting. From February 2010 until now, the relativity has continued to decline and now stands at just under 15%.

It may be concluded, therefore, that the weighting of the UK and Eurozone listed sector is currently below its "normalized" level.

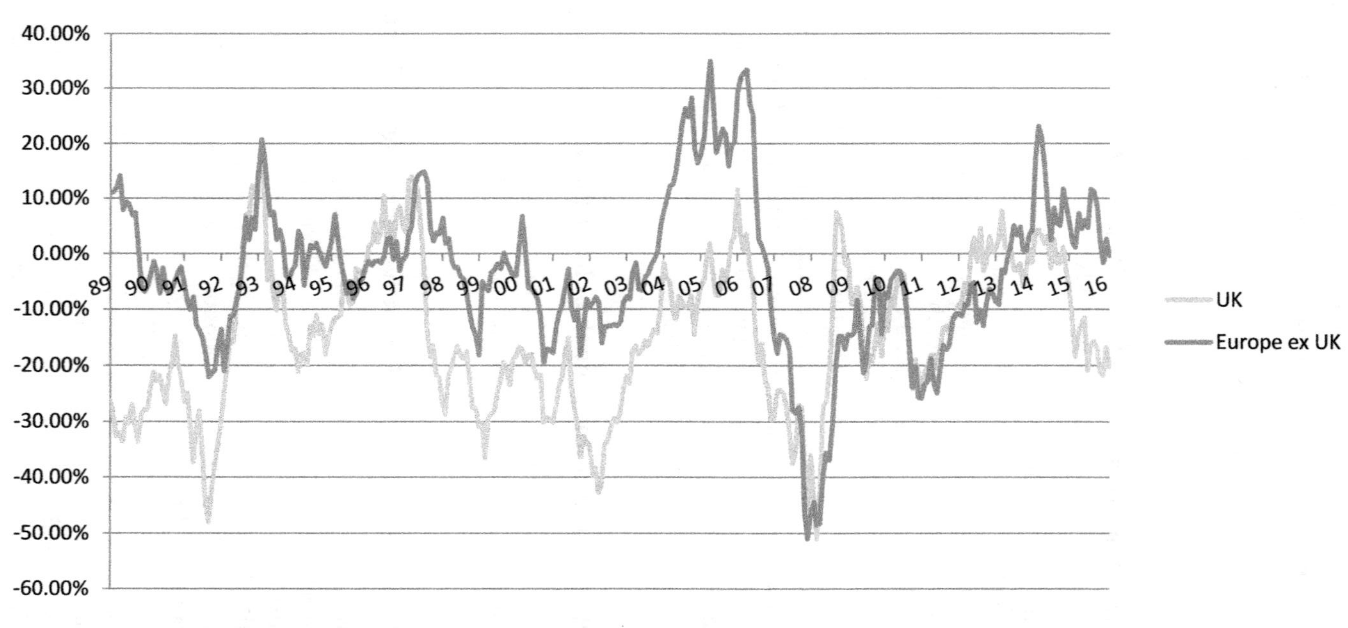

Figure 10.2 Discount/Premium to NAV for UK and European Real Estate Stocks

Source: EPRA/Author

10.4.2 What has the change been in absolute terms?

While the relative weighting of the UK and Eurozone listed sector has been driven down by the dynamic growth of Asia as well as by trends in the US, has the European sector also declined in absolute terms?

The peak, in terms of market capitalization, was a total US$189bn in March 2007. What is interesting and perhaps surprising for some commentators to note is that, from a trough of US$48bn in February 2009, the combined market cap is now US$102bn, being back to where it was in October 2005. This is a sharp contrast to indicators of value in the direct market. In the UK, where valuation data is most frequent, values have started to decline again and are currently not higher than in March 2000. In Europe, valuation declines looked much shallower with the annual IPD France Index, for example, suggesting that values are now back at the levels of end-2006.

A key difference between the US and Europe is that US REITs have continued to issue equity from March 2009 to the present day, whereas European REITs had one round of "rescue" rights issue in 1H.2009 and little else since.

In absolute terms, therefore, it may be concluded that the European listed sector has been more robust and healthier than its global weighting might suggest, but has failed to continue to recapitalize with equity issuance after the bottom of the GFC.

10.4.3 Are valuations to blame?

One of the key elements behind shifts in relative market capitalization are share valuation methodologies. In Europe, a discount to NAV methodology is commonplace whereas, in the US, a price multiple of Adjusted Funds from Operations (AFFO) is the convention.

One of the causes for the decline in weighting may be an expansion of the US rating and a contraction of the European rating. This would translate, for Europe, into stocks trading at a wider discount to NAV than their long-term average (i.e. below normalized levels). A second cause may be the ability of REITs to raise equity capital, which is partly a function of valuation and partly a function of mechanics (including pre-emption rights).

Figure 10.2 shows the long-term average discount for European and UK stocks, including the pre-REIT period where capital gains tax was included within the NAV figure and the CGT liability averaged c. 20-30% of the NAV per share.

For the UK, current valuations are of concern. Prima facie, valuations post-REIT conversion close to the average for pre-REIT status might imply that the benefits of a REIT structure to investors – elimination of capital gains tax and a higher dividend pay-out – have been zero. However, there are more plausible explanations. First, the implied level of property value changes contained in this discount can be assessed. Current pricing implies a fall in property values of around 12% over the next 12 months. Using a loan to value ratio of 40%, this equates to a fall of 20% in NAV per share. Given the uncertain economic outlook post the EU referendum and brokers property market forecasts of around -8% to -10% during the period of Brexit, this pricing does not seem unreasonable.

Second, it is not clear that all UK management teams have fully embraced the changes in corporate strategies that REIT status were expected to bring, including lower development exposure, focus on long-term sustainable cash flow, moderated gearing and regular recourse to both debt and equity capital markets for funding. Prima facie, it seems there may be agency issues which are stopping UK REITs from achieving superior valuations to tax inefficient PropCos.

It may, therefore, be concluded that valuations are currently lower than their normalized rate and are a partial contributor to the current low relative weighting.

10.4.4 How big should the listed real estate market be?

The final question to be addressed in this section is what fraction of the total commercial real estate market could be expected to be in the listed sector? Data from EPRA, in their monthly statistical bulletin, may help to answer this question. The EPRA methodology takes the market capitalization of each country shown by the EPRA Index divided by the EPRA estimate of the size of the overall investment market in commercial real estate. The rather clumsy label "equitization" may be used here to avoid confusion with the more commonly used "securitization", which may be associated with debt rather than equity products.

The EPRA results confirm previous studies, finding that Australia has the largest listed sector, relatively, accounting for 14.9% of total real estate investment. The global average is far lower with the listed sector accounting for 4.5% of the commercial real estate market and only 2.5% of the domestic stock market capitalization. In terms of real estate "equitization", the listed sector in Europe (less so the UK) clearly looks undercapitalized.

It may be concluded, therefore, that the European listed real estate sector is undercapitalized relative to the global average and that significant growth would be required to bring it to a global norm.

10.4.5 Where might European listed sector growth come from?

If the European listed real estate sector is undercapitalized relative to the global average and significant growth is required to bring it to a global norm, where might such growth come from?

Having regard to the evidence from the explosion in the US REIT market, may parallels be drawn highlighting similarities and differences across Europe?

10.4.5.1 Is there a pool of investible stock that could form the asset base for REIT IPOs in Europe?

The first question is whether there is a base of sufficiently high quality assets in the UK and Europe to attract institutional investment via REIT vehicles. While non-performing bank debt might seem an obvious and certainly plentiful source of stock, anecdotal evidence suggests that many such assets are not institutional quality.

It may, therefore, be contended that the most relevant source of assets may be likely to be unlisted funds, many of whom are coming to the end of their investment life. There are currently 336 UK unlisted funds with a gross asset value of €103bn and 877 European Funds with an asset value of €389bn. Interestingly, allowing for the prevailing discount and accounting for leverage, it would appear that the UK unlisted sector is similar in size to the listed market, but the European sector is significantly larger with stock in unlisted funds possibly up to four times that of the listed sector. Intuitively, this makes sense as a number of countries are significantly under-represented in the listed (commercial) sector, most notably Germany, Spain and Italy.

Therefore the pool of investible assets to support REIT growth, particularly in Europe, undoubtedly exists within the unlisted funds sector.

In the US, the closure of the private debt markets was accompanied by legislation which kicked off the growth of the public commercial mortgage backed securities (CMBS) market. A similar debt freezing now prevails in the UK and Europe. Traditional bank lending is becoming harder to obtain and comes only with margins increased and low LTVs. The question is, therefore, can European REITs access alternative debt markets? The answer is a resounding yes.

As bank debt has dried up, a number of different sources of debt have become available. The most popular are corporate bonds as investors continue to chase yield, US private placements, senior debt from insurance companies who are entering the market in response to Solvency 2 and mezzanine debt from newly formed debt funds.

One aspect of the situation in the US was arguably exceptional, whereby institutions had to double their weighting in REITs to bring them up to a similar level to their holdings in equities in general. This is clearly not the case for the developed markets of the UK and France but, again, given the under representation in the listed sector, there are structural underweight positions in Germany, Spain and Italy. For growth in these markets to occur there has to be a shift, not just in the assets from the unlisted sector to the listed, but also by the investors. This will require the right assets, the right management team, the right REIT structure and, crucially, the right valuation.

In summary, therefore, there is a sufficient pool of institutional, quality, investible assets to support REIT growth in Europe. The debt markets are open, accessible and available to refinance bank debt as witnessed by bond issues by European REITs over the last three years.

Whilst the UK and French REIT models are well known and in the case of the UK recently improved, it is the Spanish, Italian and to a lesser extent German models that need to be fine-tuned if the sector is to capitalize on demand for real estate assets in a listed form at the turning point of a cycle. Unlike the US in the 1990s, investors are not structurally underweight the existing REIT markets in Europe.

10.4.5.2 Where is the growth likely to come from?

Having considered the parallel of the US REIT market growth, there are a number of possibilities as to where growth in the listed sector may come from with the currently most likely areas which meet the requisite criteria and have recent catalysts being the following:

- the UK, post the abolition of the conversion charge, with potential stock and management teams from the overleveraged and redemption heavy unlisted funds sector;
- Germany, particularly after the recent announcement that the German Government is considering legislation to phase out c reation of open-ended property funds and require that all future vehicles are closed ended. There are clear parallels here with the decline of the RELP sector in the US in the late 1980s and also the decline of the open-ended property fund structures in Australia and the Netherlands which was the precursor to growth in the listed sector. The supply of assets (from the GOEFs) and demand for an efficient vehicle (providing a steady dividend stream) would therefore appear to be in place, although issues remain to be resolved regarding the underlying property valuation methodology; and
- Italy and Spain, which have negligible listed sectors and governments/banks which are under enormous pressure to deliver. In these cases, it is the REIT structure itself which needs to be fine-tuned to become institutionally acceptable.

10.5 REITs in Europe

This chapter analyses REITs in the European region, using case studies from the UK (being a developed REIT sector in the region), Spain (being a developing REIT sector in the region) and Poland (being an emerging REIT sector in the region).

10.5.1 REITs in the UK

The UK REIT legislation was introduced in the UK with effect from 1 January 2007 by the Finance Act 2006. On 1 January 2007, nine companies elected to become REITs – a number that grew significantly within the first year of the regime but, since then, the increase has been small each year though it has increased recently due to new REIT launches and offshore property funds managed by asset managers converting to REIT status.

There are now 36 REITs in the UK, of which 23 are in the EPRA Index, representing around 4.6% of the Global REIT Index as at September 2016.

The UK REIT regime operates through a combination of legislation plus guidance. Although there have been some changes since first inception, a number of commentators believe that updated guidance to accompany the legislation and subsequent amendments is long awaited.

Amendments to the REIT rules were introduced with effect from 2012, making the UK REIT regime more attractive to both companies and investors, with the principal amendments including the following:

- entry to the REIT regime becoming cheaper, since the entry charge was abolished;
- new REITs can list on the junior market, AIM, which has more relaxed reporting requirements than the main market as was initially required; and
- a three-year grace period for the share ownership of REITs to become widely held and not "close" (i.e. controlled by five shareholders or less). Certain institutions are encouraged to invest in REITs given their shareholdings in a REIT will be treated as widely held.

More recently, in 2013 and 2014, the UK Government has introduced further amendments in relation to UK REITs investing in other UK REITs. The measures allow the income from UK REITs investing in other UK REITs to be treated as income of the investing REIT's tax-exempt property rental business and REITs shareholders to be ignored when considering "close" status.

10.5.1.1 Requirements of UK REIT legislation

The key requirements of the UK REIT legislation are as follows:

- shareholder requirements:
  - cannot be a close company (i.e. controlled by five or less shareholders);
  - no corporate shareholder should hold 10% or more of the shares or voting rights;
  - there are no restrictions on foreign shareholders;
- listing requirements:
  - unlike US REITs, UK REITs must be listed on the London Stock Exchange (LSE) or any other "recognized" stock exchange;
- asset level tests:
  - at least 75% of a UK REIT's net profits must be derived from the property rental business;
  - at least 75% of a UK REIT's assets must be used in the property rental business;
  - a UK REIT must hold at least three separate assets;

    - no one asset may exceed 40% of the total assets;
    - a UK REIT may invest in real estate outside the UK;

- leverage:

    - the property profits must be at least 1.25 times the property financing costs. Where income cover is less than 1.25 times, a tax charge will arise based on the amount of the property financing costs that cause the ratio to fall below 1.25 times;
    - as the test looks only at the relationship between rental income and interest costs, a sudden unexpected increase in interest rates or a drop in income may result in a tax penalty. HMRC has the power to waive this penalty charge if the UK REIT is in severe financial difficulty, the ratio is breached due to unexpected circumstances and the UK REIT could not reasonably have taken action to avoid the ratio falling below 1.25 times income cover;

- profit distribution obligations:

    - a distribution out of the property rental business of the REIT (rental income and capital gains) is called a Property Income Distribution – a "PID";
    - a UK REIT must distribute 90% of property rental profits;
    - a UK REIT must distribute 100% of PID from other REITs;
    - capital gains are not included in the distribution obligation;

- tax treatment at the REIT level:

    - income from the property rental business is not subject to UK corporation tax;
    - non-rental business income (residual income) is taxable in the ordinary manner at the highest rate of corporation tax which is currently 20%;
    - capital gains or losses that arise on disposal of property used in a UK REIT's property rental business are not chargeable to tax;
    - the sale of "developed properties" may be subject to tax if they are disposed of within three years of the completion of any development activities conducted by the UK REIT;

- entry to the REIT regime is now cheaper, since the entry charge has been abolished;
- new REITs can list on the junior market, AIM, which has more relaxed reporting requirements than the main market as was initially required; and
- a three-year grace period for the share ownership of REITs to become widely held and not "close" (i.e. controlled by five shareholders or less). Certain institutions are encouraged to invest in REITs given their shareholdings in a REIT will be treated as widely held.

10.5.1.2 Market commentary and issues

The UK market was already well established by the time REIT legislation came into force. Indeed, before the advent of the modern REIT era, the UK had a larger listed real estate market than the US. Now it represents around 10% of the value of the US market. Therefore the new legislation was not expected to stimulate a new market, but rather make an existing market more tax-efficient and comparable to its European neighbours. However, there were a number of expectations at the time of launch that are worth re-visiting a decade later to determine how accurate they proved.

RESIDENTIAL REITS

The original legislation was designed by a Labour Government with a view to stimulating investment in the private rented residential sector. This has not occurred, although institutional interest in this sector has increased significantly over the period. It remains to be seen whether a residential REIT could achieve critical mass in the way that the multifamily and apartment REITs have in the US.

LOWER DEVELOPMENT ACTIVITY

One of the key predictions was that companies would engage in a lower level of development activity post-REIT status than before. The rationale being that because non-income-producing assets cannot contribute to dividend payments they would be an inefficient deployment of capital resources. In practice this has not proven to be the case, the reason being that during the global financial crisis companies were not able to acquire distressed assets at the trough of the market (as they had no capital) so the most economic use of resources was to invest in redeveloping existing assets. This led to the construction of the iconic "Walkie-Talkie" and "Cheesegrater" office towers in the City of London by Land Securities and British Land.

INCREASED ISSUANCE OF EQUITY CAPITAL

As a REIT has to distribute 90% of its taxable profits as income it typically accesses debt and equity capital markets on a regular basis to provide the capital for expansion. Leaving aside the distressed rights issues of 1Q.2009, this has not been the case, particularly when compared to the US experience. There are two explanations for this. First, shareholders are loathe to subscribe to new issues at a material discount to NAV and, second, because of rising values and falling interest rates courtesy of quantitative easing, debt issuance has become a more favoured option.

EQUITY MARKET VALUATIONS WILL BE AT AROUND NAV

Because of the disruption caused by the global financial crisis and by the EU referendum in 2016, this has not happened (see Figure 10.2), although without the contingent CGT it is noticeable that they trade at a narrower discount than previously.

INCREASED LEVEL OF SPECIALIZATION

It is hard to know whether it is due to REIT status, or market forces, but in broad terms there does appear to have been a greater emphasis on sector specialization from REITS.

TAX-EFFICIENT CURRENCY FOR ACQUISITIONS

Although the UK REITs do not have an UPREIT structure similar to the US, they do have a tax advantage when offering shares as whole or part consideration. However, this advantage has not been utilized in a meaningful way with the notable exception being the acquisition of the Trafford Centre by Intu Properties (then Capital Shopping Centres).

CONCLUSIONS

Overall, it can be seen that the introduction of REIT status has been beneficial to the UK listed sector, although the companies themselves have not changed their corporate strategies significantly and have not issued materially more equity than during pre-REIT status.

10.5.1.3 Summary – REITs in the UK

The case study of REITs in the UK, being a developed REIT sector in the European region, is particularly interesting in that the UK started its REIT era with an existing and substantial property company and unlisted fund market, acknowledged through transition legislation which was then amended as REITs became established and transition arrangements were no longer required.

Unlike the developing REIT market of Spain, the UK retained limitations on shareholder holdings and constraints on leverage while maintaining the usual requirements for distribution and tax benefits commonly found in REITs internationally.

While REIT markets are well-established internationally and may, therefore, be expected to have an element of predictability, the focus on residential property, lower levels of development and increased equity issuance anticipated at the establishment of REITs in the UK did not transpire, which provides an interesting insight into the dynamics of individual markets relative to common collective experiences that may prove informative for emerging markets such as Poland.

10.5.2 REITs in Spain

This chapter analyses REITs in the European region, using case studies from the UK (being a developed REIT sector in the region), Spain (being a developing REIT sector in the region) and Poland (being an emerging REIT sector in the region).

In Spain, the original REIT (or "Sociedades Anónimas Cotizadas de Inversión en el Mercado Inmobiliario" – "SOCIMI") legislation was introduced in 2009. However the changes required to the taxation element of the legislation, to make Spanish REITs comparable to other European REITs, were not made until December 2012. Hence, the "modern REIT era" for Spain began in January 2013.

There are currently five Spanish REITs listed on the main market, of which four are in the EPRA Global REIT Index, accounting for around 0.5% of the Global weighting. There are also 23 listed in the Mercado Alternativo Bursátil, ("MAB"), but these are relatively small and are not in the EPRA Index.

10.5.2.1 Requirements of Spanish REIT legislation

The key requirements of the Spanish REIT legislation are as follows:

- shareholder requirements:
    - there are no threshold limits for ownership percentages;
- listing requirement:
    - Spanish REITs must be listed;

- asset level tests:
  - at least 80% of their assets must be invested in the following: (a) urban real estate (acquired or developed) for rental, or (b) other SOCIMIs, or (c) foreign REITs or (d) Spanish or foreign qualifying subsidiaries ("Sub-SOCIMI") or real estate collective investment schemes;
  - in terms of their revenues at least 80% of the SOCIMI's revenue must derive from (a) the leases of qualifying assets, and/or (b) the dividends distributed by qualifying subsidiaries. There is no asset diversification rule and SOCIMIs are entitled to hold a single property asset;
- leverage:
  - there are no restrictions on leverage;
- profit distribution obligations:
  - operating income – at least 80% of the operating income of the SOCIMI must come from rental and ancillary activities. However it is compulsory to distribute 100% of profits stemming from dividends distributed by qualifying entities;
  - capital gains – at least 50% of the profit corresponding to income derived from the transfer (where the holding period has been met) of real estate assets and qualifying holdings must be distributed. The other 50% of that profit must be reinvested in Qualifying Assets in a period of three years or, otherwise, distributed to the SOCIMI's shareholders;
- tax treatment at the REIT level:
  - current income – as per 1 January 2013, all the income received by a SOCIMI will be taxed under CIT at a 0% rate. Nevertheless, rental income stemming from Qualifying Assets being sold prior to the end of the minimum holding period (three years) would be subject to the standard CIT rate (i.e. 25%); and
  - capital gain – as a general rule, a SOCIMI will be taxed under CIT with a 0% flat rate being applicable. Nevertheless, a SOCIMI will be taxed at the standard CIT rate of 25% if the relevant asset has been sold prior to the end of the minimum holding period (three years).

10.5.2.2 Market commentary and issues

The Spanish REIT market is a classic example of how the structure can help transfer high-quality assets from unwilling holders (the banks) to willing holders (equity investors). The initial REITs were set up as "cashboxes" (i.e. equity-funded vehicles with little initial assets) in an attempt to rebuild the Spanish listed sector. In the first instance, since they were opportunistic, there was little to differentiate them in terms of strategy or execution capabilities. The key to new shareholders was the ability to enter into a recovering market, with no legacy issues, so there was a distinct first mover advantage to this phase of Spanish REIT development. This success of commercial REITs is expected to lead to the IPO of several Spanish house building companies, a situation that would have been unthinkable several years ago.

Some of the distinguishing features of Spanish REITs include the following:

- private equity style remuneration structures – because of their first mover advantage, a number of the initial IPOs were set up with private equity style remuneration structures for external management teams. This is contrary to the perceived best practice which is that internal management is preferable. One of the key aspects which drew attention was the level of performance fees, based on an absolute rather than relative benchmark. Merlin Properties was a good example of this initial remuneration structure. However, as the company's shareholder base changed, it became apparent that there was resistance to the scheme. It was, therefore, changed to reflect recommendations by both American and European regulators;
- finite life vehicle – typically a listed REIT is regarded as having perpetual capital. However, adopting another practice from private equity, Hispania has announced that it will return capital to shareholders within seven years of IPO by disposing of the assets of the company. This alleviates any problem equity investors may have had regarding the opportunistic and non-specialist nature of the assets acquired and again reflects the first mover advantage in terms of management teams being able to dictate structure;
- residential and tourism assets in REITs – whereas the asset base of developed REIT markets tends to focus on office, retail and industrial properties, Spain is notable for having an element of residential and hotels in the portfolios of its REITs reflecting the importance that tourism plays in the domestic economy; and
- developing specialization – although starting life as "cashboxes", all the IPOs have subsequently developed an asset specialization and management style. It is only possible to do this against a background of a resurgent direct property market and a shortage of investible listed opportunities.

In conclusion, Spain exhibited all the key characteristics that were essential in driving the growth of REITs in the US into the Modern REIT Era in 1992–1993, namely;

- legislative change (in this case to the REIT structure);
- a debt-fuelled commercial real estate bust preceding it; and
- structurally underweight institutions (the listed sector had been decimated by the global financial crisis).

Spain is now moving into the second phase of REIT sector development, with corporate growth driven less by acquisitions and more by improving property fundamentals.

10.5.2.3 Summary – REITs in Spain

The case study of REITs in Spain, being a developing REIT sector in the European region, shows how quickly a market may develop when the legislative structure allows considerable flexibility. The absence of limits for ownership percentages and restrictions on leverage provide scope for growth, while maintaining the conventional distribution of income parameters and tax benefits that are a familiar feature of REIT legislation internationally, resulting in the successful migration of property assets from banks to equity investors.

It is particularly interesting that the Spanish REIT sector started with a prevalence of private equity remuneration structures for external management teams, only to revert to generally accepted international bases over time. While starting with private equity–style remuneration structures may be attractive to those seeking to establish a REIT sector in an emerging market such as Poland, it would be prudent to heed the lessons from Spain.

10.5.3 REITs in Poland

This chapter analyses REITs in the European region, using case studies from the UK (being a developed REIT sector in the region), Spain (being a developing REIT sector in the region) and Poland (being an emerging REIT sector in the region).

Poland does not currently have a REIT regime, but it does tick most of the boxes for the successful introduction of a REIT regime. It is the most developed financial market across Central and Eastern Europe with the largest stock exchange in the region. It has a healthy provision of assets with prime properties accounting for around 5–7% of stock. There is also an underutilized potential of sourcing financial assets from Polish households amounting to more than €400 billion.

In this context, a Polish REIT regime is long overdue. It would give a considerable boost to the market, activating local capital and drawing in foreign investors. In addition, Polish REITs could be an attractive tool for commercial real estate projects in the country.

On 14 October 2016, the Polish Ministry of Finance published the first draft of a bill implementing the Real Estate Investment Trust (REIT) as a new corporate income tax (CIT) exempt vehicle for certain real estate investments. This follows the Ministry's announcement earlier in the year.

According to the Ministry, the proposal is aimed at attracting investors who to date have not entered the Polish real estate market or, due to lack of sufficient local incentives, operated via other investment structures. The REIT is also envisaged as a platform to allow smaller investors to participate in the commercial real estate market, despite high natural entry thresholds in this market.

10.5.3.1 Expected key features

The key features of the Polish REIT market are expected to be as follows:

- shareholder requirements:
  - minimum capitalization of PLN60m (approximately US$15m);
- listing requirement:
  - stock to be listed on Warsaw Stock Exchange (minimum free float threshold not set yet);
- asset level tests:
  - at least 70% of assets are to be comprised of real estate or shares in real estate subsidiaries (that must also meet certain criteria);
  - at least 70% of revenue to arise from real estate lease or sale;
  - revenue earned by a REIT must be generated from at least three properties;
- leverage:
  - liabilities cannot exceed 70% of total assets;

- profit distribution obligations:
  - at least 90% of profit to be derived from real estate lease or sale, sale of stock of other REITs or dividends from real estate subsidiaries that must be repatriated as dividend. Alternatively, such profit can be used to acquire real estate or shares/stock (at least 95%) of a joint stock company, limited liability company or a joint stock partnership where at least 70% of assets are real estate; and
- tax treatment at the REIT level:
  - according to the bill, a REIT is exempt from CIT on income from the lease and sale of real estate as well as the sale of shares of other REITs and so-called real estate subsidiaries (that must also meet certain criteria). Real estate subsidiaries that meet such criteria and are 95% held by REITs can also enjoy a tax exemption on certain real estate-related sources of income.

The bill demonstrates a positive intention to offer new incentives for organized forms of real estate investment. At this stage, the bill remains silent on certain aspects which would be desirable for more comprehensive regulation.

10.5.3.2 Market commentary and issues

Poland has a number of issues to address in order for it to be able to create a successful REIT market, including:

- improving property fundamentals – unlike the Spanish or Irish market, the Polish REIT market is being launched at a time when yields are at historically low rather than high levels. As a result investors will need confidence in the path of rental growth and the impact of potential yield expansion. In addition there are a number of concerns regarding the level of over-supply in Warsaw offices;
- developing a strong domestic investor base – historically, a lot of the investment activity in Poland has been from overseas investors. This is regarded as increasing the risk to capital values, if capital flows were to reverse; and
- developing Warsaw as a capital market – investors have previously accessed Polish exposure by cross border companies, be they Austrian, French or Israeli.

Whilst Poland does have a number of positive characteristics, it faces a number of challenges in attracting institutional capital if it is to develop critical mass in the REIT market.

10.5.3.3 Summary – REITs in Poland

The case study of REITs in Poland, being an emerging REIT sector in the European region, serves to illustrate how the basic pre-requisites for a REIT market are now so well established internationally than the suitability of an emerging market can be easily determined.

A well-established equity market, with access to both assets and capital and nascent legislation that broadly accords with the features of legislation in other REIT jurisdictions around the world, tempered with a prudent leverage limitation, should provide a sound basis upon which a REIT market may emerge in Poland.

10.6 Summary and conclusion

Part II includes six chapters each analysing REITs in a region of the world, using case studies from a developed, developing and emerging REIT sector in the region.

Chapter 8 analysed REITs in the North American region, using case studies from the USA, Canada and Mexico, with Chapter 9 analysing REITs in the Latin American region, using case studies from Brazil, Argentina and Uruguay for the developed, developing and emerging REIT sectors in the region, respectively.

This chapter analysed REITs in the European region, using case studies from the UK (being a developed REIT sector in the region), Spain (being a developing REIT sector in the region) and Poland (being an emerging REIT sector in the region).

Europe has a long history of successful REIT markets and, in the UK and France, the REIT structure is the dominant one. Moss (2012) highlights the relationship between a good track record in developing new REIT markets and prior banking crises, the most recent examples being Spain and Ireland. Future growth is expected to come from both new markets, such as Italy and Poland, as well as the development of existing markets in Germany, rather than the mature markets of the UK and France.

While the UK REIT sector is classified as developed, it is only just over a decade old, having achieved the listing of 36 REITs. In similar relativity, the developing REIT sector in Spain is only 5 years old and already has five major REITs listed on the main market, with many more on the secondary market. Accordingly, for an emerging market like Poland, the trajectory for growth is potentially steep if suitable conditions prevail.

While Spain benefitted from good timing, with a debt-fuelled commercial real estate bust preceding the introduction of enabling REIT legislation at a time when institutions were also structurally underweight equities due to the GFC and the UK already had a substantial property company and unlisted fund market upon which to build, Poland currently lacks either as the foundation for growth at the current time. Economies and markets do, however, often change unexpectedly rapidly.

Given the size of the existing property company and unlisted fund market pre-REITs in the UK, it is interesting that REITs do not appear to have gone through the initial diversified portfolio phase, commonly found in other regions, but have moved swiftly into sector-specific REITs such that further broadening into other sectors may be anticipated as the market evolves. Similarly, it is in interesting that Spain included residential and hotel properties in diversified portfolios from the commencement of the REIT sector rather than simply holding the traditional office, retail and industrial properties. Such precedents give Poland the choice of either the traditional diversified route or a sector specializing route or both as it's REIT market emerges.

Chapter 11 will analyse REITs in the South East Asian region, using case studies from Singapore, Malaysia and Thailand with Chapter 12 analysing REITs in the North Asian region, using case studies from Japan, Hong Kong and China. Chapter 13 will analyse REITs in the Oceania region, using case studies from Australia, South Africa and India and Chapter 14 will conclude Part II by considering *Directions for the Future of International REITs*.

References

Devaney, S., Xiao, Q. and Clacy-Jones, M. 2013, *Listed and direct real estate investment: A European analysis*, 20th Annual European Real Estate Society Conference, Vienna.

EPRA 2016, *Global REIT survey September 2016*, EPRA, Brussels.

EPRA 2017, *Statistical bulletin February 2017*, EPRA, Brussels.

EPRA and INREV 2016, *Real estate in the real economy*, EPRA, Brussels.

Hoesli, M. and Oikarinen, E. 2012, 'Are REITs real estate? Evidence from international sector level data', *Journal of International Money and Finance*, Vol. 31, No. 7, pp. 1823–1850.

Kroencke, T., Schindler, F. and Steinenger, B.I. 2015, *Are REITs real estate or stocks? Dissecting REIT returns in an asset pricing model*, EPRA, Brussels.

Lee, C.L. and Lee, M.L. 2014, 'Do European real estate stocks hedge inflation? Evidence from developed and emerging markets', *International Journal of Strategic Property Management*, Vol. 18, No. 2, pp. 178–197.

Moss, A. 2012, *Growing the European listed real estate market: What are the reasons behind the relative decline of the sector, and can we learn any lessons from the factors behind the explosion of the US REIT sector in the 1990s*, EPRA, Brussels.

Moss, A. and Baum, A. 2013a, *Are listed real estate stocks managed as part of the real estate allocation?* EPRA, Brussels.

Moss, A. and Baum, A. 2013b, *The use of listed real estate securities in asset management*, EPRA, Brussels.

11
SOUTH EAST ASIA

Liow Kim Hiang and Huang Yuting

11.1 Introduction

This book aims to identify key areas for research in the REIT discipline for the next five to ten years by surveying the current state of the REIT discipline around the world and identifying emerging and cutting edge research areas through a thematic review of current contextual issues and a regional analysis based on case studies.

This book comprises two parts, the first part being a thematic review of emerging and cutting edge global research into current contextual issues in REITs internationally and the second part being a regional analysis of REITs around the world, each written by authoritative academic authors from the world's leading Universities and REIT industry experts.

Part I included six chapters each reviewing a current theme of REIT evolution through the lens of contemporary research. Chapter 1 focused on critical contextual issues in international REITs while Chapter 2, the *Post-Modern REIT Era*, examined the evolution of the global REIT market, what is deemed to be best current market practice and how the market is expected to change going forward and Chapter 3, *Emerging Sector REITs*, investigated such sectors as residential, health care, self-storage, timber, infrastructure and data centres generally and then specifically through a case study of US REITs.

Chapter 4, *Sustainable REITs*, analysed REIT environmental performance and the cost of equity and Chapter 5, *Islamic REITs*, examined the evolution of global Islamic REITs through a study of the Malaysian market. Chapter 6, *Behavioural Risk in REITs*, addressed the management of behavioural risk in global REITs and Part I concluded with Chapter 7 which reviewed recent research into *REIT Asset Allocation*.

Part II includes six chapters each analysing REITs in a region of the world, using case studies from a developed, developing and emerging REIT sector in the region. Chapter 8 analysed REITs in the North American region, using case studies from the US, Canada and Mexico, with Chapter 9 analysing REITs in the Latin American region, using case studies from Brazil, Argentina and Uruguay and Chapter 10 analysed REITs in the European region, using case studies from the UK, Spain and Poland for the developed, developing and emerging REIT sectors in the region, respectively.

This chapter analyses REITs in the South East Asian region, using case studies from Singapore (being a developed REIT sector in the region), Malaysia (being a developing REIT sector in the region) and Thailand (being an emerging REIT sector in the region) as well as analysing their relationships with major REIT markets in an international environment.

This chapter goes on to analyse the investment dynamics of the Singapore (SGREIT) and Malaysia (MAREIT) markets in the context of modern portfolio theory, as well as globalization and integration of financial markets. The Thai-REIT market is excluded from this part of study due to insufficient data observations required for MGARCH analyses. The key research question considered is whether a globalization and integration process is taking place in global REIT markets. This question is particularly relevant to policymakers and regulators because increasing integration between local REIT markets and regional/international REIT markets, as well as increasing integration between local REIT markets and local/regional/global stock markets, is a pre-condition for REIT market growth due to the internationalization of investment capital. This is why the return co-movement, volatility spillover and return causality of the two REIT markets (SGREIT and MAREIT) with the US, Australia and Japan will be investigated using a MGARCH and asymmetric dynamic conditional correlation (ADCC) framework. The chapter also assesses to what extent the two South East Asian REIT markets are integrated with the regional/global stock markets of the US, Japan and Australia using the same MGARCH framework.

Chapter 12 then analyses REITs in the North Asian region, using case studies from Japan, Hong Kong and China with Chapter 13 analysing REITs in the Oceania region, using case studies from Australia, South Africa and India, before Chapter 14 concludes Part II by considering *Directions for the Future of International REITs*.

11.2 Singapore REITs (SGREITs)

This chapter analyses REITs in the South East Asian region, using case studies from Singapore (being a developed REIT sector in the region), Malaysia (being a developing REIT sector in the region) and Thailand (being an emerging REIT sector in the region).

In just 14 years since its inception in 2002, the Singapore REIT (SGREIT) industry now ranks third in the Asia-Pacific region, behind Japan and Australia. The sector has grown to 43 REITs with a total market capitalization of about US$50.3 billion. It is generally accepted that Singapore's clear regulatory regime and support from the government are among the key factors contributing to the sector's growth.

Many of the SGREITs undertake active asset management, improvement and enhancement in all sectors of the property investment market including retail, office and industrial. SGREIT property holdings comprise retail (30%), industrial (25%), office (16%), diversified (13%), hotel (9%), health care (4%) and residential (3%) (Atchison, 2014). About 30% of the properties held by SGREITs are located outside Singapore.

The idea of creating a REIT in Singapore was first mooted by the Property Market Consultation Committee in 1986. In July 2002, CapitaMall Trust (CMT) was successfully launched by CapitaLand after an abortive attempt earlier. The first attempt by CapitaLand to launch an IPO of SingMall Property Trust (SPT), in November 2001, was unsuccessful. This saw SGREITs established as the second REIT market in Asia, after REITs in Japan. With the regulatory body, the Monetary Authority of Singapore (MAS)'s continuing efforts to encourage the growth and attractiveness of REITs through the implementation of various supportive and conducive regulatory changes, the SGREIT market has grown in breadth and depth.

> The pro-business and supportive regulatory regime in Singapore have laid a solid foundation for the REIT industry and created an environment conducive to CMT's

growth. The regulator's continual efforts to refine and enhance the SGREIT regime in Singapore has further helped improve the professionalism of the sector and promote the attractiveness of SGREITs. The results: a growing local and international investor base and an expanding pool of industry talent.

(*Mr. Wilson Tan, CEO of CMT Management, Business Times, 31 January 2017*)

11.2.1 SGREIT market structure and key requirements

The following comprise the requirements for the SGREIT market structure, principal activities, restrictions, regulation and taxation (EPRA, 2016b):

- SGX-listed SGREITs must be constituted as a trust and have a minimum initial capital of at least SG$300 million. For SGD-dominated REITs listed on the SGX, at least 25% of the capital must be held by at least 500 public unit holders;
- a SGREIT may be managed internally or externally. As of December 2016, all are externally managed;
- SGREITs may invest in real estate (freehold or leasehold, in or outside Singapore), real estate-related assets, debt securities (listed or unlisted) and listed shares of non-real estate companies/ government securities and cash/cash equivalent assets. They may invest in real estate by way of direct ownership or a shareholding in an unlisted real estate special purpose vehicle (SPV) and joint-venture. The SPV can take the form of a company, trust or partnership;
- all SGREITs are subject to a leverage limit of 45%;
- SGREITs are also subject to some restrictions on their investment activities, with at least 75% of the property portfolio to be income-producing real estate and no more than 10% property development activities or investing in unlisted property development companies is to be undertaken – although this 10% limit of the SGREIT's property can be increased to 25% subject to the SGREIT meeting certain conditions;
- SGREITs need not be listed, but only a REIT that is listed on the SGX is eligible for tax concessions. All eligible rental income is exempted from tax at the REIT level. No tax is imposed on capital gains. However, gains would be taxed at the REIT level at the prevailing corporate tax rate if the REIT is trading in properties. Additionally, foreign-sourced income of SGREITs may qualify for tax exemption. However, no foreign withholding tax refunds are allowed in respect of tax-exempted income;
- at the individual unitholder level, all distributions made by SGREITs are exempt from Singapore income tax. Gains realized on the sale of SGREIT units by an individual unit holders are not taxable. Finally, distributions to domestic unit holders are not subject to withholding tax; and
- although there are no legal or regulatory requirements for an SGREIT to distribute any pre-determined percentage of its income as distributions for a given financial year, in order to enjoy tax transparency treatment, for Singapore properties, a SGREIT is required to distribute at least 90% of its "tax transparent income" in cash or in the form of units, subject to meeting certain financial conditions.

11.2.2 Principal active SGREITs

Table 11.1 lists the top eight SGREITs as at December 2016 with regard to their listing date, market cap and key performance indicators. These REITs cover three major property sectors in

Table 11.1 Principal Active S-REITS: December 2016

Name	*Nature*	*Listing Date*	*Mkt Cap (USD m)*	*Dividend Yield*	*Return on Equity(%)*	*P/E Ratio*
CapitaMall Trust	Retail	7/17/2002	4622.87	5.90	7.02	11.10
Ascendas REIT	Indutrial	11/19/2002	4480.17	4.29	6.61	18.50
CapitaCommercial Trust	Office	5/11/2004	3035.88	5.88	5.01	15.40
Suntec REIT	Office, Retail	12/9/2004	2897.12	6.16	4.73	12.10
Keppel REIT	Office	4/28/2006	2323.97	6.44	5.35	8.20
Mapletree Commercial Trust	Retail	4/27/2011	2770.99	5.27	11.33	10.00
Mapletree Logistic Trust	Industrial	7/28/2005	1765.10	7.24	7.29	16.50
SPH REIT	Retail	7/24/2013	1677.42	5.79	5.33	10.89

Sources: EPRA (2016a) and Various S-REIT Websites and Annual Reports

Singapore (office, retail and industrial) and their portfolios include properties from Singapore, China, Malaysia, Australia, Hong Kong, Japan, South Korea and Vietnam.

11.2.3 Summary – SGREITs

The Singapore approach to REIT development includes a significant role for government in regularly improving the operating and regulatory environment for SGREITs, so that they can continue to grow while at the same time operating with good corporate governance and embracing the best industry practices. The supportive regulatory structure, legislation and taxation have contributed to the initial and current success of SGREITs, which have great potential for future development.

In this approach, SGREITs have provided an opportunity for small investors to access real estate in a liquid and diversified form. This has seen the SGREIT market cap increase from US$0.71 billion in 2002 to US$50.3 billion in 2016 (Figure 11.1).

However, what is the next stage for SGREITs? As highlighted by Brounen and de Koning (2012), one option is for SGREITs to expand their engagement in property management and development. In this regard, the Tax Modernization Act of 1986 paved the way for the large-scale adoption of REITs by institutional investors. Another direction would be for the MAS to introduce an alternative REIT structure whereby an SGREIT does not need to own properties, similar to an umbrella REIT (UPREIT) structure in the US.

A challenge for the MAS is to gradually introduce more supportive regulations to facilitate the transformation of some local real estate entities from a "property company" corporate structure to a "REIT" trust form, so that the local securitized real estate industry is able to move more cohesively in the "REIT" direction and develop the popularity of the trust structure in the Singapore securitized real estate sector.

Overall, SGREITs must continue to grow horizontally, as well as vertically, in order to maintain and improve competitive advantage over other regional markets and allow Singapore to position itself as the pre-eminent South East Asian hub for REIT listing, thus advancing the goal of establishing Singapore as a fund management and asset management hub in the region.

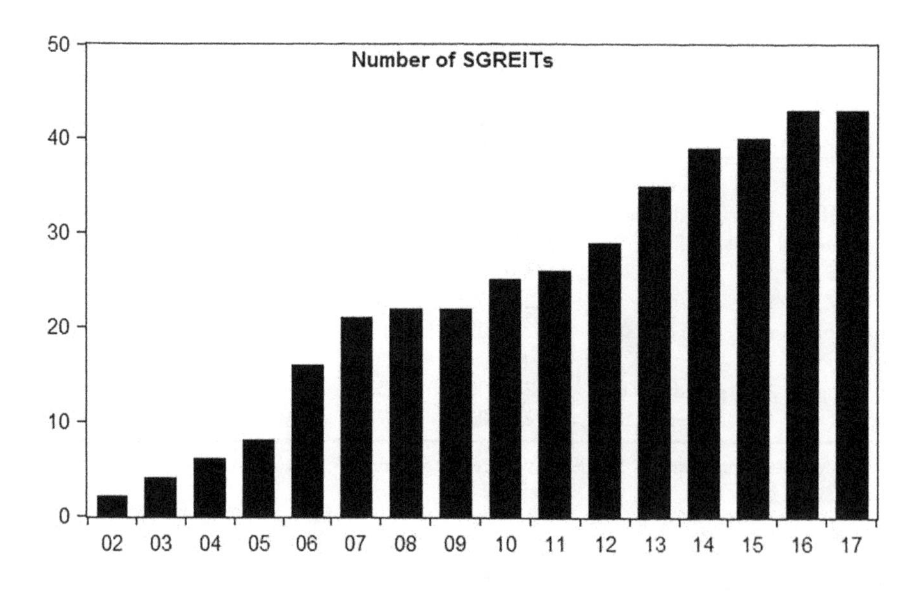

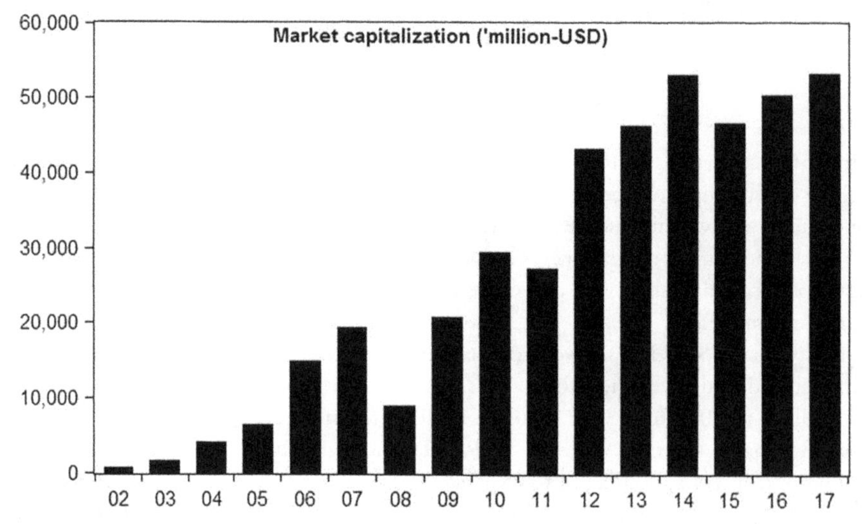

Figure 11.1 Development of SGREITs

Source: Author Analysis from Datastream S&P REITs

11.3 Malaysian REITs (MAREITs)

This chapter analyses REITs in the South East Asian region, using case studies from Singapore (being a developed REIT sector in the region), Malaysia (being a developing REIT sector in the region) and Thailand (being an emerging REIT sector in the region).

REITs are developing as an important investment vehicle in the Malaysian economy, with the Malaysian REIT (MAREIT) market providing a common platform for the existence of both conventional and Islamic REITs (Wong, 2016). Prior to 2005, REITs existed in the form of listed property trusts (LPTs) in Malaysia (Newell and Osmadi, 2010). Since the debut of the first MAREIT (Axis REIT) in August 2005, the total market capitalization of the MAREIT sector has grown from about RM1.8 billion at the end of 2005 to RM23.5 billion as of December 2016, being around a 13-fold increase and positioning the MAREIT market as one of the leading and developing Asian REIT markets.

Further, the MAREIT market has matured over the years, especially since the global financial crisis (GFC) of 2008–2009, with 16 MAREITs listed on the Kuala Lumpur Stock Exchange (KLSE) having a combined market capitalization of about US$5.76 billion at December 2016.

11.3.1 MAREIT market structure and key requirements

The following comprise the requirements for the MAREIT market structure, principal activities, restrictions, regulation and taxation (EPRA, 2016a):

- MAREITs take the form of a unit trust fund which must be registered in Malaysia and approved by the Securities Commission (SC), with a minimum fund size of RM100 million;
- the trust must be managed and administered by a management company approved by the SC and foreigners can hold up to 70% of the equity of the management company. Furthermore, real estate held by the MAREIT must be managed by a qualified property manager approved by the trustee;
- in addition to the main components of a conventional MAREIT, Wong (2016) reports that an Islamic REIT in Malaysia must maintain Shariah advisors who comprise learned scholars in Shariah law and custom to provide Shariah-related advice to the REIT manager on Islamic asset management activities in order for the REIT to be Shariah compliant;
- MAREITs may only invest in real estate, real estate-related assets, non-real estate-related assets, single-purpose real estate owing companies and cash, deposits and money market instruments. At least 50% of the MAREIT's total asset value must be invested in real estate and/or single-purpose real estate companies, with the remaining two types (non-real estate and cash etc.) not exceeding 25% of the REIT's total asset value. All REITs may acquire foreign real estate assets where it is viewed as a viable investment;
- for Islamic REITs, although they are permitted to acquire real estate, the fund manager must ensure that the rental income from non-permissible activities under Shariah Law does not exceed 20% of the total turnover of the Islamic REIT;
- borrowing may not exceed 50% of the total asset value, with MAREITs permitted to issue debentures;
- MAREITs are not required to make any minimum distribution of income but they will only benefit from a tax exemption if at least 90% of their total income for the year of assessment is distributed to investors; and
- MAREITS are tax-exempt at the level of the MAREIT if 90% of total income is distributed. There is no capital gains tax in Malaysia, except for real property gains tax (RPGT) on disposal of real properties or shares in real property companies. Further, the disposal of REIT units is not subject to RPGT.

11.3.2 Summary – Malaysian REITs

Over the decade since 2005, the MAREIT market has grown significantly, which clearly endorses the popularity of MAREITs and their potential in Malaysia (Wong, 2016) (Figure 11.2).

In addition to the market issues for conventional MAREITs in Malaysia, there are also issues arising from the co-existence of conventional and Islamic MAREITs since there are some

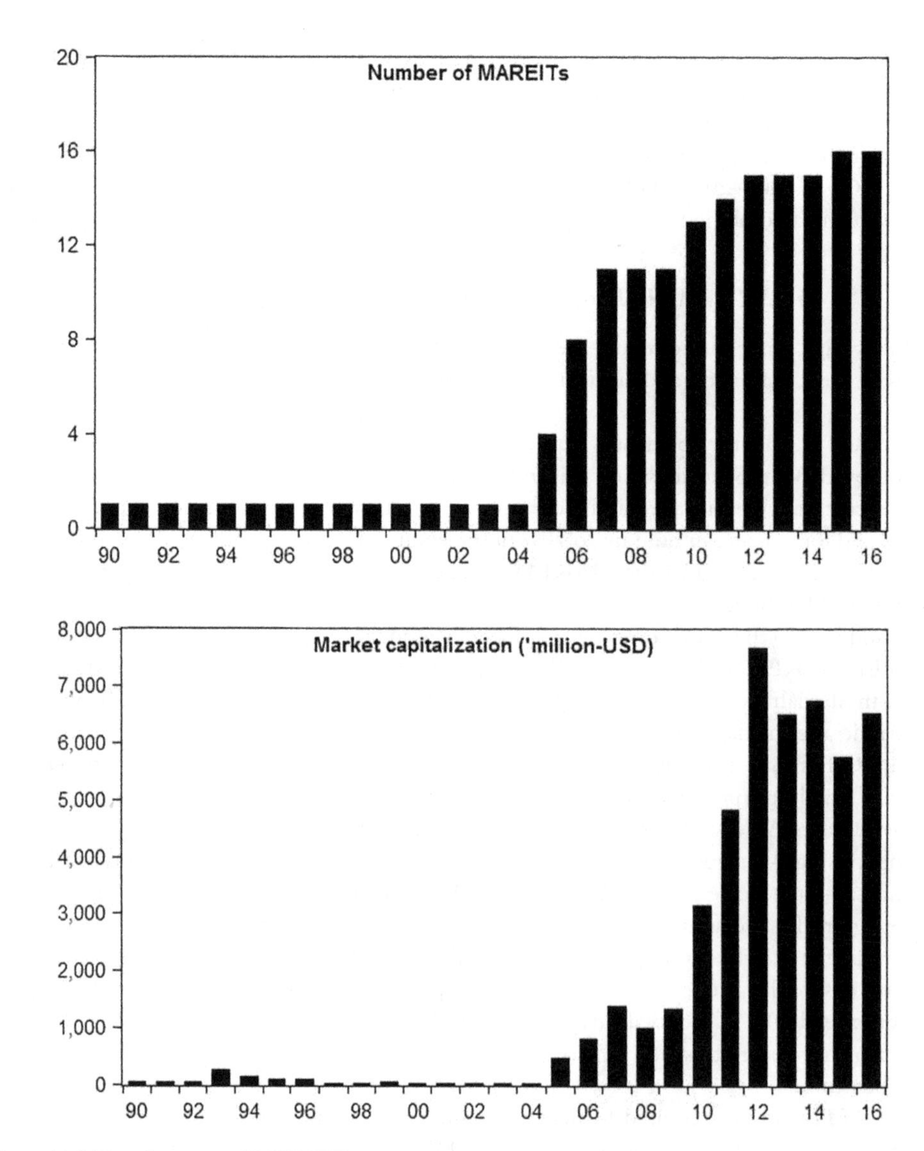

Figure 11.2 Development of MAREITs

Source: Author Analysis from Datastream S&P REITs

fundamental differences between the MAREIT types. One example of such difference is that conventional REITs are more sensitive to changes to long-term yields than Islamic REITs, which are more sensitive to changes in medium-term yields (Wong, 2016).

11.4 Thailand REITs (Thai-REITs)

This chapter analyses REITs in the South East Asian region, using case studies from Singapore (being a developed REIT sector in the region), Malaysia (being a developing REIT sector in the region) and Thailand (being an emerging REIT sector in the region).

In Thailand, REITs (Thai-REITs) were originally known as a Listed Property Fund for Public Offering (PFPO), with the first Thai-REIT, UOB Apartment Property 1 Leasehold, listed on the Thailand Stock Exchange (SET) in 2003. Pham (2011) notes that the PFPO regulatory code was more rigid than other REIT regulations overseas, with Thai-REIT investment confined to properties in Thailand only as well as many restrictions regarding leverage and institutional holding. An important change, according to Pham (2011), is that the maximum borrowing was raised to 50% of NAV, as opposed to 10% under the PFPO code.

On 11 October 2010, a new REIT regulatory framework was adopted by the Securities and Exchange Commission of Thailand (SEC) which follows international practices and provides greater flexibility than the PFPO code, being expected to make Thai-REITs more attractive than under the PFPO regime. Further, no new PFPO can be set up from 1 January 2014, neither will an existing PFPO be allowed to extend its size thereafter.

11.4.1 Thai-REIT market structure and key requirements

The following comprise the requirements for the Thai-REIT market structure, principal activities, restrictions, regulation and taxation (EPRA, 2016c):

- Thai-REIT is a trust structure which can invest in overseas real estate and can borrow up to 60% of total assets if rated as investment grade. Minimum capital of Baht 500 million is required with listing on the SET mandatory;
- Thai-REIT can be established and managed by a REIT manger (RM), which may be an asset management company (AMC) or a qualified person, through a public offering (PO). The trustee is responsible to monitor the activities of the RM;
- the minimum number of unit holders for Thai-REITs is 250 for an IPO and 35 after listing. Additionally, former property owners and related persons shall not acquire more than 50% of the total units sold in each tranche. Further, at least 15% of the units must be held by public investors in each tranche;
- a Thai-REIT must invest at least 75% of net asset value (NAV) in income-producing properties. Additionally, the fund is allowed to invest in projects under construction (green field projects) up to 10% of the NAV. Further, no restriction on type of real estate investment is imposed for investment overseas. Acquisition and disposal prices must be based on the appraised value with properties revalued every two years;
- borrowing by a Thai-REIT is allowed but not more than 35% of its total assets, extended to 60% if rated as investment grade;
- for operative income, at least 75% of the total income must be generated from the rental income. At least 90% of the net profit must be distributed to unit holders within 90 days after the end of each annual accounting period. Also, a Thai-REIT must distribute at least 90% of capital gains and only keep a maximum of 10% of the net profit as retained earnings; and
- a Thai-REIT is not subject to income tax, but a 12.5% land and building tax on the rental income from immovable properties will be imposed. A Thai-REIT is not subject to withholding tax and is tax-exempt for capital gains.

11.4.2 Summary – Thai-REITS

Figure 11.3 shows development in the Thai-REIT market following changes in the regulatory structure, with in-depth analysis similar to that for SGREITs and MAREITs possible as Thai-REITs mature further in years to come.

With Thai-REITs moving towards internationally accepted practice with a less restrictive regime since late 2012, it is expected that they will offer new opportunities and challenges for international investors (Pham, 2011), with Thai-REITS becoming more popular in portfolios since the global financial crisis.

Interestingly, Thailand's Impact Growth REIT, controlled by property developer Bangkok Land, announced on 18 September 2014 that its retail offering hand been fully subscribed in

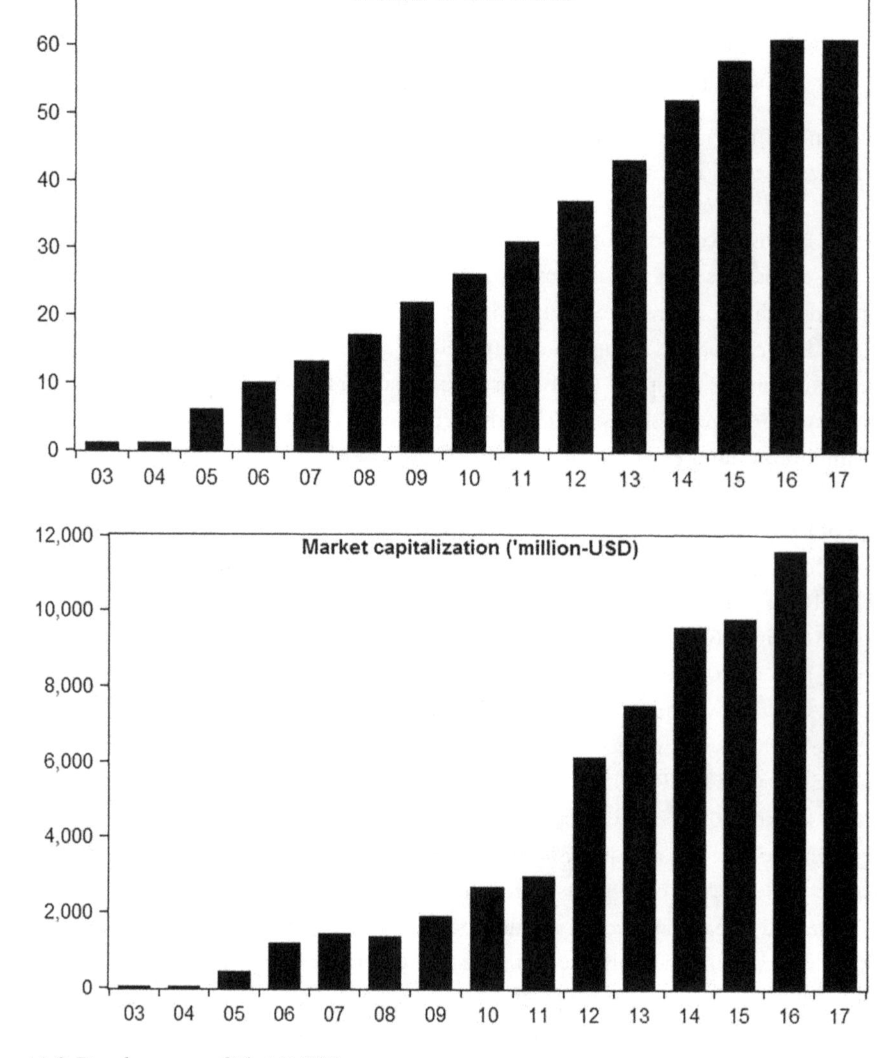

Figure 11.3 Development of Thai-REITs

Source: Author Analysis from Datastream S&P REITs

just one day and was ultimately 4.1 times oversubscribed. The Baht 15.7 billion (US$488 million) offering was priced at Baht 10.60, the top of the original estimated range. It was the largest REIT initial public offering in Asia outside Japan in 2014. Impact Growth REIT owns Bangkok's Muang Thong Thani complex, the largest indoor exhibition and convention centre in Southeast Asia.

11.5 Country-level analysis: relative performance and investment dynamics analysis

The risk-adjusted performance of the three South East Asian REIT markets may be assessed by comparison to their domestic stock, property stock and bond market performance. Accordingly, the monthly returns for local Standard and Poor (S&P) REITs, S&P real estate stocks, S&P general stocks and ten-year government bonds and a risk-free rate (three-month Treasury-Bill discount rate) over the relevant period were extracted from Datastream. The start of the analysis for each market was chosen based on the first available month of the S&P REIT series concerned.

The time series commenced from August 2003 for S-REITs, December 2006 for M-REITs and September 2012 for Thai-REITs and the end month for the analysis is November 2016. All data are collected in local currency terms to maintain the consistency and equivalency of the analysis. Figure 11.4 provides some usual descriptive statistics for the monthly returns for three countries' REITs.

The performance analysis covers the measurement of both return and risk. Risk-adjusted returns are assessed using the coefficient of variation (CV = risk/return) and Sharp ratio (SR) (excess return per unit of risk). Portfolio diversification benefits are assessed by the correlation of REITs, shares, property companies and bonds. Finally, for SGREITs and MAREITs, the analysis was repeated for the global financial crisis period, defined as from August 2007 to March 2009,

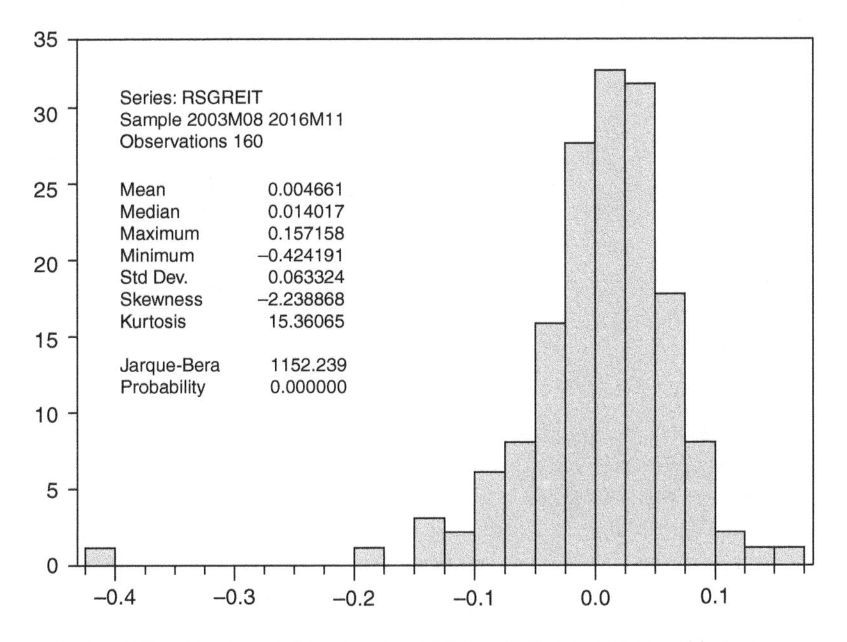

Figure 11.4 Statistical Distribution of SGREIT, MAREIT and Thai-REIT Monthly Returns

Source: Author Analysis from Datastream S&P REITs

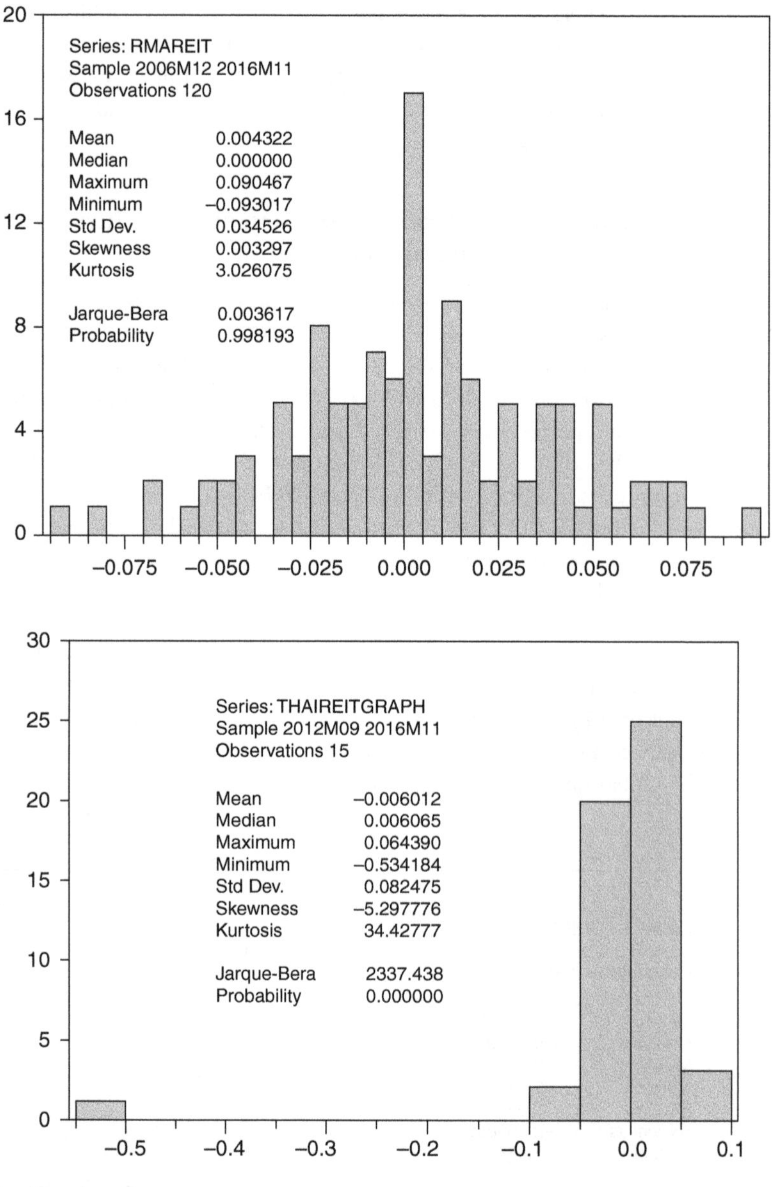

Figure 11.4 (Continued)

in order to assess the dynamic interactions and effects of the global financial crisis on the performance of the respective investment sectors.

The adoption of a global financial crisis period from August 2007 to March 2009 follows that defined by the Federal Reserve Board of St Louis (Federal Reserve, 2009), which identified the first three phases of the global financial crisis from 2 August 2007 to 31 March 2009. Phase 4 of the global financial crisis period (1 April 2009 to 4 November 2009 – "stabilization and tentative signs of recovery") is not included as it indicated a tentative recovery period for the global economy. Finally, Thailand is excluded from the global financial crisis analysis due to lack of Thai-REIT return data.

Excluding cases of negative returns, the four investment types are ranked from 1 (best performance) to 4 (worst performance) for each indicator, which are then totalled to derive the overall ranking. Table 11.2 reports the performance analysis results for SGREITs, property companies, shares and bonds from 8/2003 to 11/2016 (full period), as well as for the GFC period. During the full study period (Panel A), SGREITs achieved higher monthly returns (0.47%) than bonds (0.25%) and shares (0.32%), but were below that seen for property companies (0.52%). SGREITs have a lower risk level (6.33% per month) compared to that of property stocks (6.80%). On a risk-adjusted basis, both CV and SR indicate that S-REITs outperformed shares, but underperformed property stocks and bonds. Based on the four criteria, it may be concluded that SGREITs ranked third over the full study period (i.e. they outperformed shares but underperformed property stocks and bonds).

With the exception of bonds, the effect of the global financial crisis was evident across the other three asset classes (Panel B). For SGREITs, the average monthly return went down from 0.47% to -5.04% with the risk level rising by about 86%, which is higher than that of property stocks (69% increase) and common stocks (85% increase).

Panel C provides the unconditional correlation matrices for the four asset types over the full period (below the diagonal) and during the global financial crisis period (above the diagonal).

Table 11.2 SGREIT Performance Analysis

Panel A: Full period (08 2003 11/2016)				
	SREITs	*Property stocks*	*General stocks*	*Bonds*
Average monthly returns	0.47% (2)	0.52% (1)	0.32% (3)	0.25% (4)
Monthly risk	6.33% (3)	6.80% (4)	5.90% (2)	0.05%(1)
Coefficient of variation	13.585 (3)	13.001 (2)	18.707 (4)	0.215(1)
Sharpe ratio	0.058 (3)	0.062 (2)	0.037 (4)	2.161 (1)
Total ranking index	11	9	13	7
Panel B: GFC period (08/2007–03/2009)				
	SREITs	*Property stocks*	*General stocks*	*Bonds*
Average monthly returns	-5.04%	-5.49%	-4.01%	0.22%
Monthly risk	11.76%	11.50%	10.91%	0.06%
Coefficient of variation	-2.322	-2.096	-2.783	0.444
Sharpe ratio	-0.437	-0.485	-0.374	3.101
Panel C: Correlation matrices: (08/2003–11/2016)/(08 2007–03/2009)				
	SREITs	*Property stocks*	*General stocks*	*Bonds*
SREITs	1.000	0.896***	0.872***	-0.0.404*
Property stocks	0.873***	1.000	0.869***	-0.406*
General stocks	0.834***	0.907***	1.000	-0.379*
Bonds	-0.047	-0.035	-0.046	1.000

Notes: In Panel A: numbers in brackets indicate relative ranking, where one (1) indicates the best performance and four (4) indicates the worst performance. In Panel C, numbers below the diagonal are the respective full period (8/2003–11/2016) correlations; numbers above the diagonal are the respective GFC period (8/2007–11/2016) correlation. ***, **, * – indicates statistical significance at the 1%, 5% and 10% levels.
Source: Authors

Over the full period, it may be observed that SGREITs were strongly correlated with property stocks (correlation = 0.873) and shares (correlation = 0.834), implying relatively little diversification benefit for both SGREITs and property companies with the local stock market over this period.

In contrast, the results indicate that bonds offered diversification benefits with SGREITs, property stocks and common stocks over the same period as the respective correlation estimates were insignificantly negative. During the global financial crisis period, the correlations between SGREITs and property stocks, as well as between SGREITs and common stocks, rose by 2.6% and 4.6%, respectively, indicating a further reduction (although not significant) in diversification benefits in the Singapore investment markets during the crisis period. At the same time, bonds become more negatively correlated with SGREITs, property stock and stock markets.

Table 11.3 provides the equivalent analysis for Malaysia, where MAREITs emerged as the second-best performer after bonds over the full period from 12/2006 to 11/2016. During

Table 11.3 MAREIT Performance Analysis

Panel A: Full period (12/2006–11/2006)

	M-REITs	*Property stocks*	*General stocks*	*Bonds*
Average monthly returns	0.43% (1)	0.39% (2)	0.39% (2)	0.32% (3)
Monthly risk	3.45% (2)	6.16% (4)	4.36% (3)	0.03% (1)
Coefficient of variation	7.988 (2)	15.695 (4)	11.235(2)	0.080(1)
Sharpe ratio	0.049 (2)	0.021(4)	0.028 (3)	2.105(1)
Total ranking index	7	14	10	6

Panel B: GFC period (08/2007–03/2009)

	M-REITs	*Property stocks*	*General stocks*	*Bonds*
Average monthly returns	-1.91%	-3.31%	-2.36%	0.33%
Monthly risk	3.46%	7.83%	6.05%	0.04%
Coefficient of variation	-1.776	-2.368	-2.565	0.120
Sharpe ratio	-0.651	-0.460	-0.439	0.678

Panel C: Correlation matrices: (12/2006–11/2016)/(08/2007–03/2009)

	M-REITs	*Property stocks*	*General stocks*	*Bonds*
M-REITs	1.000	0.361*	0.422**	-0.063
Property stocks	0.578***	1.000	0.593***	-0.371*
General stocks	0.538***	0.802***	1.000	-0.415*
Bonds	-0.153*	-0.228*	-0.230*	1.000

Notes: In Panel A: numbers in brackets indicate relative ranking, where one (1) indicates the best performance and four (4) indicates the worst performance. In Panel C, numbers below the diagonal are the respective full period (8/2003–11/2016) correlations; numbers above the diagonal are the respective GFC period (8/2007–11/2016) correlations. ***, **, * – indicates statistical significance at the 1%, 5% and 10% levels.
Source: Authors

Table 11.4 Thai-REIT Performance Analysis

Panel A: Full period (09/2012–11/2006)

	Thai-REITs	*Property stocks*	*General stocks*	*Bonds*
Average monthly returns	-0.60% (4)	0.48% (1)	0.38% (2)	0.26% (3)
Monthly risk	8.25% (4)	6.16% (3)	4.62% (2)	0.06% (1)
Coefficient of variation	-13.719(4)	12.754 (3)	12.273 (2)	0.221 (1)
Sharpe ratio	-0.093 (4)	0.052 (3)	0.046 (2)	1.600(1)
Total ranking index	16	10	8	6

Panel B: Correlation matrix: (09/2012–11/2016)

	Thai-REITs	*Property stocks*	*General stocks*	*Bonds*
Thai-REITs	1.000			
Property stocks	0.620***	1.000		
General stocks	0.571***	0.880***	1.000	
Bonds	-0.281**	0.037	0.040	1.000

Notes: In Panel A: numbers in brackets indicate relative ranking, where one (1) indicates the best performance and four (4) indicates the worst performance. In Panel B, numbers below the diagonal are the respective full period correlations. ***, **, * – indicates statistical significance at the 1%, 5% and 10% levels.
Source: Authors

the global financial crisis period, MAREITs were also not spared, however it appears that the adverse global financial crisis effects on MAREITs were less severe than those on SGREITs. Moreover, unlike SGREITs, MREITs were less (although still significant) correlated with property stock and common stock markets over the full period and global financial crisis period.

Finally, over the short period from 9/2012 to 11/2016, Thai-REITs provided the lowest monthly returns of -0.6% per month; they significantly underperformed property stocks (0.48%), common stock (0.38%) and bonds (0.26%). On a risk-adjusted basis, Thai-REITs emerged as the worst asset class, with a negative coefficient of variation value (-13.72) and a negative Sharpe ratio (-0.093). Similar to MAREITs, the correlation estimates show that Thai-REITs were moderately linked to property stocks (correlation of 0.62) and less to common stocks (correlation of 0.571). Finally, bonds were negatively correlated to Thai-REITs (-0.28), reflecting some defensive characteristics of Thai-REITs.

11.6 A multivariate analysis of international REIT markets

Volatility and correlation of asset returns are two significant inputs into the calculation of risk in modern portfolio theory, risk management and asset allocation. This section considers the issue of the globalization of REITs by evaluating their market integration in the context of return co-movements and volatility spillovers among the key global REIT markets from the US, Japan and Australia, as well as Singapore and Malaysia from 9 November 2006 to 24 November 2016 (Thailand is excluded from the analysis as it has only price data from July 2009 onward) using weekly returns.

The increasing economic globalization and greater correlation of financial markets, as reported in many stock market studies, raises a question as to whether a similar globalization process is taking place in international REIT markets, with Figure 11.5 displaying the time-series movement of five REIT indices over the full-study period.

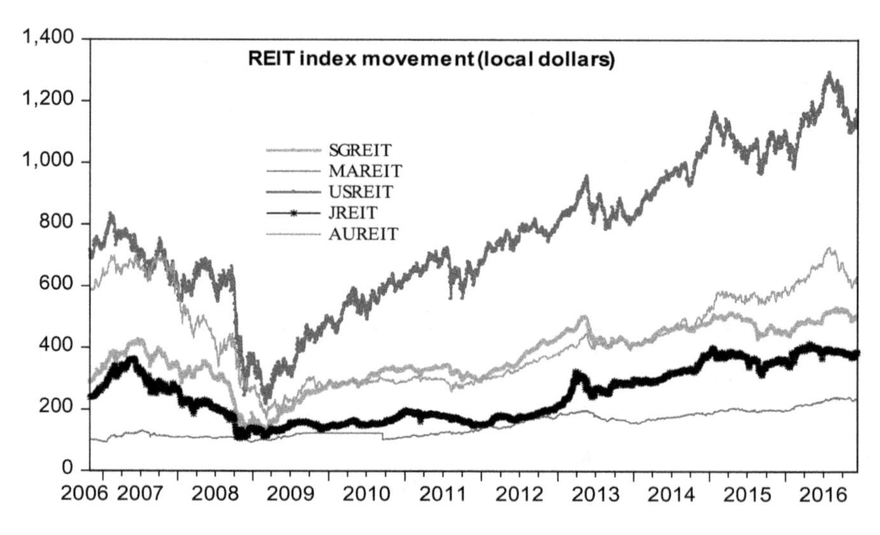

Figure 11.5 REIT Index Movement

Source: **Authors Analysis from Datastream S&P REITs**

In this section, a multivariate model with GJR-GARCH (Glosten, Jagannathan and Runkle, 1993) and asymmetric dynamic conditional correlation (ADCC) (Engle, 2002; Cappiello, Engle and Sheppard, 2006) is used for the five REIT markets' weekly indices return. First, it is hoped to capture the time-varying conditional correlations from a multivariate AR (1) -GJR-GARCH-ADCC model that takes into account that conditional variance is an asymmetric function of past innovations (leverage effect), which increases proportionately more during market declines. Liow et al. (2009) study international correlation and volatility dynamics of five major securitized real estate markets using monthly returns from 1984 to 2006. In the second stage of estimation, we employ the ADCC. An attractive feature of the ADCC is that it keeps the simple interpretation of the univariate GARCH models, as well as providing consistent estimates of the dynamic correlation matrix. Moreover, the model is useful to take into account the leverage effects of return volatility.

Table 11.5 reports the estimation results of the GJR-GARCH-ADCC model, including the univariate GJR-GARCH for each REIT market. In addition, most of the estimated ARCH and GARCH are statistically significant, which implies that the basic GARCH (1, 1) (i.e. without the GJR term) may be able to describe the weekly REIT return behaviour adequately. The DCC parameter (b) is statistically significant. Therefore, the assumption of constant conditional correlation is not supported empirically. Finally, the multivariate asymmetrical correlation term (g) is statistically insignificant for the five-REIT model, implying a symmetrical DCC model may be sufficient to characterize the correlation dynamics examined.

Figure 11.6 describes the evolution of the estimated DCC dynamics among seven pairwise REIT markets over the full period. As observed, all the correlation series shows a fairly stable trend. The mean correlation values are 0.199 (SG/MA). 0.385 (SG/US), 0.440 (SG/JP), 0.452 (SG/AU), 0.108 (MA/US), 0.143 (MA/JP) and 0.182 (MA/AU), implying that the market interdependence between the two SEA REIT markets with the three global REIT markets can be described as weak (MAREIT) and moderate (SGREIT). The most immediate implication of the DCC model in this case is that there is a moderate (weak) interaction between the returns of SGREIT (MAREIT) indices that highlights the time-varying nature of conditional variances and co-variances. The results further indicate there is little evidence that the two SEA REIT markets

Table 11.5 Estimation Results of AR (1) – GARCH-ADCC Model

	SGREITs	*MAREITs*	*USREITs*	*JREITs*	*AUREITs*
Univariate GJR-GARCH estimates					
GIR	0.0428	0.6506	0.0994	0.0195	0.04339*
	(t = 1. 11)	(t = 1.42)	(t - 1.44)	(t = 0.21)	(t = 1 .89)
ARCH	0.0667*	0.1199	0.1236**	0.1777**	0.0699***
	(t = 1.66)	(t = 1 .49)	(t = 2.45)	(t = 2.34)	(t = 2.65)
GARCH	0.9214***	0.7223***	0.8591***	0.8517***	0.9114***
	(t = 37.76)	(t=6.03)	(t = 22.1)	(t = 12.93)	(t = 45.33)
Multivariate equation					
ADCC model					
a	0	t = 0			
b	0.9475***	t = 73.15			
g	0.0269	t = 1.40			

Source: GIR-GARCH-ADCC Model, Ox Metrics 7.0

have become more integrated with the three global markets in recent years. It may be concluded that, although a similar globalization process may be taking place in the global REIT markets, it is much slower as many national REIT markets are still small and need longer time to develop.

Table 11.6 provides the correlation estimates according to four sub-periods (pre-crises, global financial crisis, EDC and post-crises). As noted, the St. Louise Fed Reserve Board (2009) and the Bank for International Settlements, BIS 79th Annual Report, 2009 official timelines are followed. Based on these guides, the four sub-periods are designated as follows:

(a) pre-crises: 11/09/2006–08/02/2007;
(b) GFC: 8/2/2007–11/05/2009;
(c) EDC: 11/05/2009–12/01/2011; and
(d) post-crises: 12/01/2011–11/24/2016.

Focusing on the global financial crisis period when REIT markets are more volatile, the results indicate the seven REIT market-pairs experienced an increase of cross-market linkage from 12.53% (SG-US), 12.83% (SG-AU), 20.3% (SG-JP), 32.65% (MA-US), 33.13% (MA-AU), 57.82% (SG-MA) and 78.79% (MA-AU), implying possible contagion effect among some REIT markets. Despite this increase in correlation, the average market integration level is still only between 0.13 (MA-US) and 0.486 (SG-JP). In recent years (post-crises periods), the results indicate the REIT market integration level was between 0% (SG-US) and 38.4% (MA-JP) higher that in the pre-crises periods, resulting from the two crises (GFC and EDC) and globalization of financial markets.

Based on Diebold and Yilmaz (2012)'s generalized spillover methodology, volatility spillover results are shown in Table 11.7. They are based on vector auto-regressions of lag length one and generalized variance decompositions of the 20-week ahead forecast errors. The volatility spillover index for the full sample is 44.6%, meaning that less than one-half of the total variance of the forecast errors for the five REIT markets can be explained by the spillover of return volatility shocks across countries. The remaining 55.4% of the volatility movement is caused by a purely domestic factor (i.e. the idiosyncratic dynamics of the domestic REIT volatilities in the past).

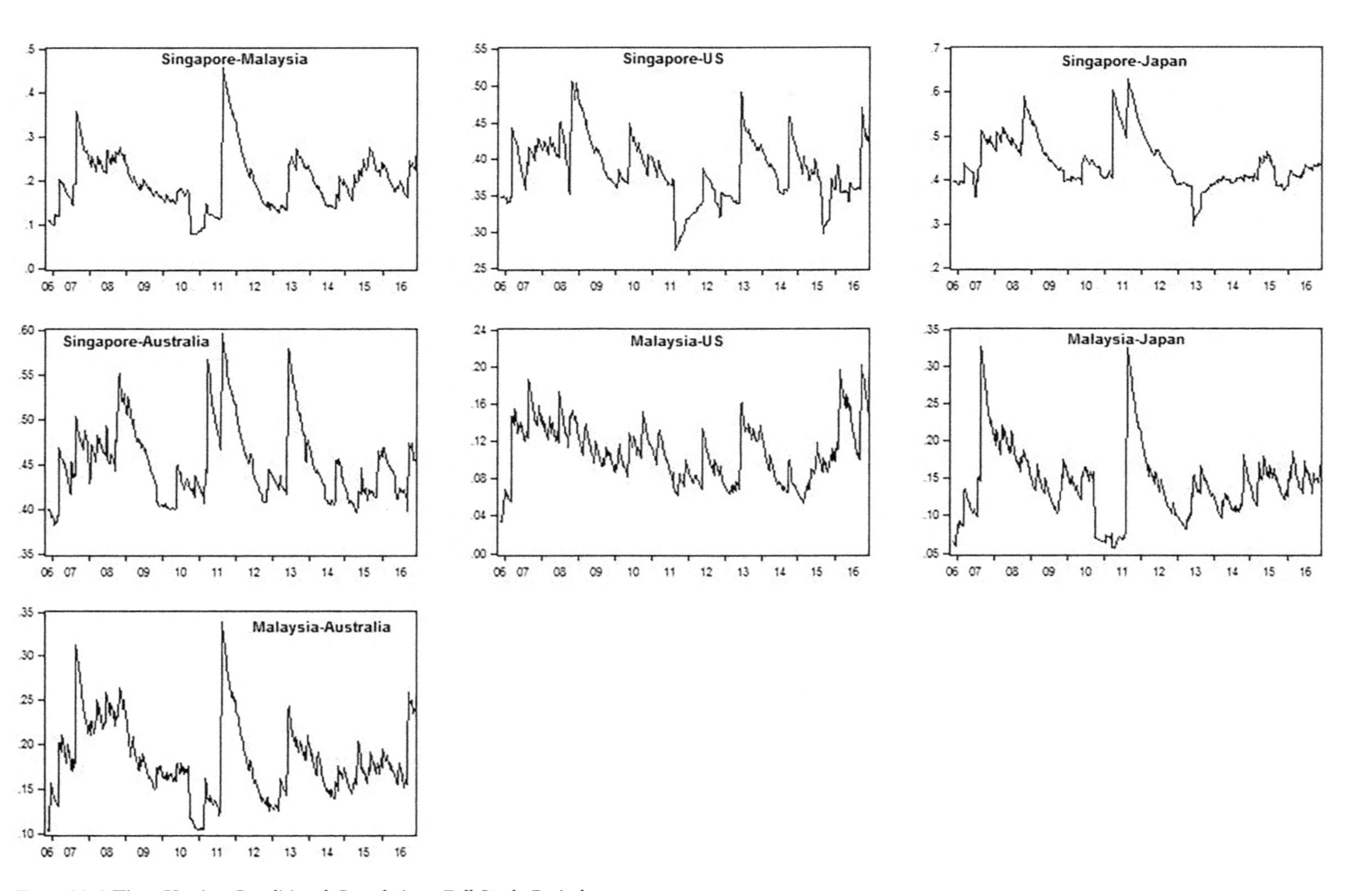

Figure 11.6 Time-Varying Conditional Correlations: Full Study Period

Source: Derived from AR (1)-GJR-ADCC Model, Ox Metrics 7.0

Table 11.6 REITs' Correlations at Different Sub-Periods

	SG-MA	*SG-US*	*SG-JP*	*SG-AU*	*MA-US*	*MA-JP*	*MA-AU*
PRE-CRISIS (PRE)	0.147	0.375	0.404	0.421	0.098	0.099	0.163
GFC	0.232	0.422	0.486	0.475	0.13	0.177	0.217
GFC vs PRE	57.82%	12.53%	20.30%	12.83%	32.65%	78.79%	33.13%
EDC	0.174	0.372	0.466	0.456	0.103	0.132	0.167
POST-CRISIS	0.201	0.375	0.413	0.445	0.102	0.137	0.175
POST vs PRE	36.73%	0.00%	2.23%	5.70%	4.08%	38.38%	7.36%

Based on these guides, four sub-periods were designated as follow: (a) pre-crises: 09/11/2006–02/08/2007, (b) GFC: 02/08/2007–05/11/2009, (c) EDC: 05/11/2009–01/12/2011 and (d) post-crises: 01/12/2011–24/11/2016.

Source: Following the St. Louise Fed Reserve Board (2009) and the Bank for International Settlements, BIS 79th Annual Report, 2009 Official Timelines

Table 11.7 Generalized Volatility Spillover Table, REITs, VAR (1), Forecast Step: 20-weeks 2006:11:06 to 2016:11:24

	SG	*MA*	*US*	*JP*	*AU*	*From*
SG	30.8	0.2	8.8	30.8	29.4	69
MA	1.4	95.8	0.8	1.9	0.1	4
US	12.1	0	36.4	33.1	18.3	64
JP	12.9	0.3	9.3	57.2	20.2	43
AU	15.4	0	2.8	24.9	56.8	43
TO	42	1	22	91	68	223
FROM	69	4	64	43	43	
NET	-27	-3	-42	48	25	
TOTAL	73	96	58	148	125	**44.60%**

Sources: Diebold and Yilmaz (2012) and Authors' Own Estimates

Moreover, own market's volatility spillover explains the highest share of forecast error variance since all diagonal elements receive the highest values (between 30.8% for SGREITs and 95.8% for MAREITs) compared to off-diagonal elements. Thus, volatility persistence is relatively dominant in all five REIT markets with the SGREIT market being the most endogenous market registering the lowest own-market volatility spillover of about 46% over the study period, whereas the MAREIT market is most exogenous during the same period.

Figure 11.7 shows the dynamics of the total rolling volatility spillover index based on 104-week rolling window estimations. The average spillover level is about 44.7%, with the plot indicating the highly fluctuating nature of volatility spillover with two major spikes happening at 27 November 2008 (index level: 80, GFC) and 30 September 2010 (European debt crisis, EDC: index level: 72). Thus, similar to stock markets, there is significant and pronounced volatility spillover across the REIT markets during the two financial crises (global financial crisis and EDC), indicating possible contagion effects.

Thereafter, the spillover index fluctuated, reaching its third highest level of about 59 around mid-May 2015 and then ending at a low of 30% on 24 November 2016, indicating a decline of volatility linkages among the five REIT markets examined. Additionally, volatility co-movements across the five REIT markets were most pronounced during the global financial

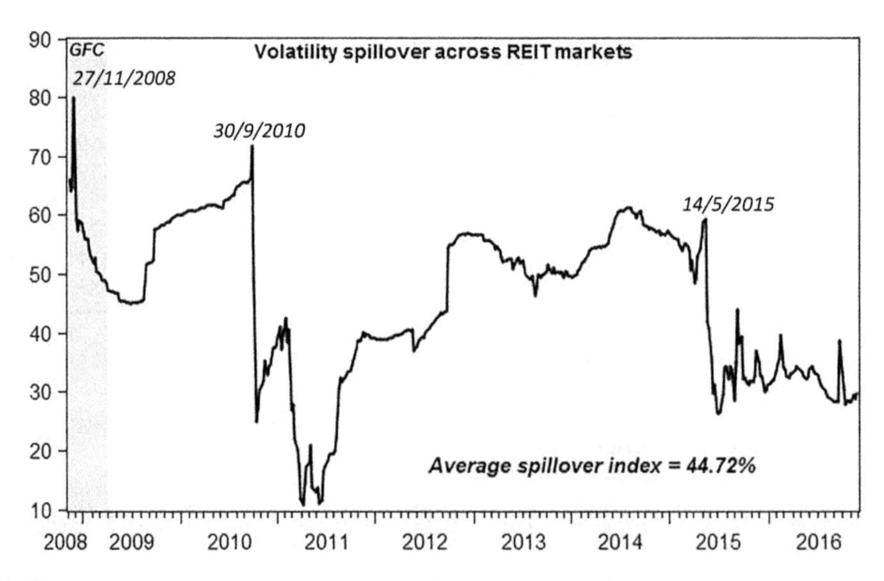

Figure 11.7 Generalized Conditional Volatility Spillovers, Five REIT Markets, 104-week Rolling Windows, 20-step Forecast Horizons, VAR (1), Full Study Period

Notes: The five REIT markets in the VAR system are the following: SGREITs MAREITs, USREITs, JREITs and AUREITs.

Sources: Diebold and Yilmaz (2012) and Authors' Own Estimates

crisis period. Specifically, the volatility spillover index during the global financial crisis period was 52.179%, which is about 14.13% increase in volatility interaction over the full period.

Additionally, two other indicators of interest are the directional total spillover TO and FROM each market, as well as the net volatility spillover index. For brevity, the four plots for SGREIT and MAREIT markets in Figure 11.8 are considered, where it is observed that there are substantial variations in the directional gross and net spillover effect. Overall, SREITs and MAREITs are net volatility absorbers relative to the REIT markets in the US, JP and AU during the full sample period.

One other observation is that, during the global financial crisis period, SREITs become more open and interactive with other REIT markets. With the "FROM" (transmission) directional spillover index increased from 41.07% to 59.69%, compared to a smaller increase (from 61.09% to 66.72%) in the "TO" (absorption) directional spillover index, this sees SREITs net directional index improved from -20.01% to -7.02%. In contrast, MAREITs see their net directional spillover index experienced only slight change (from -24.48% to -24.83%), although both their "FROM" and "TO" registered almost same magnitude of decline (between 6–7% from the full sample period).

Finally, based on the bivariate net spillover index estimates, Figure 11.9 reveals that SGREITs were a net volatility exporter of MAREIT REITs (net spillover value = 2.90). However, both SGREITs and MAREITs were largely pairwise receivers of volatility shocks from the US, Japan and Australia. These results indicate the three largest global REIT markets have the most dominant volatility impacts on the two SEA REIT markets over the full sample period. One immediate implication from these volatility transmission findings is that SGREITs need to become the volatility leader over JREITs in the next few years if they hope to become the REIT hub in Asia.

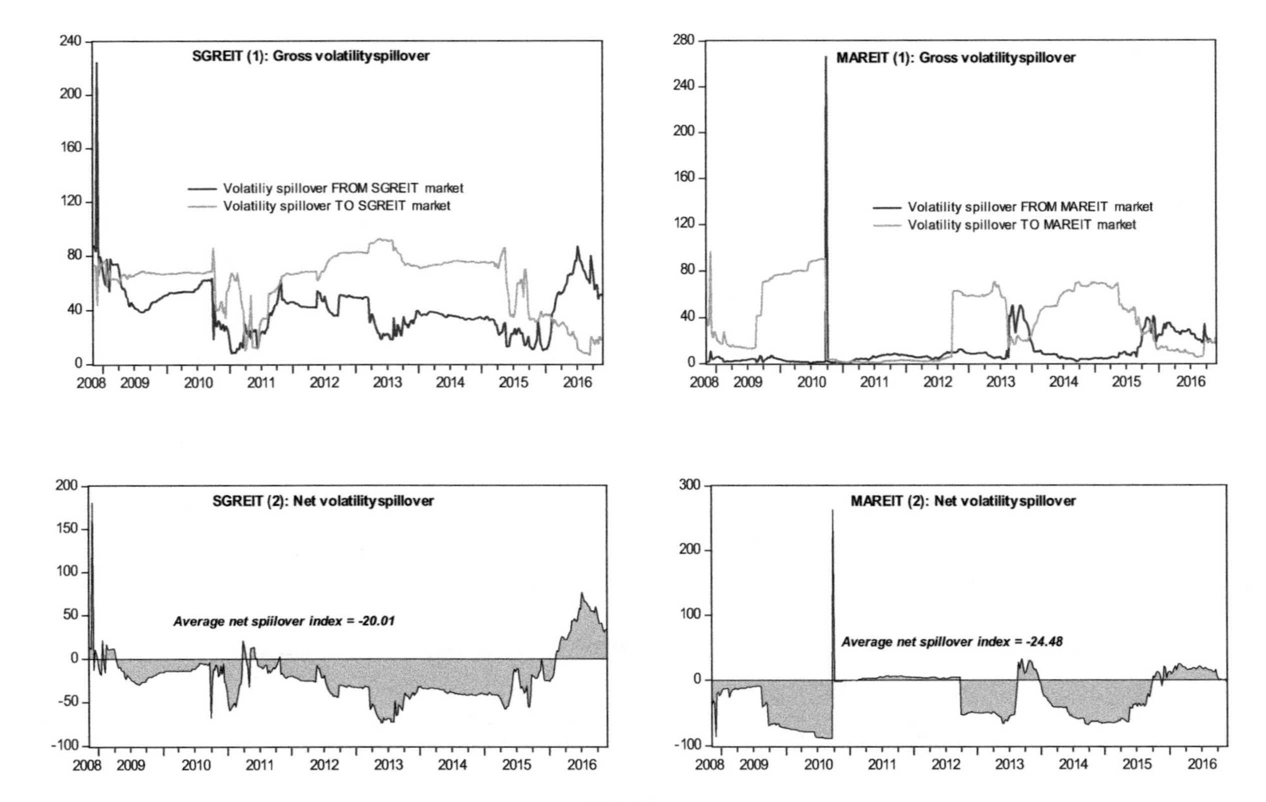

Figure 11.8 Generalized REIT Volatility Spillover (Multivariate): Total From, Total To and Net Spillover (Multivariate), 104-week Rolling Windows SGREITs and MAREITs

Sources: Diebold and Yilmaz (2012), and Authors' Own Estimates

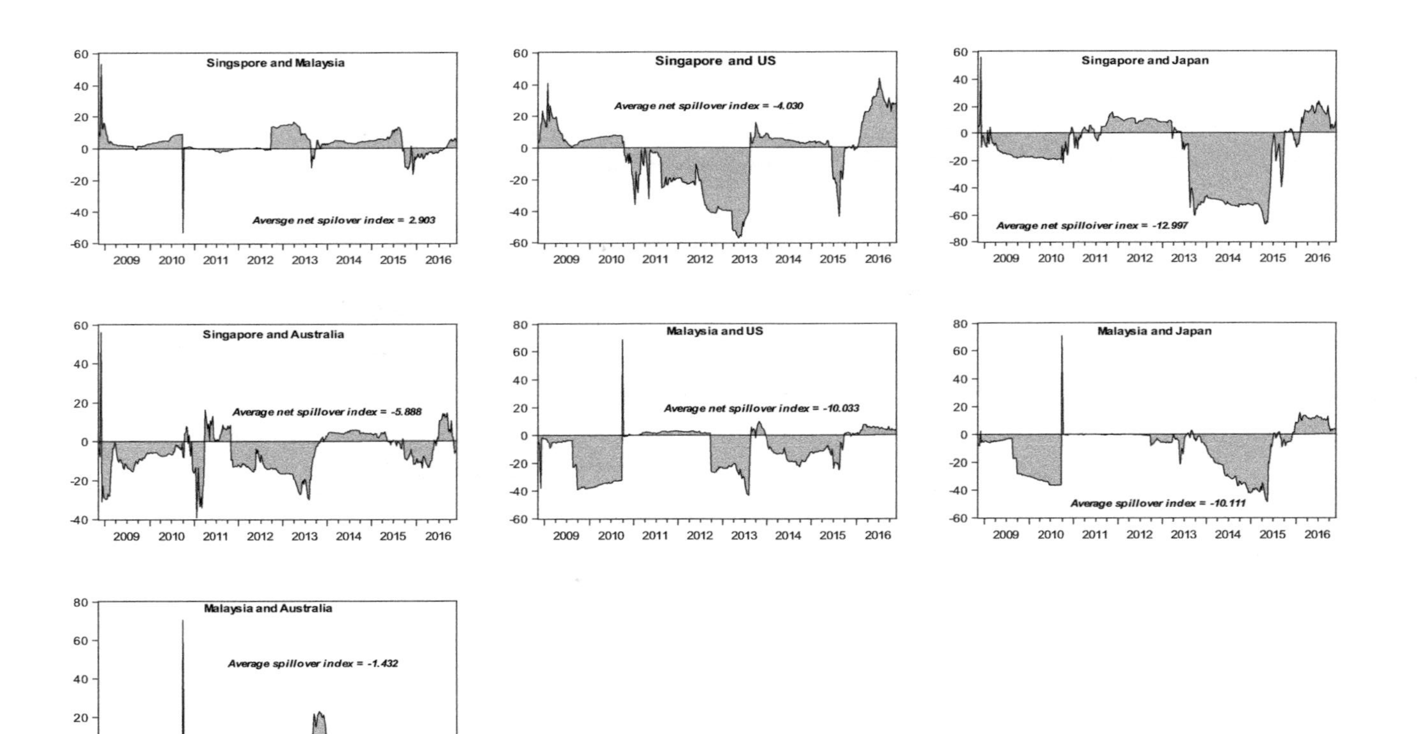

Figure 11.9 Generalized Net Volatility Spillover (Bivariate): Between SGREITs (MAREITs) and USREITs, JREITs and AUREITs, 104-Week Rolling Windows

Sources: Diebold and Yilmaz (2012), and Authors' Own Estimates

11.7 Causal relationship between SGREITs, MAREITs and global stock markets

The notion of causality is related to the study of market integration/segmentation across the entire economy (Okunev, Wilson and Zurbruegg, 2000). A domestic stock market which is highly developed will be able to better support the development of a REIT market. It may be argued that, in order for SGREIT market to be more developed, it is important to understand to what extent the SGREIT market is integrated with the three global stock markets of the US, Australia and Japan. In this section, it is proposed to examine whether any significant causal relationship exists between SGREIT, MAREIT and the US, Japan and Australia stock markets.

Linear and nonlinear Granger causality tests are employed. In this case, Granger (1969) causality tries to test whether knowing the current and lagged values of (say) REIT returns improves the forecast of future values of stock returns, and vice versa. The linear causality for two stationary series REIT returns (R) and stock returns (S) is based on the bivariate representation:

$$R_t = a_0 + \sum_{1=1}^{p} a_i R_{t-i} + \sum_{j=1}^{q} a_j S_{t-j} + \varepsilon_{H,t}$$

$$S_t = c_0 + \sum_{1=1}^{p} a_i R_{t-i} + \sum_{j=1}^{q} a_j S_{t-j} + \varepsilon_{S,t}$$

where R and S are stationary REIT and stock return series, p and q are the lag lengths of R and S, t is the number of observations. The null hypothesis in the Granger causality test is that S does not cause R, which is represented by H0: $a_1 = \ldots\ldots = a_q = 0$, and the alternative hypothesis is H1: $a_0 \neq 0$ for at least one j in the equation, with a standard F-distribution with (p, t-p-q-1) degree of freedom.

The nonlinear Granger causality test proposed by Hiemstra and Jonathan (HJ) (Hiemstra and Jonathan, 1994) is also employed. The HJ test is a modified version of the Baek and Brock (1992) test for conditional independence, with critical values based on asymptotic theory. Unlike the linear test, this non-parametric test is able to accommodate nonlinearity in the dynamics of REIT and stock prices. As in Okunev, Wilson and Zurbruegg (2000), this non-linear Granger causality test is applied to the estimated residual series from the VAR model which can remove any linear predictive power. Specifically, for two weekly dependent time series $\{X_t\}$ and $\{Y_t\}$, $\{X_t^m\}$ denotes the m-length lead vector. The L_x length lag vector of X_t is denoted by $X_{t-L_x}^{L_x} = \left(X_{t-L_x}, X_{t-L_x+1}, \ldots, X_{t-1}\right)$, where $L_x = 1, 2, \ldots, t = L_x + 1, L_x + 2, \ldots$ Then for given values of m, L_x and $L_y \geq 1$ and for some $e \geq 0$, Y is said to not strictly Granger cause X if:

$$\Pr(\| X_t^m - X_s^m \| < e \| \| X_{t-L_x}^{L_x} - X_{s-L_x}^{L_x} \| < e, \| Y_{t-L_y}^{L_y} - X_{s-L_y}^{L_y} \| < e) = \Pr(\| X_t^m - X_s^m \| < e \| \| X_{t-L_x}^{L_x} - X_{s-L_x}^{L_x} \| < e)$$

Table 11.8 reports the linear causality results. There seems to be evidence of strong bidirectional relationship between SGREITs and local stocks, between SGREITs and the US stocks, as well as between SGREITs and Australian stocks. Additionally, there is some evidence of strong one-way directional causality, at the 5% significance level, running from SGREITs to Japan stocks. In contrast, there is complete absence of return causality between MAREITs and all stock markets. These results imply that SGREITs are integrated with the world stock markets, which may be conducive for their future development.

Table 11.8 Linear Granger Causality Results: SGREITs (MAREITs) and Regional/International Stocks, Full Period

Optimal ff of lags	*Results*	*Null Hypothsis*	*F_stat*	*P_value*	*Null Hypothsis*	*F_stat*	*P_value*
SINGAPORE REITS							
3	**Bilateral**	SG_REITS does not cause SG_STOCKS	**8.5827**	**0.000**	SG_STOCKS does not cause SG_REITS	**3.2746**	**0.021**
4	**Bilateral**	SG_REITS does not cause US_STOCKS	**8.9985**	**0.000**	US_STOCKS does not cause SG_REITS	**13.2940**	**0.000**
3	**REITS to Stock**	SG_REITS does not cause JP_STOCKS	**2.6861**	**0.046**	JP_STOCKS does not cause SG_REITS	0.8080	0.490
3	**Bilateral**	SG_REITS does not cause AU_STOCKS	**5.7287**	**0.001**	AU_STOCKS does not cause SG_REITS	**4.5829**	**0.004**
MALAYSIA REITS							
1	**No Causality**	MA_REITS does not cause MA_STOCKS	**1.1060**	**0.293**	MA_STOCKS does not cause MA_REITS	0.5957	0.441
2	**No Causality**	MA_REITS does not cause US_STOCKS	**0.2260**	**0.798**	US_STOCKS does not cause MA_REITS	0.5612	0.571
1	**No Causality**	MA_REITS does not cause JP_STOCKS	**0.2122**	**0.645**	JP_STOCKS does not cause MA_REITS	1.2900	0.257
1	**No Causality**	MA_REITS does not cause AU_STOCKS	**0.0357**	**0.850**	AU_ST0CKS does not cause MA_REITS	0.0051	0.943

Notes: Highlighted in bold indicates rejection of the null hypothesis of the linear Granger non-causality test.

Source: Authors' Own Estimates

Table 11.9 Nonlinear Granger Causality Results: SGREITs (MAREITs) and Regional /International Stocks, Full Period

Panel A: SINGAPORE REITS								
	HO: SG_REITS does not cause SG_STOCKS		*HO: SG_REITS does not cause US_STOCKS*		*HO: SG_REITS does not cause JP_STOCKS*		*HO: SG_REITS does not cause AU_STOCKS*	
	Prob(HJ)	*T(HJ)*	*Prob(HJ)*	*T(HJ)*	*Prob(HJ)*	*T(HJ)*	*Prob(HJ)*	*T(HJ)*
IX=IY=1	**0.000**	**3.3998**	**0.003**	**2.7694**	**0.003**	**2.7764**	**0.002**	**2.8623**
IX=IY=2	**0.000**	**3.5489**	**0.007**	**2.4753**	**0.001**	**3.0645**	**0.001**	**3.0337**
IX=IY=3	**0.000**	**3.6354**	**0.003**	**2.7399**	**0.002**	**2.9167**	**0.001**	**3.0495**
IX=IY=4	**0.001**	**3.2345**	**0.009**	**2.3736**	**0.005**	**2.5982**	**0.007**	**2.4649**
IX=IY=5	**0.007**	**2.4761**	**0.032**	**1.8527**	**0.010**	**2.3409**	**0.030**	**1.8827**
IX=IY=6	**0.008**	**2.4205**	**0.059**	**1.5649**	**0.040**	**1.7502**	**0.087**	**1.3578**
IX=IY=7	**0.028**	**1.9053**	0.195	0.8601	0.108	1.2380	0.447	0.1337

IX=IY=8	**0.016**	**2.1477**	0.231	0.7356	0.205	0.8234	0.631	-0.3338

	HO: SG_STOCKS does not cause SG_ REITS		*HO: US_STOCKS does not cause SG_ REITS*		*HO:JP_STOCKS does not cause SG_ REITS*		*HO:AU_STOCKS does not cause SG_ REITS*	
	Prob(HJ)	*T(HJ)*	*Prob(HJ)*	*T(HJ)*	*Prob(HJ)*	*T(HJ)*	*Prob(HJ)*	*T(HJ)*
IX=IY=1	**0.003**	**2.7419**	**0.003**	**2.6970**	**0.023**	**1.9971**	**0.002**	**2.9336**
IX=IY=2	**0.004**	**2.6878**	**0.030**	**1.8749**	0.187	0.8901	**0.002**	**2.8417**
IX=IY=3	**0.003**	**2.7474**	**0.085**	**1.3740**	0.214	0.7939	**0.005**	**2.5987**
IX=IY=4	**0.032**	**1.8518**	**0.094**	**1.3160**	0.580	-0.2014	**0.025**	**1.9539**
IX=lY=5	0.178	0.9248	0.104	1.2598	0.685	-0.4829	0.115	1.1982
IX=IY=6	0.138	1.0880	0.261	0.6413	0.733	-0.6213	0.167	0.9661
IX=IY=7	0.358	0.3634	0.508	-0.0196	0.823	-0.9281	0.371	0.3305
IX=lY=8	0.338	0.4180	0.694	-0.5066	0.846	-1.0182	0.657	-0.4047

Panel B: MALAYSIA REITS

	HO: MA_REITS does not cause MA_ STOCKS		*H0: MA_REITS does not cause US_ STOCKS*		*HO: MA_REITS does not cause JP_STOCKS*		*HO: MA_REITS does not cause AU_ STOCKS*	
	Prob(HJ)	*T(HJ)*	*Prob(HJ)*	*T(HJ)*	*Prob(HJ)*	*T(HJ)*	*Prob(HJ)*	*T(HJ)*
IX=IY=1	0.142	1.0728	0.881	-1.1781	0.392	0.2731	0.900	-1.2809
IX=IY=2	0.275	0.5969	0.967	-1.8448	0.498	0.0050	0.940	-1.5522
IX=IY=3	0.634	-0.3435	0.956	-1.7065	0.632	-0.3369	0.924	-1.4321
IX=IY=4	0.632	-0.3382	0.905	-1.3116	0.743	-0.6529	0.925	-1.4387
IX=IY=5	0.681	-0.4700	0.906	-1.3159	0.919	-1.3981	0.952	-1.6661
IX=IY=6	0.614	-0.2910	0.864	-1.0984	0.884	-1.1951	0.880	-1.1758
IX=IY=7	0.720	-0.5824	0.897	-1.2619	0.569	-0.1728	0.890	-1.2272
IX=IY=8	0.656	-0.4003	0.803	-0.8517	0.432	0.1705	0.809	-0.8737

	HO: MA_STOCKS does not cause MA_ REITS		*HO: US_STOCKS does not cause MA_ REITS*		*HO:JP_STOCKS does not cause MA_ REITS*		*HO:AU_STOCKS does not cause MA_ REITS*	
	Prob(HJ)	*T(HJ)*	*Prob(HJ)*	*T(HJ)*	*Prob(HJ)*	*T(HJ)*	*Prob(HJ)*	*T(HJ)*
IX=IY=1	0.259	0.6472	0.291	0.5491	0.866	-1.1091	0.920	-1.4071
IX=IY=2	0.121	1.1696	0.375	0.3181	0.828	-0.9461	0.757	-0.6951
IX=IY=3	0.082	1.3898	0.142	1.0722	0.833	-0.9671	0.539	-0.0984
IX=IY=4	**0.085**	**1.3692**	0.216	0.7849	0.747	-0.6662	0.617	-0.2973
IX=IY=5	**0.214**	**0.7925**	0.161	0.9893	0.563	-0.1580	0.520	-0.0497
IX=IY=6	0.290	0.5529	0.137	1.0952	0.556	-0.1398	0.494	0.0160
IX=IY=7	0.325	0.4539	0.346	0.3949	0.579	-0.1997	0.556	-0.1397
IX=IY=8	0.433	0.1689	0.475	0.0616	0.555	-0.1393	0.619	-0.3018

Notes: This table reports the results of the modified Baek and Brock (1992) nonlinear causality test results (as adapted by Hiemstra and Jonathan, 1994) based on the VAR filtered return residual series. Lx = Ly denote the number of lags on the residual series for length scales of 1.5. Prob (HJ) and T (HJ) reefer to significance level and t statistic of the test. Highlighted in bold indicates rejection of the null hypothesis of nonlinear Granger non-causality at least at the 10% level.

Source: Authors' Own Estimates

Table 11.9 reports the nonlinear Granger causality test results, from employing the modified Baek and Brock test procedure to the residuals of the respective VAR models. For SGREITs, its respective significant two-way causal relationships with local stocks, US stocks and Australian stocks, seems to be apparent for one, two, three and four lags, indicating that SGREITs follow and chase up to changes occurring in the local and international stock markets, and vice versa.

Similarly, there is some evidence of bilateral relationship, at the 5% significant level, between SGREITs and Japan stocks for one-week lags. In contrast, some evidence of uni-directional causality, at the 10% significance level can only be detected, running from MA stocks to MAREITs for three- and four-week lags.

11.8 Summary

Part II includes six chapters each analysing REITs in a region of the world, using case studies from a developed, developing and emerging REIT sector in the region.

Chapter 8 analysed REITs in the North American region, using case studies from the USA, Canada and Mexico, with Chapter 9 analysing REITs in the Latin American region, using case studies from Brazil, Argentina and Uruguay and Chapter 10 analysed REITs in the European region, using case studies from the UK, Spain and Poland for the developed, developing and emerging REIT sectors in the region, respectively.

This chapter analysed REITs in the South East Asian region, using case studies from Singapore (being a developed REIT sector in the region), Malaysia (being a developing REIT sector in the region) and Thailand (being an emerging REIT sector in the region).

Since their introduction, SGREITs, MAREITs and Thai-REITs have become an important investment vehicle for the respective countries. Further, they provide additional investor choice for listed property exposure on the respective stock markets (Newell, Pham and Ooi, 2015).

This chapter has reviewed the development and significance, as well as market structure and key regulations, of the three SEA REIT markets. One general conclusion arising from this review is that, although the three markets are in their different stages of development, they offer different opportunities and challenges for institutional investors:

> SGREITs have the potential to match or even overtake those in Japan in the next few years . . . Managing the tax, regulatory and financial regimes have supported SGREITs over the past 15 years is key to the sector's success.
>
> (Business Times, 01/31/201)

> The MAREIT market has been recognized as one of the leading Asian markets. . . . . . . The MAREIT at this stage holds great potential in helping the local financial market as a whole. . . . . For policymakers, the lack of integration between MAREIT and the regional REIT markets is a cause for concern if we wish to continue to maintain the growth momentum of the REIT market in Malaysia.
>
> (Wong, 2016)

> As Thai-REITs are moving towards international practices with a less restrictive regime, they offer opportunities and challenges for international investors. . . . The new changes in regulatory structure might also lead to the Thai-REITs exposure to higher levels of risk.
>
> (Pham, 2011)

Second, applying the mainstream finance theories and methodologies indicates that, while SGREITs needs to improve their current return integration level and volatility connectedness with the regional/global developed REIT markets, the low level of return integration between MAREITs and the regional/global players is an obstacle against the market growth.

Together with the findings that MAREITs are not causally linked to the regional/global stock markets suggests that policies and guidelines should be revised or enacted with a clear objective to synchronize the best practices among the regional/global REIT and stock markets in order to increase their degree of market integration over time (Wong, 2016). Finally, with further data for Thai-REITs over a longer time frame, more-in depth analysis can be done similar to the present study.

Chapter 12 will analyse REITs in the North Asian region, using case studies from Japan, Hong Kong and China, with Chapter 13 analysing REITs in the Oceania region, using case studies from Australia, South Africa and India and Chapter 14 will conclude Part II by considering *Directions for the Future of International REITs*.

References

Atchison, K. 2014, *The impact of REITs on Asian economies*, Asia Pacific Real Estate Association Limited, Singapore.

Baek, E. and Brock, W., 1992. A general test for nonlinear Granger causality: Bivariate model. *Iowa State University and University of Wisconsin at Madison Working Paper.*

Brounen, D. and de Koning, S. 2012, '50 years of real estate investment trusts: An international examination of the rise and performance of REITs', *Journal of Real Estate Literature*, Vol. 20, No. 2, pp. 197–223.

Cappiello, L., Engle, R. and Sheppard, K. 2006, 'Asymmetric dynamics in the correlations of global equity and bond returns', *Journal of Financial Econometrics*, Vol. 4, No. 4, pp. 37–572.

Diebold, X. and Yilmaz, K. 2012, 'Better to give than to receive: Predictive directional measurement of volatility spillovers', *International Journal of Forecasting*, Vol. 28, No. 1, pp. 57–66.

Engle, R. 2002, 'Dynamic conditional correlation: A simple class of multivariate generalized autoregressive conditional heteroskedasticity models', *Journal of Business & Economic Statistics*, Vol. 20, pp. 339–350.

EPRA 2016a, *Global REIT survey: Malaysia-REIT*, European Public Real Estate Association, Brussels.

EPRA 2016b, *Global REIT survey: Singapore-REIT*, European Public Real Estate Association, Brussels.

EPRA 2016c, *Global REIT survey: Thailand-REIT*, European Public Real Estate Association, Brussels.

Federal Reserve Board of St. Louis 2009, *The financial crisis: A timeline of events and policy actions*, Federal Reserve Board of St. Louis, St. Louis.

Glosten, L.R., Jagannathan, R. and Runkle, D.E. 1993, 'On the relation between the expected value and the volatility of the nominal excess return on stocks', *The Journal of Finance*, Vol. 48, No. 5, pp. 1779–1801.

Granger, J. 1969, 'Investigating causal relations by econometric models and cross-spectral methods', *Econometrica: Journal of the Econometric Society*, pp. 424–438.

Hiemstra, C. and Jonathan, J. 1994, 'Testing for linear and nonlinear Granger causality in the stock price-volume relation', *The Journal of Finance*, Vol. 49, No. 5, pp. 1639–1664.

Liow, K.H., Ho, K.H.D., Ibrahim, M.F. and Chen, Z. 2009, 'Correlation and volatility dynamics in international real estate securities markets', *The Journal of Real Estate Finance and Economics*, Vol. 39, No. 2, pp. 202–223.

Newell, G. and Osmadi, A. 2010, 'Assessing the importance of factors influencing the future development of REITs in Malaysia', *Pacific Rim Property Research Journal*, Vol. 16, No. 3, pp. 358–374.

Newell, G., Pham, A.K. and Ooi, J. 2015, 'The significance and performance of Singapore REITs in a mixed-asset portfolio', *Journal of Property Investment & Finance*, Vol. 33, No. 1, pp. 45–65.

Okunev, J., Wilson, P. and Zurbruegg, R. 2000, 'The causal relationship between real estate and stock markets', *The Journal of Real Estate Finance and Economics*, Vol. 21, No. 3, pp. 251–261.

Pham, A.K. 2011, 'The performance of Thai-REITs in a mixed-asset portfolio', *Pacific Rim Property Research Journal*, Vol. 17, No. 2, pp. 197–214.

Wong, Y.M. 2016, 'Malaysia REIT: First decade development and returns characteristics', *International Real Estate Review*, Vol. 19, No. 3, pp. 371–409.

12
NORTH ASIA

Joseph TL Ooi and Woei-Chyuan Wong

12.1 Introduction

This book aims to identify key areas for research in the REIT discipline for the next five to ten years by surveying the current state of the REIT discipline around the world and identifying emerging and cutting edge research areas through a thematic review of current contextual issues and a regional analysis based on case studies.

This book comprises two parts, the first part being a thematic review of emerging and cutting edge global research into current contextual issues in REITs internationally and the second part being a regional analysis of REITs around the world, each written by authoritative academic authors from the world's leading Universities and REIT industry experts.

Part I included six chapters each reviewing a current theme of REIT evolution through the lens of contemporary research. Chapter 1 focused on critical contextual issues in international REITs while Chapter 2, the *Post-Modern REIT Era*, examined the evolution of the global REIT market, what is deemed to be best current market practice and how the market is expected to change going forward and Chapter 3, *Emerging Sector REITs*, investigated such sectors as residential, health care, self-storage, timber, infrastructure and data centres generally and then specifically through a case study of US REITs.

Chapter 4, *Sustainable REITs*, analysed REIT environmental performance and the cost of equity and Chapter 5, *Islamic REITs*, examined the evolution of global Islamic REITs through a study of the Malaysian market. Chapter 6, *Behavioural Risk in REITs*, addressed the management of behavioural risk in global REITs and Part I concluded with Chapter 7 which reviewed recent research into *REIT Asset Allocation*.

Part II includes six chapters each analysing REITs in a region of the world, using case studies from a developed, developing and emerging REIT sector in the region. Chapter 8 analysed REITs in the North American region, using case studies from the USA, Canada and Mexico, with Chapter 9 analysing REITs in the Latin American region, using case studies from Brazil, Argentina and Uruguay for the developed, developing and emerging REIT sectors in the region, respectively.

Chapter 10 analysed REITs in the European region, using case studies from the UK, Spain and Poland, with Chapter 11 analysing REITs in the South East Asian region, using case studies from Singapore, Malaysia and Thailand for the developed, developing and emerging REIT sectors in the region, respectively.

This chapter analyses REITs in the North Asian region, using case studies from Japan (being a developed REIT sector in the region), Hong Kong (being a developing REIT sector in the region) and China (being an emerging REIT sector in the region).

The two largest Real Estate Investment Trust (REIT) markets in North Asia are Japan and Hong Kong. Japan paved the way for REITs in Asia with the simultaneous listing of two REITs in September 2001. The two REITs, namely Nippon Building Fund Inc. of Japan and the Japan Real Estate Investment Corporation, were launched by Nikko Salomon Smith Barney and Nomura Securities, respectively. They carry a face value of around US$4,000 per unit and annual dividends of 4–5%, as compared with near-zero interest rates for bank deposits in Japan (Ooi, Newell and Sing, 2006).

Hong Kong saw the successful launch of its maiden REIT in November 2005, with the listing of Link REIT which was the largest REIT IPO in the world (US$2.6 billion). Two further Hong Kong REIT IPOs followed soon after with the listing of Prosperity REIT and GZI China REIT in December 2005.

Since their debut, both the Japanese REITs and Hong Kong REITs have recorded impressive growth, reaching a market capitalization of around US$103.5 billion and US$25.6 billion, respectively, at December 2016. Appendix 12.1 and Appendix 12.2 list the names and market capitalization of 56 Japanese REITs and 10 Hong Kong REITs, respectively.

Other North Asian economies that have seen the introduction of REITs include Taiwan and South Korea, but their growth is stunted because of restrictive guidelines. Taiwan REITs are restricted from offering secondary offerings. In South Korea, corporate restructuring REITs (CR-REITs) dominate the REIT market. They are recession-centric instruments that allow investors to participate in real estate held by corporate and financial institutions for a finite holding period. With the recent introduction of several quasi-REIT products, the prospect for REITs in China is becoming more optimistic (RICS, 2016).

This chapter traces the growth and performance of REITs listed in Japan and Hong Kong. It will also address the prospect of a vibrant REIT market emerging in China. Figure 12.1 shows the market capitalization and total number of REITs listed in Japan and Hong Kong from 2001

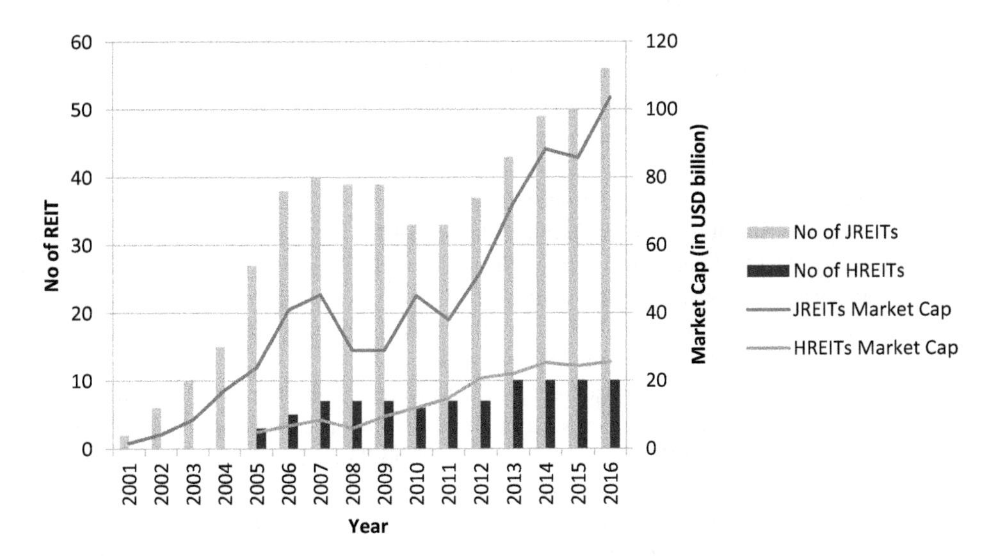

Figure 12.1 The Growth of REIT Markets in Japan and Hong Kong

Source: Author's Analysis Based on Datastream

to 2016. The total market capitalization of J-REITs increased exponentially from US$1.6 billion in 2001 to US$103.5 billion in 2016. The number of publicly listed J-REITs also grew from 2 in 2001 to 56 by end of 2016. In addition to having more J-REITs listed, the average size of REITs in Japan has also increased from US$0.8 billion to US$1.8 billion over the past 15 years.

The lines on Figure 12.1 indicate total market capitalization in US$ billion for REITs in Japan and Hong Kong from 2001 to 2016, while the vertical bars show the total number of REITs in the corresponding period.

Chapter 13 then analyses REITs in the Oceania region, using case studies from Australia, South Africa and India for the developed, developing and emerging REIT sectors in the region, respectively, with Chapter 14 concluding Part II by considering *Directions for the Future of International REITs*.

12.2 REITs in Japan

This chapter analyses REITs in the North Asian region, using case studies from Japan (being a developed REIT sector in the region), Hong Kong (being a developing REIT sector in the region) and China (being an emerging REIT sector in the region).

The Japanese REIT (J-REIT) market is now the second largest in the world, but it is still a long way from the US REIT market which had a market capitalization in excess of US$1 trillion in 2016. The growth trajectory of J-REITs has also not been smooth sailing. During the height of the global recession (2008) and Euro crisis (2011), the sector's market capitalization dropped by 36% and 16%, respectively. The Japanese REIT sector consolidated with nine financially distressed REITs delisted and merged with other surviving REITs. The delisted REITs include LCP Investment Corporation, Nippon Residential Investment, Advance Residence Investment Corporation, LaSalle Japan REIT, New City Residence Investment Corporation, Prospect Retail Investment Corporation, Japan-Single-Residence REIT, Nippon Commercial Investment Corporation and Ichigo Real Estate Investment Corporation.

For example, New City Residence Investment Corporation could not raise money to repay outstanding debt and had to file for court protection from its creditors in 2008. This episode underscores the importance, even in a developed market, of maintaining a healthy capital structure to provide sufficient debt headroom for the REIT to pursue growth as well as to provide sufficient buffer against a credit crisis. Ooi, Wong and Ong (2012) document that in a tight credit market, the primary concern of most REITs is the ability to access capital and maintain adequate liquidity.

The quantitative easing policy implemented by the Japanese government from 2012, which involved buying J-REIT stocks by the Bank of Japan (BOJ), has expanded the overall market capitalization of J-REITs. Fueled by a flurry of new REIT IPOs as well as aggressive expansion through property acquisitions, the post-financial crisis era saw a second wave of growth recorded by J-REITs. In a brief period of only six months, from September 2016 to February 2017, J-REITs acquired US$5.7 billion worth of properties.

It should be noted that half of the current J-REITs were only listed post-2011. The younger J-REITs contributed to the sector's expansion through their insatiable appetite for property acquisitions, fueled by the BOJ's low interest rate regime. The historically low interest rates, however, imposed pressures on investors seeking higher yield investments. Figure 12.2 shows the sector's average yield spread, which is defined as a Japanese or Hong Kong REIT's dividend yield minus the prevailing interest rate on treasury bills in the two markets. Except for the crisis period (2008–2009) when the dividend yield of J-REITs shot up to 12.7% on the back

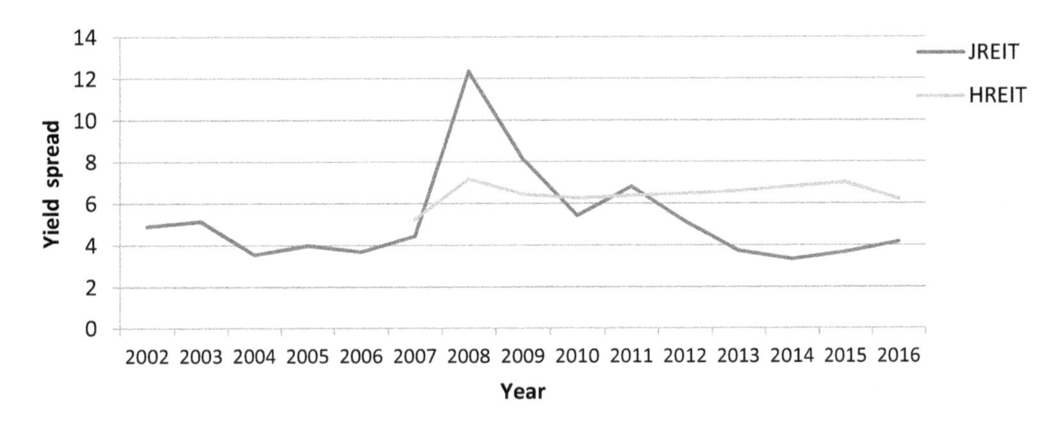

Figure 12.2 Dividend Yield Spread (2001–2016)

Note: Yield spread is defined as the difference between dividend yield of the REIT stocks and the Treasury-Bill rate.

Source: Author's Analysis Based on Datastream

of declining stock prices, J-REIT stocks have been trading at between 4.0 and 5.0 percentage points above treasury-bill rate.

12.2.1 Summary – Japanese REITs

The development of the REIT sector in Japan is a global success story, becoming the second largest REIT market in the world in less than 20 years, despite the impact of the global financial crisis of 2007–2009 and the Euro crisis of 2011. It is particularly interesting that the growth in the sector was contributed to by the Bank of Japan including buying J-REITs as part of its quantitative easing programme.

12.3 REITs in Hong Kong

This chapter analyses REITs in the North Asian region, using case studies from Japan (being a developed REIT sector in the region), Hong Kong (being a developing REIT sector in the region) and China (being an emerging REIT sector in the region).

For the developing REIT sector of Hong Kong, Figures 12.3 and 12.4 show that the price-to-book and price-to-earnings ratios of Hong Kong REITs (HK-REITs) are significantly lower than J-REITs, which doubled between 2011 and 2016. Since 2011, the stock price of a typical HK-REIT has consistently traded below its book value by some 40%.

As at the end of 2016, there were 10 HK-REITs with an aggregate market capitalization of US$25.6 billion. This constitutes around 5.8% of the total market capitalization of the listed property sector in Hong Kong, which includes both the traditional property companies and HK-REITs. In contrast, REITs constitute 44.2% and 44.3% of the quoted property sector in Japan and Singapore, respectively. While REITs are dominant players in commercial property acquisitions in Japan, the opposite is true in Hong Kong. For example, property companies were behind all the US$8.8 billion of property acquisitions transacted between September 2016 and February 2017 in Hong Kong.

Figure 12.3 shows that the HK-REIT units have been trading below their book value. Two major factors have hampered the development and growth of the REIT market in Hong Kong. First, in addition to being subject to a 15% property tax, HK-REITs do not enjoy corporate

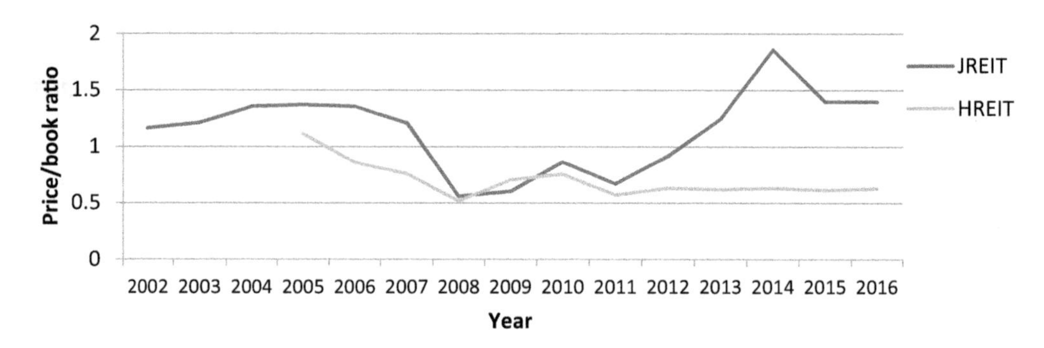

Figure 12.3 Price-to-Book Ratio (2001–2016)

Source: Author's Analysis Based on Datastream

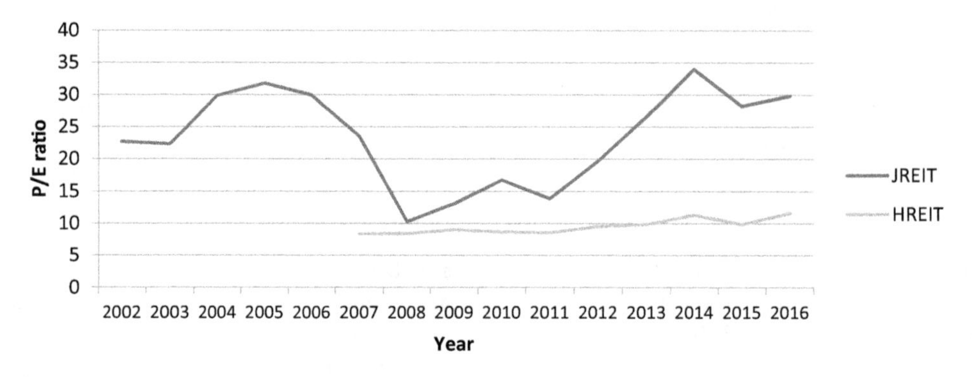

Figure 12.4 Price-to-Earnings Ratio (2002–2016)

Source: Author's Analysis Based on Datastream

tax exemption. Second, the image of HK-REITs has been impaired by sponsors who artificially inflated the HK-REITs' distribution yields through clever financial engineering. Ooi and Neo (2010) identify various tactics adopted by HK-REIT sponsors and managers to enhance the distribution yields of HK-REITs, such as the following:

- accepting lower property values;
- transferring lower quality or higher risk properties into the HK-REIT;
- providing temporary income support in the form of rental guarantees;
- waiving of dividend rights for stock owned by the sponsors;
- paying of management fees in units with the purpose of preserving cash for distribution to the unitholders; and
- leveraging with cheap debt.

An example of a heavily financially engineered HK-REIT structure is the US$348 million Sunlight REIT IPO in December 2006. Even though the yield of the 20 properties in the HK-REIT portfolio ranged from 2.8% to 3.0%, the sponsor was able to promise an initial yield of 8.49% with Ooi and Neo (2010) estimating that 65% of the distribution yield was financially engineered. Table 12.7 shows that Sunlight REIT closed at 6.54% below the offer price on the first day of trading, indicating that excessive use of financial engineering to enhance distribution yield was not favourably viewed by the stock market.

12.3.1 Summary – Hong Kong REITs

As a developing REIT sector, the Hong Kong market might be expected to be seeking growth but it is structurally challenged in such aspirations through its unique combination of a long-standing and well-understood competing property company sector, heavily financially engineered REIT structures and the absence of tax transparency in its REIT legislation and regulation.

12.4 REITs in China

This chapter analyses REITs in the North Asian region, using case studies from Japan (being a developed REIT sector in the region), Hong Kong (being a developing REIT sector in the region) and China (being an emerging REIT sector in the region).

In a strict sense, there is currently no REIT market in China (Roxburgh, 2015). Nevertheless, Chinese properties have been packaged into REITs and listed offshore, namely in Hong Kong and Singapore. For example, Yuexiu REIT, which owns five commercial properties in Guangzhou, was listed on the Hong Kong Stock Exchange in December 2005. Similarly, CapitaLand China Retail Real Estate Trust was listed on the main board of the Singapore Exchange in December 2006, starting with seven shopping malls in China.

Table 12.1 summarizes three recent quasi-REIT products that have been introduced in China. In April 2014, CITIC Securities launched the CITIC Sailing Special Asset Management Plan which, according to RICS (2016), is China's first equity REIT product. The three- to five-year term securities, allocated through private placement, are listed on the Shenzhen Stock Exchange. The underlying assets include two office buildings owned by CITIC. In October 2014, CITIC Securities and Suning Appliance launched the Suning Appliance Private Investment Fund which involved securitizing 11 Suning retail stores. The securities are tranched, with the 18-year Class A securities tranche offering a yield between 7% to 8.5%, and the 3+1 year Class B securities tranche offering a yield of 8% to 9.5%.

Table 12.1 Three Recent Quasi-REITs in China

	CITIC Sailing Special Asset Management Plan	*CITIC Suning Yunchuang Private Investment Fund*	*Penghua Qianhai Vanke REITs*
Date:	April 2014	October 2014	June 2015
Issuer:	CITIC Securities	CITIC Securities & Suning Appliance	Vanke Fund & Penghua Fund
Duration:	3–5 years	Class A: 18 years Class B: 3+1 years	10 years
Min. Investment:	1 million yuan	1 million yuan	100, 000 yuan
Return:	7%	Class A: 7.0–8.5% Class B: 8.0%-9.5%	8%
Underlying Assets:	2 office buildings owned by CITIC in Beijing and Shenzhen	11 Suning stores	Right to rental income ofVanke Qianhai Enterprise Dream Park office building.

Source: RICS (2016)

In June 2015, the Penghua Qianhai Vanke REIT Closed-End Hybrid Securities Investment Fund was launched, raising CN¥ 3 billion on the initial listing. The securities, which have an investment life of 10 years, are publicly listed in Shenzen and available to retail investors. The underlying assets are the rental income rights of Vanke's Qianhai Enterprise Dream Park office building in Qianhai, a special economic zone in Shenzhen. Although it is dubbed as China's first public REIT, Roxburgh (2015) argues that it is, in reality, a closed-end fund.

A survey conducted by RICS in 2016 suggests that the prospects for REITs in China are bright, particularly with the potential of REITs as an additional source of liquidity for developers who are caught with a high level of unsold inventory in a tight funding environment (RICS, 2016). There are, however, limitations that may impede the development of a REIT market in China, including restrictive government policy, a controlled finance system and lack of people talent. Due to concerns over loss of tax revenue, not every government has been willing to grant REITs tax transparency status which is a key element for REIT market growth. As is evident in Hong Kong, the absence of preferential tax treatment has been cited as a reason for the HK-REIT market developing more slowly than that of Singapore.

12.4.1 Summary – Chinese REITs

It is challenging to identify a REIT sector in the North Asia region that would generate greater REIT investor demand than China when the sector finally emerges. Given the popularity of a range of debt and equity property fund products based on Chinese property and already launched, it may be anticipated that, when the Chinese Government legislates and regulates for a REIT sector, its emergence will be rapid. The rate of growth, however, may be determined by the extent of resemblance of the Chinese REIT sector structure to other international REIT sector structures with, as Hong Kong has experienced, tax transparency being critical for a massive and rapid rate of growth in the sector.

12.5 J-REITs vs HK-REITs

The typical property sectors preferred by J-REITs are offices, industrial premises, retail malls and residential, which collectively accounted for 65.2 % of total market capitalization of J-REITs as at February 2017. Table 12.2 shows that approximately one in three J-REITs have a diversified portfolio. In addition, there are three J-REITs that each specialize in the health care and hospitality sectors, respectively.

Table 12.2 Classification of J-REITs by Property Types (Feb 2017)

	Market Capitalization	*No. of REITs*
Office	US$34,918 million	13
Industrial	US$12,570 million	6
Retail	US$11,047 million	5
Residential	US$10,809 million	7
Hotel and Resort	US$3,528 million	3
Health care	US$395 million	3
Diversified	US$33,107 million	20
All	**US$106.37 billion**	**57**

Source: Author's Analysis Based on S&P Capital IQ

In Hong Kong, the two largest retail REITs accounted for 58.3% of the total market capitalization of HK-REITs. Link REIT, which is the largest REIT in Hong Kong, singularly constituted 55% of the market capitalization of the HK-REIT sector. A further three office REITs accounted for 16.8% of the sector's market capitalization, while two hospitality HK-REITs made up 4.7% of the market capitalization. Finally, three HK-REITs held diversified portfolios, constituting 20.2% of the sector's market capitalization, as shown in Table 12.3.

In term of the geographical location, 65.5% of the revenue of HK-REITs is generated domestically in Hong Kong. Table 12.4 shows that the remaining 34.5% of portfolio revenue is derived from real estate assets in mainland China. J-REITs, on the other hand, have invested predominantly in the domestic market with almost all of the property assets owned by J-REITs located in Japan. Indeed, Yanagihara and Kim (2016) noted that nearly 75% of the assets held by J-REITs are located in the Tokyo Metropolitan Area, which consists of the 5 Central Wards of Tokyo and the 23 Wards of Tokyo and Kanto Area.

Although J-REITs could, theoretically, invest overseas since 2008, the first and only overseas property acquisition by a J-REIT occurred in June 2014 when AEON REIT acquired 18% of a Malaysian shopping mall (Taman University Shopping Centre) that is co-owned and master-leased by business units linked to the J-REIT sponsor, with Yanagihara and Kim (2016) providing a detailed case study on this unique acquisition by a J-REIT.

Table 12.5 contrasts the ownership structure of J-REITs and HK-REITs. The major owners of HK-REITs are corporations, either privately held or publicly listed, which collectively own 38.8% of the units issued by HK-REITs. They are most likely the sponsors of the HK-REITs and may continue to exert influence through their control of the listed HK-REITs as well as through the external management of the HK-REITs. For example, Champion REIT is 59% owned by its sponsor, Great Eagle Holdings. In addition, the manager, Eagle Asset Management, is a wholly owned subsidiary of Great Eagle Holdings.

Further, Table 12.5 shows that HK-REITs are more captive than J-REITs, which generally have a lower proportion of their shares owned by their sponsors. Overall, corporate shareholders

Table 12.3 Classification of HK-REITs by Property Types (Feb 2017)

	Market Capitalization	*No. of REITs*
Retail	US$15,084 million	2
Office	US$4,334 million	3
Hotel and Resort	US$1,215 million	2
Diversified	US$5,226 million	3
All	**US$25.86 billion**	**10**

Source: Author's Analysis Based on S&P Capital IQ

Table 12.4 Geographical Asset Allocation by HK-REITs (February 2017)

Market	*Total Assets*	*Revenue*
Hong Kong	US$36,569 million (70.6%)	US$1,744 million (65.6%)
Mainland China	US$15,260 million (29.4%)	US$916 million (34.4%)
Total	**US$51,829 million**	**US$2,660 million**

Source: Author's Analysis Based on S&P Capital IQ

Table 12.5 Ownership Structure of HK-REITs and J-REITs (Dec 2016)

	J-REITs	*HK-REITs*
Institutions	46.6%	19.0%
Corporations (Public)	3.7%	24.7%
Corporations (Private)	2.2%	14.1%
Individuals/Insiders	0.2%	1.8%
Public and Other	47.4%	40.4%
Total	100%	100%

Source: Author's Analysis Based on S&P Capital IQ

Table 12.6 Financial performance of J-REITs and HK-REITs (Dec 2016)

	J-REITs	*HK-REITs*
Profitability		
Return on Capital	2.0%	2.3%
Return on Equity	4.8%	7.6%
EBITDA/Total Revenue	64.9%	67.6%
Net Income/Total revenue	24.4%	29.5%
Long-Term Solvency		
Total Liabilities/Total Assets	0.493	0.303
EBIT/Interest Expenses	7.3x	6.2x
EBITDA/Interest Expenses	10.2x	7.0x
Three-Year Compound Annual Growth		
Total Revenue	5.1%	8.5%
EBITDA	5.3%	10.9%
Total Assets	3.6%	8.4%

Source: Author's Analysis Based on S&P Capital IQ

own only 5.9% of J-REIT units. Institutional shareholders, on the other hand, own 46.6% of J-REIT shares. For example, Nippon Building Fund was sponsored by Mitsui Fudosan who held only 3.4% equity interest in the J-REIT while institutional investors collectively owned 55.1% of the J-REIT in June 2016. The data shows that 86.7% of the institutional investors of J-REITs are mutual funds, followed by banks (9.4%) and insurance companies (3.2%). In contrast, institutional shareholders accounted for only 19% of total units issued by HK-REITs.

Table 12.6 presents the financial performance of J-REITs and HK-REITs. It is interesting to note that while the returns on capital employed (ROCE) of HK-REITs are only marginally higher than J-REITs (2.3% vs 2.0%), their returns on equity (ROE) are significantly different being 4.8% for J-REITs vs 7.6% for HK-REITs. While this could be attributed to the effect of financial leveraging, Table 12.6 shows that HK-REITs actually employ less debt in their capital structure relative to J-REITs.

While there is no restriction on the amount of debt that can be employed by J-REITs, the regulatory debt limit for HK-REITs is 45% of their total assets. This suggests that HK-REITs may engage in greater financial engineering activities to boost distribution yield to their unitholders. It is interesting to note that, despite the higher level of debt employed by J-REITs, their interest coverage ratios (EBIT/Interest Expenses and EBITDA/Interest Expenses) are higher than HK-REITs. This is probably due to the lower interest rate regime in Japan. The Japanese treasury-bill rate was -0.41% at the end of 2016, compared to 0.67% in Hong Kong.

Although the number of HK-REITs has remained static over the past three years, their average growth rate (measured by revenue, profitability or total assets) surpassed the growth rate recorded by J-REITs over the same period. This can be attributed to the more aggressive acquisition stance adopted by HK-REITs. The price of office space in Hong Kong has escalated by 17.0% compound annual growth rate over the past three years, compared to approximately 5% in the case of J-REITs. Similarly, retail properties in Hong Kong have shown 16.7% capital growth over the last three years (2013 to 2015), compared to 6.5% in Japan.

12.6 Case study: Link REIT, Asia's only internally managed REIT

Launched in November 2005, Link REIT was the first REIT to be listed on the Hong Kong Stock Exchange. The first attempt to list the REIT, which involved the securitization of some shopping centres and car parks owned by the Hong Kong's Housing Authority, was delayed due to a legal challenge by an elderly tenant of its public housing.

Link REIT also holds the distinction of being the largest REIT IPO (US$2.6 billion) in the world as well as being the first and only internally managed REIT in Asia. From the beginning, Link REIT was structured as an independent and professionally managed REIT. To divest non-core assets from its balance sheet, the sponsor, Hong Kong Housing Authority (HKHA) – the statutory body responsible for the provision of public housing and other related services in Hong Kong – injected 180 non-residential properties (comprising retail malls and car parks) into Link REIT. Unlike normal sponsors who will retain a majority stake in the newly listed REIT, HKHA opted not to own any units in Link REIT after the IPO. Link REIT therefore avoided the potential for conflict of interests between the controlling shareholder and the minority shareholders.

The ownership of Link REIT units by corporations stood at 1.17%, which is significantly lower than the 27% units retained by the average HK-REIT sponsor. Institutional investors, on the other hand, collectively owned around 50% of Link REIT's units vs 15.2% for the average HK-REIT. The higher degree of institutional ownership of Link REIT implies that the actions of the Link REIT manager will be more closely monitored by its unitholders (Lecomte and Ooi, 2013; Xu and Ooi, 2016). By implication, Link REIT has a higher corporate governance standard. For example, independent non-executive directors constituted 77% of the board members of Link REIT, compared to 45% for the average HK-REIT. Link REIT also adopts a Board diversity policy, which stipulates that board appointments are made based on merit and diversity taking into account the skill-set, industry expertise and experience, background, ethnicity, age and gender of the board members.

As a newly listed entity, the conventional principal growth avenue for a REIT is through external acquisitions, either from third-party property owners or from their sponsors. As noted by Downs et al. (2016), REITs that are linked heavily to their sponsors tend to be more active in their acquisitions. Thus, while a close relationship with the sponsor may lead to agency issues, it offers young REITs a steady growth avenue.

Downs et al. (2016) examine related party transactions (RPTs) and their impact on firm value for a sample of Hong Kong, Malaysia and Singapore REITs. They find that the average size of related party transactions reported annually by REITs listed in these markets is 5.45% of total assets. The three main channels for related party transactions by REITs are acquisitions of real estate assets from related parties (57.4%), income earned from related parties (22.2%) and management fees paid to related parties (14.8%). They find related party transactions have a positive firm value effect which is driven by the related party transactions acquisition channel.

Link REIT only made its first property acquisition in July 2011, being five years after its IPO. Five properties, worth US$1.73 billion, were added to its portfolio between 2011 and 2015. Specifically, Link REIT engaged in two watershed deals in 2015. First, it started to venture outside of Hong Kong by acquiring retail and office properties in mainland China. Second, it forayed into property development by successfully tendering for a commercial development site on Hoi Bun Road in Kowloon, through a joint-venture with Nam Fung Development Limited. The HK$10.5 billion (US$1.35 billion) project is scheduled to be completed in 2019. Identifying property development and redevelopment activity as one of its distribution growth drivers, Link REIT stated in its 2016 annual report that the developed properties will be held for long-term investment.

In sharp contrast to other HK-REITs that mostly acquired properties (99.2%) from their sponsors, all the properties acquired by Link REIT since its IPO were previously owned by third parties. This is despite the fact that Link REIT was granted a first right of refusal to acquire certain properties owned by HKHA during the first 10 years after its IPO. These arm-length transactions avoided potential conflicts of interest associated with related party acquisitions, such as injection of low-quality and or overpriced properties in to the REITs by their IPO sponsors.

With the exception of Link REIT, all of the REITs in Asia are externally managed with their day-to-day management outsourced to an external manager, who is usually controlled by the sponsor. Acting like investment advisors to mutual funds, the external REIT managers receive compensation through two types of fees, namely a base fee, which is expressed as a percentage of the value of the REIT's asset under management (AUM), plus an incentive fee. Examining the impact of managers' compensation structure on IPO pricing of 20 REITS in Singapore and their stock's subsequent performance, Ooi (2009) observes that the market favours REITs that pay their managers a low base fee coupled with a high incentive fee that is benchmarked against a pre-determined performance level.

In the case of Link REIT, the REIT manager, namely Link Asset Management Limited, is wholly owned by the Trustee. The management fee of Link Asset Management Limited is charged on a cost recovery basis, unlike typical REIT managers' fees which are usually based on a percentage of the AUM of the REIT, with Ooi (2009) noting that a fee structure that is based on AUM tends to encourage REIT managers to pursue growth. For the financial year ended March 2016, the actual management fee of Link REIT equates to around 0.4% of its AUM which is not dissimilar to the 0.3% to 0.4% fee incurred by the externally managed Hui Xian REIT and Champion REIT.

Though the manager's fee is disconnected from AUM, Link REIT's portfolio grew from US$4.6 billion to US$21.6 billion between March 2006 and June 2016. Since its IPO, the HK-REIT has engaged in a series of 45 asset enhancement activities, such as the creation of additional floor place, reconfiguration of inefficient space, improvement of accessibility and connectivity and reduction of energy consumption. For example, Lok Fu Plaza, which is Link REIT's flagship shopping centre, completed a major asset enhancement programme in December 2010 costing US$54.7 million (HK$425m). The value of Lok Fu Plaza reportedly increased from US$214.4 million (HK$1.66 billion) during the IPO, to US$461.0 million (HK$3.58 billion) in March 2011 after the enhancement works. Not resting on its laurels, Link REIT announced in September 2016 an asset enhancement activities pipeline for 37 properties which will cost US$398.8 million (HK$3.10 billion, based on an exchange rate of 1HKD = 0.13USD as at 14 March 2017).

Table 12.7 contrasts the initial trading day returns of Link REIT against three other HK-REITs, namely Sunlight REIT, Champion REIT and Prosperity REIT. The returns data show

Table 12.7 Post-IPO returns of four HK-REITs

	IPO Date	*First-day trading returns*	*First-year holding returns*	*CAGR (IPO to Dec 2016)*
The Link	25 Nov 2005	14.56%	57.67%	11.85%
Prosperity REIT	16 Dec 2005	20.37%	-18.98%	2.04%
Champion REIT	24 May 2006	-15.69%	-10.78%	1.11%
Sunlight REIT	21 Dec 2006	-6.54%	-10.77%	7.06%

Source of Data: Author's Analysis Based on Bloomberg and Datastream

Table 12.8 Financial Attributes of Link REIT vs its Peers (Dec 2016)

	Link REIT	*Average Peer*
Market capitalization	US$15.4 billion	US$1.3 billion
Price/book value	0.9x	0.7x
P/E ratio	8.0x	12.1x
Five-year beta	0.60	0.52
EBITDA/revenue	71.2	71.7
EBIT/revenue	71.0	67.8
Corporate governance		
% of value of related party acquisitions	0%	99.2%
Corporate shareholder (%)	1.2%	27.6%
Institutional investors (%)	49.6%	15.2%
Independent directors (%)	76.9%	44.7%

Source of Data: Author's Analysis Based on S&P Capital IQ and Annual Reports

that Link REIT enjoyed a strong debut on the initial trading day, recording 14.6% initial day return as compared to an average of 0.62% for its three peer HK-REITs, which ranged from -15.7% for Champion REIT to 20.4% for Prosperity REIT. Recognizing that the success of a HK-REIT goes beyond the response to the initial IPO returns, Table 12.7 also presents the one-year holding period returns of the HK-REITs during their first year of listing as well as the compound annual growth rate (CAGR) of the individual HK-REITs. While the three peer HK-REIT stocks recorded negative returns, Link REIT recorded an impressive 57.7% return in the first year. Link REIT's CAGR of 11.85% is also impressive compared to its peers' performance, which ranged from 1.11% to 7.06% since their listing.

Table 12.8 contrasts the financial attributes of Link REIT against the average of its peers. Link REIT's market capitalization of US$15.40 billion dwarfed the size of the other nine REITs in Hong Kong, which have an average market capitalization of US$1.3 billion. Consistent with the earlier findings, Link REIT's performance is superior to its peers except for its P/E ratio, which is trading at a significantly lower 8x compared to 12.1x for the average HK-REIT. The 12.1x P/E ratio of the average peer is driven mainly by Regal REIT's P/E of 37.8x, which was artificially caused by its low EPS in 2016 (i.e. $0.05 vs the average of $0.37 during 2008–2015), with the average peer P/E ratio reduced to 7.4x if Regal REIT is excluded from the sample of peer REITs.

12.7 Summary

Part II includes six chapters each analysing REITs in a region of the world, using case studies from a developed, developing and emerging REIT sector in the region.

Chapter 8 analysed REITs in the North American region, using case studies from the USA, Canada and Mexico, with Chapter 9 analysing REITs in the Latin American region, using case studies from Brazil, Argentina and Uruguay for the developed, developing and emerging REIT sectors in the region, respectively.

Chapter 10 analysed REITs in the European region, using case studies from the UK, Spain and Poland, with Chapter 11 analysing REITs in the South East Asian region, using case studies from Singapore, Malaysia and Thailand for the developed, developing and emerging REIT sectors in the region, respectively.

This chapter analysed REITs in the North Asian region, using case studies from Japan (being a developed REIT sector in the region), Hong Kong (being a developing REIT sector in the region) and China (being an emerging REIT sector in the region).

As a developed REIT sector, Japan's REIT market is now almost 20 years old with a market capitalization of US$103.5 billion, being some distance ahead of the developing REIT sector in Hong Kong which has only been trading for just over a decade and achieved a market capitalization of US$25.6 billion. Hong Kong is unusual in that it does not provide tax transparency for REITs, effectively making REITs a trust-based equivalent of the much more familiar and longer-established Hong Kong property company. It is also surprising that China has not yet embraced all aspects of the conventional REIT structure, with the emerging REIT sector currently comprising a range of quasi-REIT products.

Though the outlook for REITs in China is optimistic, there are several hurdles that must be cleared starting with the issue of specific legislation for REITs which requires critical support from government at various levels to introduce a friendly REIT regime with favourable tax transparency status granted to the REITs. Once the legislative framework is established, the prospective sponsors will have to structure a REIT vehicle that can offer stable returns and high distributions to investors. This is not an easy task in Asia, where institutional-grade properties are generally expensive and have very low yields (Ooi, Newell and Sing, 2006).

An on-going debate that continues is the prospect of internalizing REIT managers to reduce conflict of interests between minority shareholders and the sponsors. Link REIT, which is the only internally managed REIT in Asia, offers a good case for internalization. Wong, Ong and Ooi (2013) nevertheless argue that the continued involvement of the REIT sponsor in the day-to-day management of the REIT is relevant for a number of reasons. In particular, REITs backed by reputable sponsors are underpriced less at IPO due to implicit certification of the quality of the IPO. In addition, post-listing support by the sponsor is important for the REIT in terms of future growth opportunities.

As more and more REITs are listed, the market will start to distinguish between well-managed REITs and poorly managed REITs. In particular, a poorly managed REIT would be penalized through a deep discount to its share price. Lecomte and Ooi (2013) provide empirical support for a positive correlation between good corporate governance practices and stock performance. Moving forward, adopting good corporate governance and transparency practices that are consistent with international standards will be important for REIT managers seeking to attract institutional investors.

While REIT markets in both Japan and Hong Kong have grown substantially, there is scope for REITs to grow further in North Asia, particularly with the restructuring in South Korea and their potential introduction in China. Overall, the development of a vibrant REIT market will be good for North Asia.

Chapter 13 will analyse REITs in the Oceania region, using case studies from Australia, South Africa and India for the developed, developing and emerging REIT sectors in the region, respectively, with Chapter 14 concluding Part II by considering *Directions for the Future of International REITs.*

References

Downs, D., Ooi, J.T.L., Wong, W.C. and Ong, S.E. 2016, 'Related party transactions and firm valuation: Evidence from property markets in Hong Kong, Malaysia and Singapore', *Journal of Real Estate Finance and Economics*, Vol. 52, No 4, pp. 408–427.

Lecomte, P. and Ooi, J.T.L. 2013, 'Corporate governance and performance of externally managed Singapore REITs', *Journal of Real Estate Finance and Economics*, Vol. 46, No. 4, pp. 664–684.

Ooi, J.T.L. 2009, 'The compensation structure of REIT managers: Impact on stock valuation and performance', *Journal of Property Research*, Vol. 26, pp. 309–328.

Ooi, J.T.L. and Neo, P.H. 2010, 'Asian REITs: Playing the yield game' in *Global trends in real estate finance*, ed. G. Newell and K. Sieracki, Wiley-Blackwell, Oxford.

Ooi, J.T.L., Newell, G. and Sing, T.F. 2006, 'The growth of REIT markets in Asia', *Journal of Real Estate Literature*, Vol. 14, No. 2, pp. 203–222.

Ooi, J.T.L., Wong, W.C. and Ong, S.E. 2012, 'Can bank lines of credit protect REITs against a credit crisis? *Real Estate Economics*, Vol. 40, No. 2, pp. 285–316.

RICS 2016, *REITs in China: Opportunities and challenges*, RICS Research Report, May 2016, RICS, London.

Roxburgh, H. 2015, 'Is a China REIT imminent?' *IPE Real Estate*, September/October 2015 (magazine), https://realestate.ipe.com/markets-/regions/asia-pacific/is-a-china-reit-imminent/10009922.article.

Wong, W.C., Ong, S.E. and Ooi, J.T.L. 2013, 'Sponsors backing in Asian REIT IPOs', *Journal of Real Estate Finance and Economics*, Vol. 46, No. 2, pp. 299–320.

Xu, R.R. and Ooi, J.T.L. 2016, 'Good growth, bad growth: How effective are the corporate watchdogs?' *Journal of Real Estate Finance and Economics* (forthcoming).

Yanagihara, S. and Kim, J. 2016, *Incentive for globalization of J-REITs*, 22nd Annual Pacific-Rim Real Estate Society Conference, 17–20 January 2016, Sunshine Coast, Australia.

Appendix 12.1
LIST OF J-REITS (DEC 2016)

No	*Name*	*Date Listed*	*Mkt Cap (US$ billion)*
1	Nippon Building Fund Inc.	Sep-01	7.833
2	Japan Real Estate Investment Corp.	Sep-01	7.151
3	Nomura Real Estate Master Fund Inc.	Oct-15	6.341
4	Japan Retail Fund Investment Corp.	Mar-02	5.177
5	United Urban Investment Corp.	Dec-03	4.641
6	Orix JREIT Inc.	Jun-02	3.994
7	Nippon Prologis REIT Inc.	Feb-13	3.898
8	Daiwa House REIT Investment Corp.	Mar-06	3.850
9	Advance Residence Investment Corp.	Mar-10	3.577
10	Japan Prime Realty Investment Corp.	Jun-02	3.447
11	GLP J-REIT	Dec-12	3.279
12	Activia Properties Inc.	Jun-12	3.143
13	Japan Hotel REIT Investment Corp.	Jun-06	2.535
14	Daiwa Office Investment Corp.	Oct-05	2.518
15	Mori Hills REIT Investment Corp.	Nov-06	2.370
16	Kenedix Office Investment Corp.	Jul-05	2.333
17	Frontier Real Estate Investment Corp.	Aug-04	2.124
18	Nippon Accomodation Fund Inc.	Aug-06	2.123
19	Mori Trust Sogo REIT Inc.	Feb-04	2.087
20	Japan Logistics Fund Inc.	May-05	1.859
21	Hulic REIT	Feb-14	1.757
22	Industrial & Infrastructure Fund Invest. Corp.	Oct-07	1.684
23	Invincible Investment Corp.	May-04	1.661
24	Japan Excellent Inc.	Jun-06	1.660
25	Premier Investment Corp.	Sep-02	1.561
26	Aeon REIT Investment	Nov-13	1.436
27	Tokyu REIT Inc.	Sep-03	1.241
28	Sekisui House SI Residential Investment Corp.	Jul-05	1.231
29	Fukuoka REIT Corp.	Jun-05	1.184
30	Comforia Residential REIT Inc.	Feb-13	1.156
31	Sekisui House REIT Inc.	Dec-14	1.151

No	*Name*	*Date Listed*	*Mkt Cap (US$ billion)*
32	Japan Rental Housing Investment Corp.	Jun-06	1.102
33	LaSalle LOGIPORT REIT	Feb-16	1.044
34	Ichigo Office REIT Corp.	Oct-05	0.995
35	Kenedix Retail REIT Corp.	Feb-15	0.965
36	Nippon REIT Investment Corp.	Apr-14	0.957
37	Kenedix Residential Investment Corp.	Apr-12	0.947
38	Hoshino Resorts REIT Inc.	Jul-13	0.855
39	MCUBS Midcity Investment Corp.	Aug-06	0.769
40	Hanyu REIT Inc.	Oct-05	0.748
41	Heiwa Real Estate REIT Inc.	Mar-05	0.728
42	Global One Real Estate Investment Corp.	Sep-03	0.728
43	Invesco Office J-REIT Inc.	Jun-14	0.674
44	Mitsui Fudosan Logistics Park Inc.	Aug-16	0.640
45	Starts Proceed Investment Corp.	Nov-05	0.333
46	SIA REIT Inc.	Oct-13	0.321
47	Star Asia Investment Corp.	Apr-16	0.288
48	Ichigo Hotel REIT Investment Corp.	Nov-15	0.281
49	Sakura Sogo REIT Investment	Aug-16	0.225
50	Samty Residential Investment Corp.	Jun-15	0.210
51	Tosei REIT Investment Corp	Nov-14	0.169
52	Ooedo Onsen REIT Investment Corp.	Aug-16	0.122
53	Nippon Health Care Investment Corp.	Nov-14	0.113
54	Health care & Medical Investment Corp.	Mar-15	0.108
55	Japan Senior Living Investment Corp.	Jul-15	0.103
56	Marimo Regional Revitalization REIT Inc.	Jul-16	0.063
	Total		**US$ 103.49 billion**

Appendix 12.2
LIST OF HK-REITS (DEC 2016)

No	*Name*	*Date Listed*	*Mkt Cap (US$ billion)*
1	Link REIT	Nov-05	14.449
2	Champion REIT	May-06	3.141
3	Hui Xian REIT	Apr-11	2.468
4	Yuexiu REIT	Dec-05	1.507
5	Sunlight REIT	Dec-06	0.943
6	Regal REIT	Mar-07	0.866
7	Langham Hospitality Investment	May-13	0.835
8	Prosperity REIT	Dec-05	0.574
9	Spring REIT	Dec-13	0.470
10	New Century REIT	Jul-13	0.306
	Total		**US$ 25.56 billion**

13
OCEANIA

Chyi Lin Lee

13.1 Introduction

This book aims to identify key areas for research in the REIT discipline for the next five to ten years by surveying the current state of the REIT discipline around the world and identifying emerging and cutting-edge research areas through a thematic review of current contextual issues and a regional analysis based on case studies.

This book comprises two parts, the first part being a thematic review of emerging and cutting edge global research into current contextual issues in REITs internationally and the second part being a regional analysis of REITs around the world, each written by authoritative academic authors from the world's leading Universities and REIT industry experts.

Part I included six chapters each reviewing a current theme of REIT evolution through the lens of contemporary research. Chapter 1 focused on critical contextual issues in international REITs while Chapter 2, the *Post-Modern REIT Era*, examined the evolution of the global REIT market, what is deemed to be best current market practice and how the market is expected to change going forward and Chapter 3, *Emerging Sector REITs*, investigated such sectors as residential, health care, self-storage, timber, infrastructure and data centres generally and then specifically through a case study of US REITs.

Chapter 4, *Sustainable REITs*, analysed REIT environmental performance and the cost of equity and Chapter 5, *Islamic REITs*, examined the evolution of global Islamic REITs through a study of the Malaysian market. Chapter 6, *Behavioural Risk in REITs*, addressed the management of behavioural risk in global REITs and Part I concluded with Chapter 7 which reviewed recent research into *REIT Asset Allocation*.

Part II includes six chapters each analysing REITs in a region of the world, using case studies from a developed, developing and emerging REIT sector in the region. Chapter 8 analysed REITs in the North American region, using case studies from the USA, Canada and Mexico, with Chapter 9 analysing REITs in the Latin American region, using case studies from Brazil, Argentina and Uruguay for the developed, developing and emerging REIT sectors in the region, respectively.

Chapter 10 analysed REITs in the European region, using case studies from the UK, Spain and Poland, with Chapter 11 analysing REITs in the South East Asian region, using case studies from Singapore, Malaysia and Thailand and Chapter 12 analysed REITs in the North Asian region, using case studies from Japan, Hong Kong and China for the developed, developing and emerging REIT sectors in the region, respectively.

This chapter analyses REITs in the Oceania region, using case studies from Australia (being a developed REIT sector in the region), South Africa (being a developing REIT sector in the region) and India (being an emerging REIT sector in the region).

Australian REITs (A-REITs) are the largest REIT market in Oceania and one of the longest established and largest REIT markets in the world, making a significant contribution to the global REIT market. While the South African REIT (SA-REIT) market structure was only formally introduced in 2013, it has increased noticeably and the Indian REIT market is now finally emerging.

Finally, Chapter 14 will conclude Part II and consider *Directions for the Future of International REITs*.

13.2 Australia

REITs in Australia were formerly known as Listed Property Trusts then renamed in 2008 as Australian Real Estate Investment Trusts (A-REITs), being a more globally adopted terminology (Lee, 2010). This section will first examine the significance of the A-REIT market, then the profile of AREITs, their key features and their characteristics before considering A-REIT futures.

13.2.1 Significance of A-REITs in the global real estate market

REITs play a vital role in the global securitized real estate market. As highlighted in Figure 13.1, almost 75% of the total market capitalization of global real estate securities was contributed by REITs (EPRA, 2016a). In Australia, A-REITs have been one of the most successful indirect property investment vehicles. In fact, the success of REITs in Australia and the US has contributed to the proliferation of REIT and REIT-like vehicles globally (Lee, 2010).

Although the structure of A-REITs (formerly Listed Property Trusts [LPTs]) was introduced in 1959 in Australia, no LPT was established in the 1960s. The first LPT, General Property Trust, launched an initial public offering (IPO) in 1971. There was then no further new LPTs until the listing of Westfield and Stockland in 1982 (Sing and Ling, 2003; Lee, 2010).

The major growth of A-REITs did not occur till the early 1990s, when the failure of unlisted property trusts and the downturn of the commercial property market drove investors,

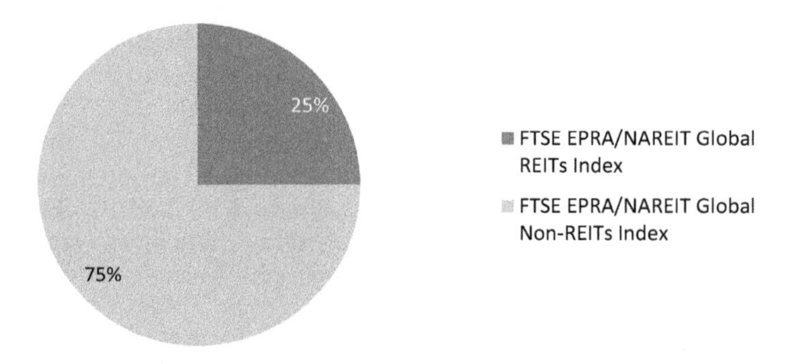

Figure 13.1 Global REITs/Non-REITs Index-Breakdown by Market Cap as of 30 October 2016

Source: EPRA (2016b)

particularly institutional investors, to seek an effective property investment vehicle with a rental income-focus, which principally involved prime commercial property investments while offering high liquidity (Stringer, 2001). A-REITs were widely seen as an effective real estate product that offered these features. Therefore, the increasing popularity of A-REITs was in response to their emerging as a way for institutional investors to hold property in a liquid form.

However, the traditional rental income-focused A-REIT structure changed significantly in the 2000s when more aggressive growth strategies were adopted by numerous A-REITs. These strategies include increased levels of gearing, increased levels of international property exposure, the use of stapled structures and investing in emerging property sectors (e.g. retirement, health care, leisure, child care and others) (Newell, 2006).

The global financial crisis (GFC) of 2007–2009 had a significant negative impact on the A-REIT market. Specifically, the crisis highlighted the significant loss of the traditional "low risk" defensive characteristics of A-REITs in response to the significant increased volatility of A-REITs (Lee and Lee, 2012). Importantly, the adopted aggressive growth strategies such as high debt levels, international property exposure and stapled structures were argued to have had an adverse impact on A-REIT performance.

In particular, high debt levels were identified as the most critical factor (Newell and Peng, 2009). Consequently, numerous A-REITs refocused on their core activities, including reducing debt levels by undergoing a major recapitalization process and focusing on Australian prime commercial properties (for example, Westfield Group separated its Australian and New Zealand operations from its international business; MorningStar, 2017). The consolidation process in the A-REIT sector then led to a recovery of A-REITs from the GFC and was seen as a key strategy to enhance returns, with around US$6.8 billion in merger and acquisition activities occurring in the Australian REIT sector in 2015 (Ernst and Young, 2016).

Today, the A-REIT market is one of the largest REIT markets in the world. As of 31 January 2017, the A-REIT market was ranked as the third largest REIT market in the world, accounting for approximately 7% of the global REIT market (EPRA, 2017). Figure 13.2 compares the size (market capitalization) of A-REITs with four of the largest REIT markets in the world. As highlighted in Figure 13.2, as of 31 January 2017, the US REIT market was the largest market in the world with a market capitalization of A$75 billion. It was almost nine times larger than the Japanese REIT market, which was the second largest REIT market having a total market capitalization of A$110 billion. The A-REIT market was slightly smaller than the Japanese

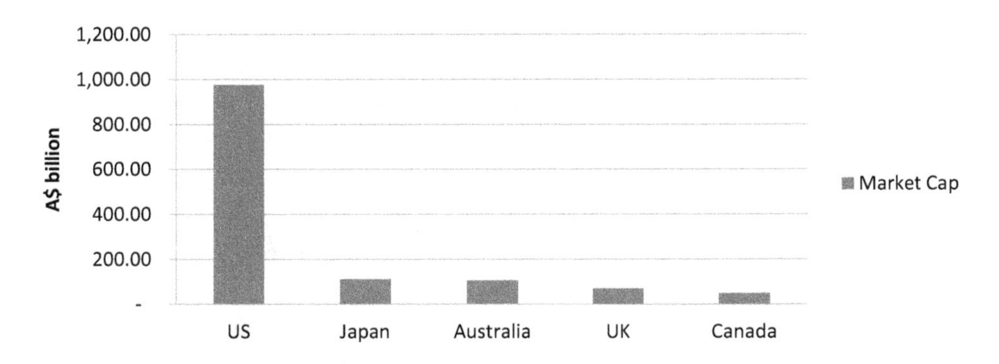

Figure 13.2 Size of Top Five REIT Markets: January 2017 (A$ billion)

Note: US$1 = A$1.30 (February 2017)

Source: EPRA (2017)

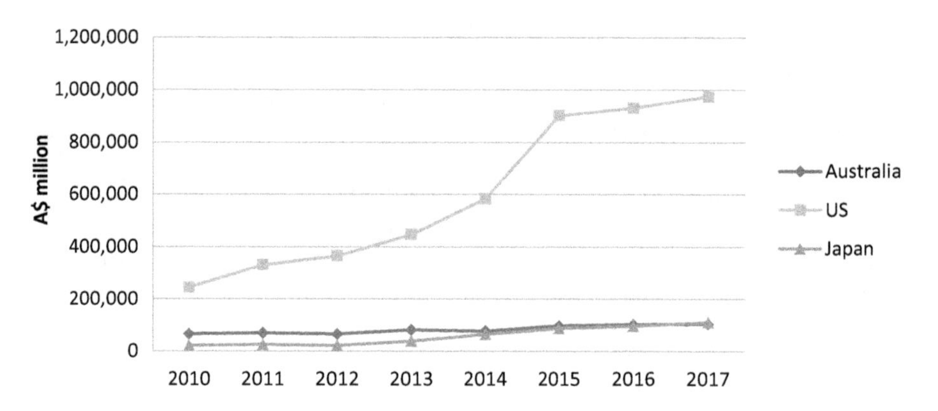

Figure 13.3 The Growth of Top 3 Global REIT Markets

Source: Author's compilation from DataStream (2017)

REIT market with a total market capitalization of A$105 billion. These were followed by the UK and Canadian REIT markets.

Figure 13.3 compares the growth of the top three REIT markets (the US, Australia and Japan) in the world over 2010–2017. As illustrated in Figure 13.3, the US and Japanese REIT markets have recorded a significant growth rate since 2013. More specifically, the size of REIT markets in the US and Japan, on average, grew 23% pa and 40% pa, respectively, over 2013–2017. However, the size of the Australian REIT market did not increase so rapidly with an average growth rate of 11% pa over the same period, thereby losing its status of the second largest REIT market to the Japanese REIT market in 2016.

Nonetheless, the A-REIT market is still a critical real estate investment sector which has received considerable attention from Australian investors as real estate, particularly A-REITs, have been widely included as one of the major asset classes for many Australian superannuation funds (Newell, Lee and Kupke, 2015). In other words, A-REITs are a key asset class in Australia.

The involvement of institutional investors in the A-REIT market has increased the demand for high quality information which has improved real estate market transparency in Australia dramatically. The recent 2016 JLL global real estate transparency index ranked the Australian real estate market as the second most transparent property market globally (JLL, 2016). Importantly, Australia continues to hold the top spot as the most transparent real estate market in the Asia-Pacific.

Overall, A-REITs have been highly successful indirect property investment vehicles in Australia despite the significant negative impact of the GFC on the A-REIT market. Nevertheless, the A-REIT market has recovered from the GFC and continues to play a major role as a leading property investment vehicle in Australia and globally.

13.2.2 Profile of Australian REITs (A-REITs)

At June 2016, the A-REIT sector had total assets of over $151 billion, comprising over 2,000 institutional-grade commercial properties in sector-specific and diversified portfolios in Australia and overseas (PIR, 2016). A-REITs are also the largest institutional owners of commercial property in Australia, ahead of unlisted wholesale property funds and unlisted retail property funds. A-REITs accounted for over $120 billion in market capitalization, being one of the largest sectors in the Australian stock market (ASX200 List, 2017).

Table 13.1 Profile of Australian REITs: 31 December 2016

No	*A-REITs*	*ASX Code*	*Market Capitalization (A$mil)*	*Management Structure*	*Number of Properties*	*Trust Type*
1	Abacus Property Group	ABP	1,728	Stapled	93	Diversified
2	Agricultural Land Trust	AGJ	5	Units	1	Agriculture
3	Aims Property Securities Fund	APW	68	Units	0	Property Securities
4	ALE Property Group	LEP	824	Stapled	86	Leisure
5	Ante Real Estate Trust	ATT	0.5	Units	2	Leisure
6	Arena REIT	ARF	437	Stapled	201	Health care
7	Asia-Pacific Data Centre Group	AJD	183	Stapled	3	Specialized
8	Aspen Group	APZ	115	Stapled	8	Diversified
9	Astro Japan Property Group	AJA	403	Stapled	31	International Property
10	Australian Unity Office Fund	AOF	284	Stapled	8	Office
11	Aventus Retail Property Fund	AVN	931	Units	20	Retail
12	Blackwall Limited	BWF	41	Units	n/a	Diversified
13	Blackwall Property Trust	BWR	79	Units	6	Diversified
14	Brookfield Prime Property Fund	BPA	310	Units	3	Office
15	BWP Trust	BWP	1,921	Units	82	Retail
16	Carindale Property Trust	CDP	532	Units	1	Retail
17	Centuria Metropolitan REIT	CMA	256	Stapled	12.5	Office/ Industrial
18	Charter Hall Group	CHC	1,956	Stapled	296	Diversified
19	Charter Hall Long Wale REIT	CLW	829	Stapled	66	Diversified
20	Charter Hall Retail REIT	CQR	1,718	Units	74	Retail
21	Cromwell Property Group	CMW	1,733	Stapled	26	Diversified
22	Dexus Property Group	DXS	9,312	Stapled	102	Diversified
23	Elanor Retail Property Fund	ERF	174	Stapled	9	Diversified
24	Folkestone Education Trust	FET	633	Units	396	Child care
25	Garda Diversified Property Fund	GDF	119	Units	8	Diversified
26	GDI Property Group	GDI	533	Stapled	5	Office
27	Generation Health Care REIT	GHC	421	Units	17	Health care

(*Continued*)

Table 13.1 (Continued)

No	A-REITs	ASX Code	Market Capitalization (A$mil)	Management Structure	Number of Properties	Trust Type
28	Goodman Group	GMG	12,756	Stapled	33	Industrial
29	GPT Group	GPT	9,043	Stapled	69	Diversified
30	Growthpoint Properties Australia	GOZ	2,104	Stapled	49	Diversified
31	Hotel Property Investments	HPI	415	Stapled	48	Leisure
32	Industria REIT	IDR	343	Stapled	17	Industrial
33	Ingenia Communities Group	INA	476	Stapled	65	Retirement
34	Investa Office Fund	IOF	2,898	Stapled	22	Office
35	Lantern Hotel Group	LTN	93	Stapled	n/a	Leisure
36	Mirvac Group	MGR	7,892	Stapled	59	Diversified
37	National Storage REIT	NSR	752	Stapled	94	Industrial
38	RNY Property Trust	RNY	9	Units	20	International Property
39	Rural Funds Group	RFF	362	Stapled	32	Agriculture
40	Scentre Group	SCG	24,705	Stapled	40	Retail
41	Shopping Centres Australasia Property Group	SCP	1,622	Stapled	81	Retail
42	Stockland	SGP	11,015	Stapled	222	Diversified
43	US Masters Residential Property Fund	URF	716	Units	595	International Property
44	Vicinity Centres	VCX	11,836	Stapled	85	Retail
45	Viva Energy REIT	VVR	1,656	Stapled	133	Specialized
46	Westfield Corporation	WFD	19,492	Stapled	34	Retail
47	360 Capital Group	TGP	210	Stapled	6	Diversified
48	360 Capital Industrial Fund**	TIX	532	Units	37	Industrial
49	360 Capital Office Fund*	TOF	157	Units	3	Office
50	360 Capital Total Return Fund***	TOT	34	Stapled	0	N/A

Notes:
*360 Capital Office Fund changed its name to Centuria Urban REIT (CUA) on 20 January 2017.
** 360 Capital Industrial Fund was renamed as Centuria Industrial REIT (CIP).
*** 360 Capital Total Return Fund successfully completed the sale of two properties in Frenchs Forrest NSW in 2016 for A$26 million.

Sources: ASX (2017a), PIR (2016) and Morningstar (2017)

As at 31 December 2016, there were several major participants in the A-REIT sector, each having a substantial market capitalization, such as Scentre Group (A$24.7 billion), Westfield Corporation (A$19.5 billion), Goodman Group (A$12.8 billion), Vicinity Centres (A$11.8 billion), Stockland (A$11.0 billion), Dexus Property Group (A$9.3 billion), GPT Group (A$9.0 billion) and Mirvac Group (A$7.9 billion). AREITs are also one of the most important sectors in the Australian securitized real estate market, accounting for 45% of the total assets under management of all property funds in Australia in 2016 (i.e. A-REITs, unlisted wholesale funds, unlisted retail funds, property securities funds, etc.) (PIR, 2016).

There were 50 A-REITs as at 31 December 2016, investing in various sectors including office (e.g. Investa Office Fund), retail (e.g. Scentre), industrial (e.g. Goodman), diversified (e.g. Stockland), leisure (e.g. ALE Property Group), health care (e.g. Generation Health Care REIT), child care (e.g. Folkestone Education Trust), retirement village (e.g. Ingenia Communities Group) and specialized property (e.g. Viva Energy REIT which invests in petrol stations), as well as international property (e.g. US Masters Residential Property Fund). (Table 13.1.)

Generally, most A-REITs adopt a stapled REIT structure instead of a traditional unit structure. Specifically, more than two-thirds of A-REITs (i.e. 33 A-REITs) adopt a stapled REIT structure (ASX, 2017a). A stapled structure allows REITs to engage in the activities of traditional REIT management as well as property development and funds management. Notably, the stapled entities play a significant role in the overall market capitalization of A-REITs, with the top 10 A-REITs being stapled entities.

13.2.3 Key features of A-REITs

This section considers several key features of A-REITs, including organizational structure, income and activity, distribution rules, gearing and management style.

13.2.3.1 Organizational structure

Generally, an A-REIT should be in a trust structure and a managed investment scheme, being subject to the rules for Managed Investment Schemes. The trust must be managed by a Responsibility Entity or corporate trustee or fund manager (EPRA, 2016b). A-REITs can adopt either a stand-alone or a stapled structure. A stand-alone structure is a conventional trust structure, being the traditional rental income-focused A-REIT structure which only engages in passive activities by focusing on operating or managing income-producing properties. The role of the trust is to generate income, which is mainly derived from the rental income of the underlying property assets (Lee, 2010). The trust is expected to distribute its taxable income (or earnings) to its unit holders as distributions (EPRA, 2016b). To avoid any conflict of interest, the traditional trust structure must be managed externally (ASX, 2017b).

Alternatively, a stapled structure allows an A-REIT to have two components (i.e. a traditional trust unit component attached to a company share component). Therefore, the stapled structure A-REITs are allowed not only to undertake a range of activities relating to passive property holding (i.e. managing and operating the underlying properties), but also to participate in development opportunities and/or funds management (Lee, 2010). To engage in these activities effectively allows the management function of a stapled security REIT to be "internalized" or internally managed (APREA, 2014; EPRA, 2016b). While both components are traded together and simultaneously at a single price on the ASX, the active (non-rental) income is subject to a separate tax treatment (ATO, 2007).

13.2.3.2 Income and activity

A-REITs are required to satisfy the "safe harbour" test in which 75% or more of the trust's revenue should be derived from rental income. In addition, A-REITs may derive not more than 2% of their gross revenue from things other than eligible investment business (APREA, 2014). In other words, A-REITs must not directly or indirectly carry on a "trading business" (i.e. a business that does not consist wholly of an eligible investment business).

13.2.3.3 Distribution rules

There are no prescribed minimum distribution rules for A-REITs who are required to distribute their taxable income as distributions (dividends) to unit holders in order to be eligible for "flow-through" taxation treatment (tax transparency). A-REITs therefore typically distribute all of their taxable income to unitholders on a quarterly or semi-annually basis. The distribution policy of 100% of trust income is typically embedded in the trust's constitution (APREA, 2014; EPRA, 2016b).

13.2.3.4 Gearing

There is no gearing limit for A-REITs under the Australian taxation law (EPRA, 2016b). However, the thin capitalization rules (or a gearing limit) may apply if an A-REIT is foreign controlled (APREA, 2014). The thin capitalization rules have a gearing limit of 60% of gross assets of an A-REIT (EPRA, 2016b). Although there is no gearing limit, A-REIT managers adopt a conservative borrowing strategy having an indicative gearing level of 36% in 2016 (PIR, 2016).

13.2.3.5 Management style

All A-REITs should be externally managed with the only exception being A-REITs with a stapled structure which are internally managed and engage in a range of active activities including property development and/or funds management (APREA, 2014; EPRA, 2016b).

13.2.4 Characteristics of A-REITs

This section provides a general profile of A-REITs, with size, dividend yield, performance analysis, investment focus, management structure and gearing considered.

13.2.4.1 Size of A-REITs

Table 13.2 shows the size (market capitalization) of the top 20 A-REITs. As at 31 December 2016, there were 50 A-REITs with the total market capitalization of A$134.66 billion.

Scentre Group was the largest REIT in Australia with a total market capitalization of A$24.7 billion, representing around 18% of the total market capitalization of A-REITs (ASX, 2017a). Scentre Group was formally established on 30 June 2014 through the merger of Westfield Retail Trust (WRT) and Westfield Group's Australian and New Zealand management business. The merger was mainly driven by the announcement of Westfield Group in December 2013 to separate its Australian and New Zealand business from its international operations (MorningStar, 2017). Although Scentre Group is a relatively new REIT, it owns and manages 40 Westfield shopping centres in Australia and New Zealand. Westfield Corporation, the second

Table 13.2 Market Capitalization of Top 20 A-REITs: 31 December 2016

No	*A-REITs*	*ASX Code*	*Market Capitalization (A$mil)*
1	Scentre Group	SCG	24,705
2	Westfield Corporation	WFD	19,492
3	Goodman Group	GMG	12,756
4	Vicinity Centres	VCX	11,836
5	Stockland	SGP	11,015
6	Dexus Property Group	DXS	9,312
7	GPT Group	GPT	9,044
8	Mirvac Group	MGR	7,892
9	Investa Office Fund	IOF	2,898
10	Growthpoint Properties Australia	GOZ	2,104
11	Charter Hall Group	CHC	1,956
12	BWP Trust	BWP	1,921
13	Cromwell Property Group	CMW	1,733
14	Abacus Property Group	ABP	1,728
15	Charter Hall Retail REIT	CQR	1,718
16	Viva Energy REIT	VVR	1,656
17	Shopping Centres Australasia Property Group	SCP	1,622
18	Aventus Retail Property Fund	AVN	931
19	Charter Hall Long Wale REIT	CLW	829
20	ALE Property Group	LEP	824

Source: ASX (2017a)

largest A-REIT with a market capitalization of A$19.5 billion, currently manages and owns 34 Westfield shopping centres internationally, including Westfield shopping centres in London, New York, San Francisco and Los Angeles (MorningStar, 2017).

Significantly, the top five A-REITs (i.e. Scentre Group, Westfield Corporation, Goodman Group, Vicinity Centres and Stockland) have a market capitalization of more than A$10 billion, contributing around 59% of the total capitalization of A-REITs. Importantly, as at 31 January 2017, the top five A-REITs (i.e. Scentre Group [#13], Westfield Corporation [#16], Goodman Group [#30], Vicinity Centres [#31] and Stockland [#34]) were also ranked within the top 35 listed companies in Australia, accounting for nearly 5.25% of the S&P/ASX 100 Index (ASX100List, 2017) and confirming the significance of these A-REITs to the Australian stock market.

The Australian REIT market is dominated by 17 A-REITs with a market capitalization of more than $1.6 billion, accounting for nearly 91% of the total market capitalization of the A-REITs market. In contrast, there are 8 A-REITs with a market capitalization less than $100 million and comprising less than 0.2% of the total market capitalization of A-REITs.

13.2.4.2 Dividend Yield of A-REITs

Table 13.3 shows the dividend yields of the top 20 A-REITs. A-REITs, in general, distribute a consistent high dividend yield to their unit holders. As at 31 December 2016, Hotel Property Investments (HPI) distributed the highest dividend yield to unitholders, being 11.09%. Most A-REITs distributed a dividend yield between 4% and 9% to unitholders (37 A-REITs, around

Table 13.3 Dividend Yield (%) of Top 20 A-REITs: 31 December 2016

No	*A-REITs*	*ASX Code*	*Dividend Yield*
1	Hotel Property Investments	HPI	11.09%
2	Garda Diversified Property Fund	GDF	8.68%
3	360 Capital Industrial Fund**	TIX	8.63%
4	Cromwell Property Group	CMW	8.51%
5	Centuria Metropolitan REIT	CMA	8.06%
6	360 Capital Office Fund*	TOF	7.94%
7	GDI Property Group	GDI	7.83%
8	Blackwall Property Trust	BWR	7.72%
9	Industria REIT	IDR	7.61%
10	360 Capital Group	TGP	7.29%
11	Charter Hall Retail REIT	CQR	6.67%
12	Aventus Retail Property Fund	AVN	6.46%
13	360 Capital Total Return Fund	TOT	6.42%
14	Growthpoint Properties Australia	GOZ	6.37%
15	Arena REIT	ARF	6.10%
16	Asia-Pacific Data Centre Group	AJD	6.09%
17	National Storage REIT	NSR	5.95%
18	Charter Hall Group	CHC	5.91%
19	Vicinity Centres	VCX	5.89%
20	Astro Japan Property Group	AJA	5.86%

Notes:
* 360 Capital Office Fund changed its name to Centuria Urban REIT (CUA) on 20 January 2017.
** 360 Capital Industrial Fund was renamed as Centuria Industrial REIT (CIP).
Source: ASX (2017a)

74% of all A-REITs), whereas 12 A-REITs (almost a quarter of all A-REITs) distributed less than a 4% dividend yield to unitholders. As of 31 December 2016, the median dividend yield for all A-REITs was 5.35%.

13.2.4.3 Performance analysis of A-REITs

Table 13.4 compares the performance of A-REITs, direct property, shares and bonds over 1-year, 3-year, 5-year, 10-year and 15-year time periods. As can be seen from Table 13.4, A-REITs outperformed the Australian share market, direct property and the bond market substantially over a one-year time period (December 2015 to September 2016). The annualized return of A-REITs was 21.19%, markedly higher than the annualized return of direct property (12.23%), Australian stocks (12.21%) and bonds (2.40%).

Over a three-year performance measure, A-REITs again achieved the highest return (18.20%) among these assets. Comparable results are demonstrated over a five-year performance measure. Nonetheless, A-REITs underperformed the Australian shares and direct property markets over 10-year and 15-year horizons which may be attributed to the GFC. As noted earlier, the GFC had a significant, negative impact on the performance of A-REITs.

However, the performance analysis also clearly demonstrates improved returns during the post-GFC period (one-year, three-year and five-year horizons), confirming that A-REITs have

Table 13.4 Asset Class Performance Analysis: September 2016 (Annualized Return)

Assets	*1Y*	*3Y*	*5Y*	*10Y*	*15Y*
Share (MSCI Australia Equities Index)	**12.21% (3)**	**5.97% (3)**	**11.61% (2)**	**6.42% (2)**	**9.70% (2)**
Direct Property (MSCI IPD/ PCA Australia Composite Index)*	**12.23% (2)**	**11.65% (2)**	**10.77% (3)**	**9.07% (1)**	**10.49% (1)**
A-REITs (S&P/ASX A-REITs 200)	**21.19% (1)**	**18.20% (1)**	**19.96% (1)**	**4.38% (3)**	**8.72% (3)**
Bonds (Australia Commonwealth Government Bond Yield 10-Year)	**2.40% (4)**	**3.03% (4)**	**3.23% (4)**	**4.37% (4)**	**4.76% (4)**

Source: Author's calculation based on the data

Note: Parentheses show the rank and (*) direct property return is represented by the MSCI IPD/Property Council Australia Commercial Property Index. This is a smoothed index, whilst no de-smoothing methodology has been performed.
Sources: MSCI (2016) and DataStream (2017)

Table 13.5 Inter-asset Correlation Matrix: December 2001–September 2016

Assets	*A-REITs*	*Direct Property*	*Shares*	*Bonds*
A-REITs	1.00			
Direct Property	0.35	1.00		
Shares	0.65	0.24	1.00	
Bonds	-0.14	0.05	-0.03	1.00

Note: Direct Property return is represented by the MSCI IPD/Property Council Australia Commercial Property Index. This is a smoothed index, whilst no de-smoothing methodology has been performed.

Source: Author's calculation based on the data from MSCI (2016) and DataStream (2017)

recovered from the GFC and continue to play a major role as a leading property investment vehicle in Australia and globally.

Table 13.5 provides a correlation analysis to indicate the diversification benefits of A-REITs, showing that A-REITs are negatively correlated with bonds (r = -0.14) and moderately positively related with shares (r = 0.65) but lowly correlated with direct property (r = 0.35). This indicates that A-REITs offer some diversification potential to other assets, particularly direct property and bonds.

13.2.4.4 Investment focus of A-REITs

Figure 13.4 depicts the property type diversification of A-REITs, showing the retail sector to be the largest sector with more than half of A-REITs' funds invested in retail property. Scentre Group, Westfield Corporation and Vicinity Group are the largest retail-specific A-REITs. The industrial sector accounted for almost 16% of total A-REIT funds with Goodman Group as the

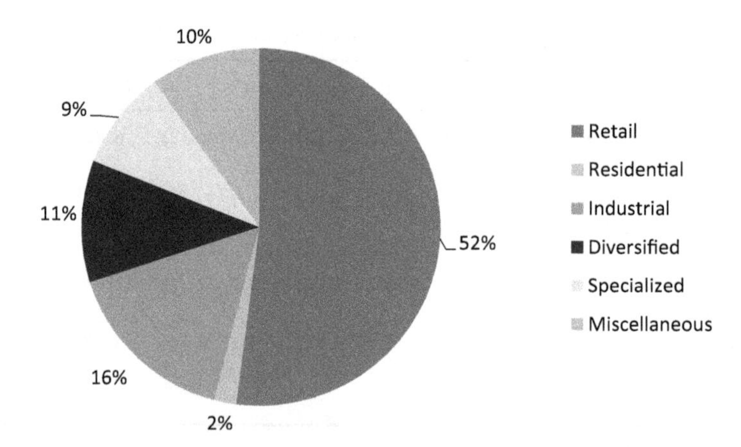

Figure 13.4 Property Type Diversification by Value: 31 December 2016

Source: ASX (2017a)

leading industrial-specific REIT in Australia. The third largest investment focus is in diversified property, contributing nearly 11% of A-REIT value. Stockland is the largest diversified REIT in Australia, with investment in a range of property sectors including retail, commercial and industrial.

Other specialized sectors such as retirement villages, child care and health care comprise around 9% of the aggregate of A-REIT values. Specialist A-REITs investing in these emerging sub-sectors include Ingenia Communities Group specializing in retirement villages, whilst Folkestone Education Trust and Generation Health Care REIT primarily focus on child care and health care, respectively.

13.2.4.5 Stapled and traditional trusts

There are two types of A-REIT structure, being stapled securities and the traditional unit trust structure. Though the first A-REIT was launched in the 1970s (i.e. GPT), there was no stapled structure A-REIT in Australia until 1988 when Stockland was established (Lee, 2010).

Given the level of maturity of the A-REIT market, an increasing popularity of stapled securities has been observed in Australia. As can be seen from Table 13.6, there is an increasing trend of A-REITs to use a stapled structure over time. In 1993, only 16% of A-REITs (six A-REITs) employed a stapled structure, while this percentage increased over time and reached 66% in 2016. Importantly, the number of stapled securities in 2016 had increased by almost threefold in comparison with the number of stapled securities in 2003.

Table 13.7 compares the profile of traditional REIT units and stapled REITs in Australia. Stapled REITs tend to have a larger market capitalization. More specifically, the average market capitalization of traditional A-REITs was A$482 million, whilst the average market capitalization of stapled A-REITs was A$3.8 billion, almost eight times larger than the average market size of A-REITs with a traditional unit structure.

In addition, the top five stapled A-REITs have a market capitalization exceeding A$11 billion, while BWP Trust, the largest A-REIT with a traditional trust structure, only has a market capitalization of A$1.9 billion confirming that stapled A-REITs play a critical role in the Australian REIT market.

Table 13.6 The Increasing Popularity of Stapled Securities

Year	*1993*	*2003*	*2016*
Number of A-REITs	38	46	50
Traditional unit REITs	32	34	17
Stapled REITs	6	12	33
% of stapled REITs/total number of A-REITs	16%	26%	66%

Source: Author's calculation based on the data from ASX (2017a) and Lee, Robinson and Reed (2008).

Table 13.7 Comparison of Traditional Units and Stapled Securities: December 2016

REITs	*Traditional Units*	*Stapled*
Number of REITs	17	33
Average Market Cap	A$482 million	A$3,832 million
Leading REITs	BWP Trust (A$1,921million) Charter Hall Retail REIT (A$1,718 million) Aventus Retail Property (A$931 million) US Masters Residential (A$716 million) Folkestone Education (A$563 million)	Scentre (A$24,705 million) Westfield (A$19,492 million) Goodman (A$12,756 million) Vicinity Centres (A$11,836 million) Stockland (A$11,016 million)

Source: Author's compilation from ASX (2017a)

13.2.4.6 Gearing levels of A-REITs

Figure 13.5 illustrates the gearing levels of A-REITs from 2010 to 2016. Gearing levels of A-REITs declined significantly after the GFC, following a significant recapitalization process including consolidation, delisting and the winding-up of numerous A-REITs over 2009–2011. A noticeable decline in gearing was evident in 2012, when the gearing level had dropped considerably from 44% in 2011 to 31% in 2012, remaining at a manageable level of 36% in 2016. This indicates that A-REIT managers have remained prudent despite the current low interest rate environment in Australia.

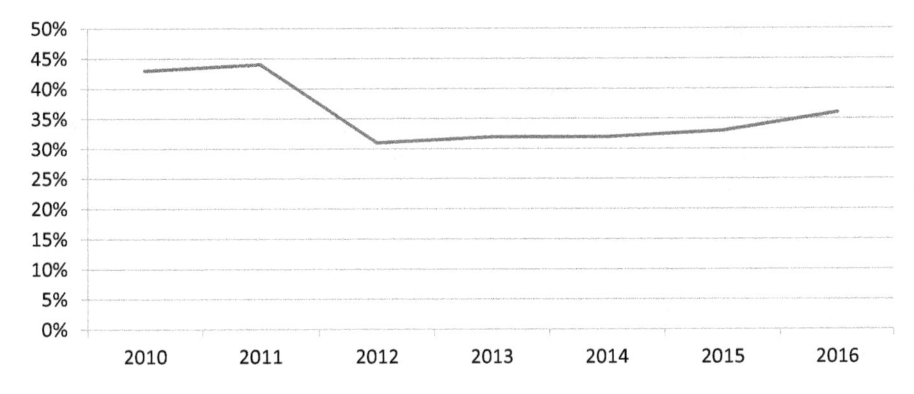

Figure 13.5 Gearing Level of A-REITs over 2010–2016

Source: Author's compilation from various PIR reports.

13.2.5 A-REIT futures

Reflecting the significance of the A-REIT market, the first REIT futures contract in the world was introduced and launched by the ASX in 2002 being the S&P/ASX 200 A-REIT Index futures contract. A futures contract is a financial instrument, being a security between two or more parties whose value is determined by its underlying asset (e.g. A-REITs) (Lee and Lee, 2012).

Importantly, the establishment of a futures market for A-REITs has not only enhanced the stature of A-REITs but also provided an efficient risk management tool for institutional investors. This risk management mechanism enables institutional investors to protect their portfolio values, as well as enhance the liquidity of their property investment (Lee, 2009; Lee and Lee, 2012; Lee, Stevenson and Lee, 2014). Subsequently, similar futures contracts written on real estate securities were also successfully established in the US (2007), Europe (2007) and Japan (2008) (Lee, Stevenson and Lee, 2014).

In Australia, the increasing popularity of A-REIT futures among property investors and fund managers was evident after the initial 2002–2004 establishment period. In particular, the A-REIT futures market has received considerable attention from property investors during the GFC period in response to the increased demand for risk sharing during volatile periods (Lee et al., 2016).

As highlighted by Lee and Lee (2012), the GFC had a strong, negative impact on the A-REIT market, with the A-REIT futures market emerging as an effective hedging tool to reduce the exposure of REIT investment portfolios to market risk. As demonstrated by Figure 13.6, the volume of A-REIT index futures trading grew sharply during the GFC (2007–2009) which could be attributed to the strong demand from institutional investors to hedge the risk of their REIT portfolios during the GFC period (Lee et al., 2016). Thereafter, the demand for A-REIT futures declined in line with the recovery process of the A-REIT market since 2010. However, a steady increase in the volume of A-REIT futures was evident in the 2012 to 2014 period.

Given the increasing popularity of A-REIT futures in Australia, numerous empirical studies have been undertaken to examine the hedging effectiveness and information transmission mechanism of A-REIT futures. Newell and Tan (2004) demonstrated how to use A-REIT futures contracts to protect the portfolio value of an A-REIT investor. Lee and Lee (2012) presented some empirical evidence that A-REIT futures are effective hedging instruments. Specifically, a risk reduction of 57% was evident for A-REITs by hedging with the S&P/ASX 200 A-REITs index futures. The hedging effectiveness increased markedly during volatile periods (i.e. the GFC), suggesting that A-REIT futures played an even more significant role during the GFC. Comparable evidence is found by Newell (2010).

In addition, Lee (2009) and Lee et al. (2016) examined the information transmission role of A-REIT futures. Lee (2009) found that A-REITs contain the critical information of A-REIT futures, whilst A-REIT futures contain little information of A-REITs. However, Lee et al. (2016) offered empirical evidence to suggest that A-REIT futures do convey some critical information of the A-REIT market during the GFC, whilst A-REITs led the A-REIT futures market after the GFC.

Overall, the onset of A-REIT futures has not only enhanced the information transmission processes of the Australian property market, but also enhanced the stature of A-REITs. Most importantly, the A-REIT futures market is an effective tool for property investors to hedge the market risk of their REIT portfolios.

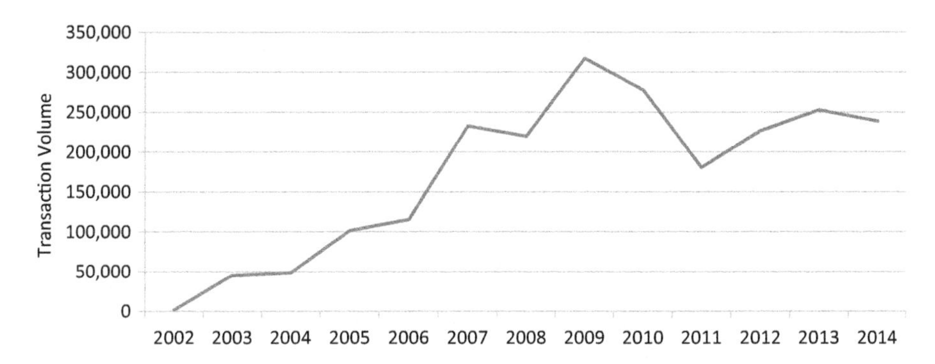

Figure 13.6 Transaction Volume of S&P/ASX 200 A-REIT Index Futures

Source: Author's compilation from DataStream (2017)

13.2.6 Australia – summary

The Australian REIT market is one of the largest and most mature REIT markets in the world, having a modernized and flexible structure. In general, those A-REITs with a larger market capitalization are more likely to employ a stapled structure. Although the GFC had a very strong, negative impact on the performance of A-REITs, numerous A-REITs have since undergone a series of major recapitalization processes. These processes included consolidation with the target of reducing gearing and refocusing on core activities such as investing in prime properties in Australia. These processes have assisted A-REITs to recover from the onset of GFC.

Performance analysis of A-REITs confirms that A-REITs offer a high dividend yield and an attractive total return to their unitholders. Specifically, the average annualized return of A-REITs is higher than stocks, bonds and direct property in recent years (one-year, three-year and five-year). Moreover, as a significant innovation, an A-REIT futures market has been established in Australia over the last 15 years. This has further enhanced the stature of A-REITs and improved the information transmission mechanisms of the property market. As a consequence, it is expected that A-REITs will continue to play a critical role in the global REIT market.

Similar to the US, the REIT sector in Australia is long established, well developed, flexible and adaptable with a deep body of academic research both following and prompting structural change in the sector. However, the introduction and successful operation of an innovative and original futures market for the REIT sector in Australia acknowledges not only the quality of the Australian REIT sector but also its forward looking attitude and risk management focus.

13.3 South Africa

This chapter analyses REITs in the Oceania region, using case studies from Australia (being a developed REIT sector in the region), South Africa (being a developing REIT sector in the region) and India (being an emerging REIT sector in the region).

13.3.1 SA-REITs history and context

The South African REIT (SA-REIT) regime started in 2013 (Ernst and Young, 2016), following the successful prior trading of very similar Property Unit Trusts (PUTs) on the Johannesburg Stock Exchange (JSE), being regulated by the South African Collective Investment Scheme

(CIS) Act. Tax transparency is provided to PUTs if PUTs distribute their taxable income to their unitholders. PUTs are allowed to invest in immoveable property and property companies, although PUTs normally invest in shares in property companies instead of immoveable property (Lee, 2010). Further, a Property Loan Stocks Company (PLS) is another type of property company in South Africa which, unlike PUTs, mainly invests in both equity and debentures while being regulated by the South African Companies Act instead of the South African CIS Act (EPRA, 2016b).

Following the introduction of the REIT structure in South Africa in 2013 through the amendment of the tax legislation and the JSE listing requirements, all PUTs and PLSs may be converted into the newly introduced REIT structure and any new listings in this sector will have to comply with the newly established JSE REIT listing requirements (EPRA, 2016b).

13.3.2 SA-REIT structure and regulation

In general, the SA-REIT regulations echo the globally used REIT structure (JSE, 2013). To qualify as a REIT in South Africa, an SA-REIT is required to have a minimum of South African Rand 300 million (A$29 million) worth of property. In addition, it is required to meet the JSE listing requirements. An SA-REIT can be a listed company REIT or a trust REIT and can be either externally or internally managed. Additionally, SA-REITs must generate 75% of their income from rental or from indirect property owned or investment income from indirect property ownership.

There is a gearing limit by which an SA-REIT must maintain its debt level below 60% of the gross asset value of its underlying assets. SA-REITs are exempted from corporation tax (tax pass-through status) if they distribute at least 75% of their taxable income to their unitholders every year. Undistributed income is subject to a tax rate of 28%. Interestingly, SA-REITs are required to have a committee to monitor the risk of the REIT and SA-REITs are prohibited from using derivative instruments that are not in the ordinary course of business (JSE, 2013; EPRA, 2016b). The key features of SA-REITs are summarized in Table 13.8.

13.3.3 SA-REIT market and performance

As at 29 July 2016, there were 43 SA-REITs with a market capitalization of A$39.39 billion, accounting for approximately 1.5% of the global REIT market (EPRA, 2016b). As summarized in Table 13.9, as at 31 December 2016, the top five SA-REITs are Growthpoint Properties with

Table 13.8 Key Features of SA-REITs

Features	*South Africa*
System	SA-REITs
Established	2013
Investment in real estate	At least 75% of its income from rental or from indirect property owned or investment income from indirect property ownership
Gearing	60% of the gross value of the underlying asset value.
Distribution of REIT income	At least 75% of its taxable income
Tax Transparency	Yes

Sources: JSE (2013) and EPRA (2016b)

Table 13.9 Top 10 SA-REITs as of 31 December 2016

Rank	*South Africa REITs*	*Market Capitalization*
1	Growthpoint Properties	A$7,406 million
2	Redefine Properties	A$5,950 million
3	Resilient Property Income Fund	A$4,640 million
4	Hyprop Investments Ltd	A$2,946 million
5	Fortress Income Fund 'A'	A$1,964 million
6	SA Corporate Real Estate Fund	A$1,373 million
7	Vukile Property	A$1,093 million
8	Investec Property Fund	A$1,094 million
9	Arrowhead Properties	A$912 million
10	Emira Property Fund	A$738 million

Source: Author's compilation from DataStream (2017)

Table 13.10 The Growth Rate of the Top 5 SA-REITs Over 2013–2016

Rank	*South Africa REITs*	*Growth Rate*
1	Growthpoint Properties	55.9%
2	Redefine Properties	96.7%
3	Resilient Property Income Fund	181.9%
4	Hyprop Investments Ltd	56.6%
5	Fortress Income Fund 'A'	268.8%
6	SA Corporate Real Estate Fund	71.9%
7	Vukile Property	55.7%
8	Investec Property Fund	109.7%
9	Arrowhead Properties	435.0%
10	Emira Property Fund	3.6%

Source: Author's calculation from the downloaded data from DataStream (2017)

a market capitalization of A$7.4 billion, Redefine Properties (A$5.9 billion), Resilient Property Income Fund (A$4.6 billion), Hyprop Investments Ltd (A$2.9 billion) and Fortress Income Fund "A" (A$1.9 billion). It may be observed that eight SA-REITs have a market capitalization of more than A$1 billion (or approximately South African Rand 10 billion), highlighting the significance of SA-REITs.

Importantly, the top SA-REITs have grown significantly since the introduction of a REIT structure in South Africa. As highlighted in Table 13.10, the market capitalization of these SA-REITs has increased by at least 55% over 2013–2016, with the only exception being Emira Property Fund with a growth rate of almost 4%. Significantly, over this study period, some SA-REITs recorded a tremendous growth rate. For instance, Arrowhead Properties, Fortress Income Fund "A" and Resilient Property Income Fund have grown 435%, 269% and 182%, respectively, over this period, suggesting that the establishment of SA-REITs, which have a globally recognized REIT investment structure, has made a significant contribution to the growth of the listed property market in South Africa.

Table 13.11 compares the performance of SA-REITs with other major investment classes in South Africa over various different time periods. SA-REITs outperformed the overall South African stock market and the South African government bond market significantly over a

Table 13.11 Asset Class Performance Analysis: December 2016 (Annual Return)

Market	*1Y*	*3Y*
Stocks (FTSE/JSE All Share Index)	**3.01% (2)**	**6.64% (2)**
SA-REITs (FTSE South Africa REIT Index)	**16.59% (1)**	**13.56% (1)**
Bonds (South Africa Government Bond Yield)	**9.03% (3)**	**8.49% (3)**

Source: Author's calculation from the downloaded data from DataStream (2017)

Table 13.12 Inter-asset Correlation Matrix for SA-REITs: January 2014–December 2016 Author's calculation from the downloaded data from DataStream (2017)

Assets	*SA-REITs*	*Shares*	*Bonds*
SA-REITs	1.00		
Shares	0.45	1.00	
Bonds	-0.11	-0.13	1.00

one-year time period (2016). The return of A-REITs was 16.59%, being significantly higher than the return of South African stocks (3.01%) and bonds (9.03%). Comparable findings are also documented over a three-year time period (2014–2016) in which SA-REITs outperformed stocks and bonds in South Africa considerably. This clearly suggests that SA-REITs are an effective property investment vehicle in South Africa.

Table 13.12 provides a correlation analysis to indicate the diversification benefits of SA-REITs, showing SA-REITs to be negatively correlated with bonds (r = -0.11) and moderately positively correlated with shares, indicating that the inclusion SA-REITs in a mixed-asset portfolio can offer some diversification benefits.

13.3.4 SA-REIT summary

Interestingly, while SA-REITs share many structural characteristics in common with other REIT regimes around the world, the 60% gearing limit and requirement for a risk monitoring committee together with a prohibition on using derivatives, other than in the ordinary course of business, provide a much stronger and more active focus on risk management than may be found elsewhere in the REIT world.

Even though the REIT structure was only formally introduced in South Africa in 2013, the SA-REIT market has grown considerably since its introduction. Furthermore, SA-REITs appear to be an effective investment vehicle, delivering superior returns when compared with the overall South African stock and bond markets. Most importantly, the establishment of SA-REITs has contributed significantly to the growth of the listed property market in South Africa.

13.4 India

This chapter analyses REITs in the Oceania region, using case studies from Australia (being a developed REIT sector in the region), South Africa (being a developing REIT sector in the region) and India (being an emerging REIT sector in the region).

13.4.1 I-REITs history and context

While the basic REIT regulations have been formulated in India, a REIT has not yet been listed in the country (Ernst and Young, 2016). Specifically, the Securities and Exchange Board of India (SEBI) introduced its draft REIT regulations in 2013, with REIT regulations officially enacted in India on 26 September 2014. SEBI has partnered with major stakeholders in India to further refine the guidelines for Indian REITs, to bring the regulatory regime governing REITs in India in line with those of other developed and developing markets especially with reference to dividend distribution policies and capital requirements (EPRA, 2016b).

In the 2016 Indian Government Budget, several relaxations of the Indian REIT regulations were announced including the exemption of dividend distribution tax on special purpose vehicles (SPVs). SPVs include a company in which a REIT holds or proposes to hold a controlling interest and an equity stake or interest of least 50% and does not engage in any other activity (PwC, 2016). Further relaxations allow Indian REITs to engage in other non-core activities (e.g. under construction properties) to 20% of the value of the REIT and remove the restriction on the SPV to invest in other SPV structures and the limit on number of sponsors (PwC, 2016; Hindustan Times, 2017).

These changes are strongly supported by the property industry. Blackstone, the world's largest private equity manager, has indicated that it plans to launch the first Indian REIT via an IPO with a total value of Indian Rupee 40 billion (A$778 million) (The Times of India, 2016). Importantly, these changes have made the Indian REIT regime more consistent with internationally followed concepts, methods and principles, particularly the tax pass-through status of REITs (tax transparency).

13.4.2 I-REITs structure and regulation

In general, Indian REITs (I-REITs) should be established in a trust form. Several key features of I-REITs include that 80% of the value of I-REITs must be invested in income-producing properties either directly or via SPVs. At least 75% of the revenue of an I-REIT should be derived from rental or leasing of assets. I-REITs are required to distribute a minimum of 90% of the net distributable cash flow of the I-REIT. The minimum value of I-REIT assets should be 500 crore Indian Rupee (A$97.5 million – crore is a widely used measurement in India, where 1 crore is equal to 10 million).

I-REITs must raise funds through the sale of units to the public and are not allowed to invest in mortgages. Furthermore, the aggregate consolidated net borrowings (i.e. net of cash and cash equivalents) and deferred payments of an I-REIT are capped at 49% of the value of an I-REIT's assets. If the gearing level of an I-REIT exceeds 25% of the value of the I-REIT assets, the I-REIT is required to provide a credit rating and seek the approval of unit holders. However, no specific conditions are imposed on those I-REITs having a gearing level below 25% of the value of the I-REIT (SEBI, 2014; EPRA, 2016b; EY, 2016; PwC, 2016). The key features of I-REITs are summarized in Table 13.13.

Apart from the I-REIT regulations, SEBI also took an innovative step and launched the regulation of infrastructure investment trusts (InvIT), together with the I-REIT regulatory framework, concurrently in 2014. Similar to I-REITs, InvITs are investment vehicles that can be used to attract private investment into the infrastructure sector. This is in line with the need for new sources of finance for stalled infrastructure projects in India (PwC, 2016). InvITs have a very similar structure to I-REITs, the only difference being that InvITs invest in infrastructure instead of real estate (Ernst and Young, 2016). InvITs are also expected to alleviate the burden on the banking system for funding the Indian infrastructure sector.

Table 13.13 Key Features of Indian REITS

Features	*India*
System	I-REITs
Established	2014; no I-REIT is listed yet (February 2017)
Investment in real estate	At least 80% in income-producing properties
Development	Yes; limited
Gearing	Up to 25% of the REIT value, no specific condition From 25% up to 49%, the REIT to be subject to the following: – credit rating is required – approval of the unitholders
Distribution of REIT income	At least 90% of its taxable income
Tax Transparency	Yes

Sources: SEBI (2014), EPRA (2016b), Ernst and Young (2016) and PwC (2016)

Overall, a regime for Indian REITs has been successfully established. Although it has undergone several steps of refinement, the revised I-REIT framework echoes the internationally recognized REIT regime. Therefore, the first Indian REIT is very likely to be successfully launched in the near future and the Indian REIT market is expected to grow very rapidly through a sound REIT regulatory framework. Importantly, the creation of a REIT structure in India is expected to increase the depth of the Indian property market and provide a greater transparency to that market over time.

13.4.3 I-REIT summary

As an emerging REIT sector in the Oceania region, India has established an I-REIT regulatory framework following the principal characteristics of other REIT markets around the world but with a very prudent gearing limit of 49% of the value of assets, together with a requirement to provide a credit rating and obtain unitholder approval if gearing is to exceed 25% of the value of assets.

Such a cautious start to the creation of an I-REIT industry in an emerging market is commendable and likely to be a significant contributor to investor confidence, engendering more rapid and substantial growth in the Indian REIT market.

13.5 Summary

Part II includes six chapters each analysing REITs in a region of the world, using case studies from a developed, developing and emerging REIT sector in the region. Chapter 8 analysed REITs in the North American region, using case studies from the USA, Canada and Mexico, with Chapter 9 analysing REITs in the Latin American region, using case studies from Brazil, Argentina and Uruguay for the developed, developing and emerging REIT sectors in the region, respectively.

Chapter 10 analysed REITs in the European region, using case studies from the UK, Spain and Poland, with Chapter 11 analysing REITs in the South East Asian region, using case studies from Singapore, Malaysia and Thailand and Chapter 12 analysed REITs in the North Asian region, using case studies from Japan, Hong Kong and China for the developed, developing and emerging REIT sectors in the region, respectively.

This chapter analysed REITs in the Oceania region, using case studies from Australia (being a developed REIT sector in the region), South Africa (being a developing REIT sector in the region) and India (being an emerging REIT sector in the region).

With the launch of its first REIT in 1971, the developed Australian REIT market is now almost 50 years old, having 50 REITs listed on the ASX with a market capitalization of A$105 billion (Figure 13.2). Comparatively, the developing South African REIT market started in 2013 and has 43 REITs listed with a market capitalization of A$39.39 billion at July 2016. The emerging REIT market of India has already established the requisite regulation and is on the verge of seeing its first REIT listed.

It is interesting, within the Oceania region, that A-REITs, as a developed market, pioneered the development and operation of REIT futures while also managing to incur massive gearing at the onset of the GFC leading to huge unit price falls and widespread recapitalization of the REIT sector. Conversely, the South African approach of a gearing limit, risk management committee and prohibition on the extraneous use of derivatives together with India's approach of requiring a credit rating and unitholder approval for gearing above 25% and a gearing cap at 49% should protect their REIT markets from the damage that can be caused by excessive debt which may otherwise be a concern in a developing and emerging market, respectively.

Part II of this book comprised an analysis of REITs in six regions of the world, using case studies from a developed, developing and emerging REIT sector in each region with the final chapter, Chapter 14, concluding Part II and considering *Directions for the Future of International REITs*.

References

APREA 2014, *Asia Pacific REITs: A comparative regulatory and tax study*, Asia Pacific Real Estate Association, Singapore.

ASX100 List 2017, *ASX top 200 companies: ASX 100 list*, ASX100List, Sydney, 1 February 2017.

ASX200 List 2017, *ASX top 200 companies: ASX 200 list*, ASX200List, Sydney, 1 February 2017.

ASX 2017a, *ASX funds (listed managed investments, funds and EFPs) monthly update – December 2016*, Australian Securities Exchange, Sydney.

ASX 2017b, *A-REITs*, Australian Securities Exchange, Sydney.

ATO 2007, *Stapled securities and capital gains tax*, Australian Taxation Office, Canberra.

DataStream 2017, *Thomson Reuters data stream*, DataStream (online database), New York.

EPRA 2016a, *EPRA research: Monthly statistical bulletin October 2016*, European Public Real Estate Association, Brussels.

EPRA 2016b, *Global REIT survey 2016*, EPRA, Brussels.

EPRA 2017, *EPRA statistics: January 2017*, European Public Real Estate Association, Brussels.

Ernst and Young 2016, *Global perspectives: 2016 REIT report*, EY, New York.

Hindustan Times 2017, *REITs opportunity in India estimated to be Rs1.25 trillion*, Hindustan Times, Mumbai.

JLL 2016, *Global real estate transparency index 2016*, Jones Lang LaSalle, London.

JSE 2013, *JSE to welcome REITs regime* [Online], Johannesburg Stock Exchange, Johannesburg, 20 January 2017.

Lee, C.L. 2009, 'Volatility transmission in Australian REIT futures', *Journal of Real Estate Portfolio Management*, Vol. 15, No. 3, pp. 221–238.

Lee, C.L. 2010, *Lower partial moment-capital asset pricing model in listed property trusts*, Lambert Academic Publishing, Saarbrucken.

Lee, C.L. and Lee, M.L. 2012, 'Hedging effectiveness of REIT futures', *Journal of Property Investment and Finance*, Vol. 30, No. 3, pp. 257–281.

Lee, C.L., Robinson, J. and Reed, R. 2008, 'Downside beta and the cross-sectional determinants of listed property trust returns', *Journal of Real Estate Portfolio Management*, Vol. 14, No. 1, pp. 49–62.

Lee, C.L., Stevenson, S. and Lee, M.L. 2014, 'Futures trading, spot price volatility and market efficiency: Evidence from European real estate securities futures', *Journal of Real Estate Finance & Economics*, Vol. 48, No. 2, pp. 299–322.

Lee, M.T., Kuo, S.H., Lee, M.L. and Lee, C.L. 2016, 'Price discovery and volatility transmission in Australian REIT cash and futures markets', *International Journal of Strategic Property Management*, Vol. 20, No. 2, pp. 113–129.

MorningStar 2017, *MorningStar DatAnalysis*, MorningStar, Sydney.

MSCI 2016, *MSCI IPD/PCA Australia composite indices*, Morgan Stanley Capital International, Sydney.

Newell, G. 2006, 'The changing risk profile of listed property trusts', *Australian Property Journal*, Vol. 39, pp. 172–180.

Newell, G. 2010, 'The effectiveness of A-REIT futures as a risk management strategy in the global financial crisis', *Pacific Rim Property Research Journal*, Vol. 16, no. 3, pp. 339–357.

Newell, G., Lee, C.L. and Kupke, V. 2015, *The opportunity of unlisted wholesale residential property funds in enhancing affordable housing supply*, AHURI Final Report No 249, Australian Housing and Urban Research Institute, Melbourne.

Newell, G. and Peng, Hw. 2009 'The impact of the global financial crisis on A-REITs', *Pacific Rim Property Research Journal*, Vol. 15, No. 4. pp. 453–469.

Newell, G. and Tan, Y.K. 2004, 'The development and performance of listed property trust futures', *Pacific Rim Property Research Journal*, Vol. 10, No. 2, pp. 132–145.

PIR 2016, *Australian property funds industry survey 2016*, Property Investment Research, Sydney.

PwC 2016, *India's new real estate and infrastructure trusts. The way forward*, PwC, Mumbai.

SEBI 2014, *Securities and exchange board of India (real estate investment trusts) regulations, 2014*, Securities and Exchange Board of India, New Delhi.

Sing, T.F. and Ling, S.C. 2003, 'The role of Singapore REITs in a downside risk asset allocation framework', *Journal of Real Estate Portfolio Management*, Vol. 9, No. 3, pp. 219–235.

Stringer, T. 2001, 'What is the best strategy for property investment: Direct, listed or both?', *Australian Property Journal*, Vol. 36, No. 5, pp. 430–432.

The Times of India 2016, *Blackstone readies India REIT listing*, The Times of India, Mumbai.

14
DIRECTIONS FOR THE FUTURE OF INTERNATIONAL REITS

David Parker

14.1 Introduction

This book aims to identify key areas for research in the REIT discipline for the next five to ten years by surveying the current state of the REIT discipline around the world and identifying emerging and cutting-edge research areas through a thematic review of current contextual issues and a regional analysis based on case studies.

This book comprises two parts, the first part being a thematic review of emerging and cutting edge global research into current contextual issues in REITs internationally and the second part being a regional analysis of REITs around the world, each written by authoritative academic authors from the world's leading Universities and REIT industry experts.

Part I included six chapters each reviewing a current theme of REIT evolution through the lens of contemporary research. Chapter 1 focused on critical contextual issues in international REITs while Chapter 2, the *Post-Modern REIT Era*, examined the evolution of the global REIT market, what is deemed to be best current market practice and how the market is expected to change going forward and Chapter 3, *Emerging Sector REITs*, investigated such sectors as residential, health care, self-storage, timber, infrastructure and data centres generally and then specifically through a case study of US REITs.

Chapter 4, *Sustainable REITs*, analysed REIT environmental performance and the cost of equity and Chapter 5, *Islamic REITs*, examined the evolution of global Islamic REITs through a study of the Malaysian market. Chapter 6, *Behavioural Risk in REITs*, addressed the management of behavioural risk in global REITs and Part I concluded with Chapter 7 which reviewed recent research into *REIT Asset Allocation*.

Part II includes six chapters each analysing REITs in a region of the world, using case studies from a developed, developing and emerging REIT sector in the region. Chapter 8 analysed REITs in the North American region, using case studies from the USA, Canada and Mexico, with Chapter 9 analysing REITs in the Latin American region, using case studies from Brazil, Argentina and Uruguay for the developed, developing and emerging REIT sectors in the region, respectively.

Chapter 10 analysed REITs in the European region, using case studies from the UK, Spain and Poland, with Chapter 11 analysing REITs in the South East Asian region, using case studies from Singapore, Malaysia and Thailand. Chapter 12 analysed REITs in the North Asian region, using case studies from Japan, Hong Kong and China and Chapter 13 analysed REITs in the

Oceania region, using case studies from Australia, South Africa and India for the developed, developing and emerging REIT sectors in the region, respectively.

This chapter concludes Part II and considers *Directions for the Future of International REITs*. With continuous change and evolution a certainty for the REIT sector around the world, the contributing chapters to this book serve to highlight five major directions for the future of international equity REITs, each of which may be expected to contribute to a change in the risk/return relativities of the REIT sector, comprising the following:

- primary focus on people not property;
- primary focus on cash flow not property;
- societal impacts on REITs;
- global flagship property REITs; and
- global regulation of REITs,

which will be considered, sequentially, in the following sections (beginning with section 14.3). However, while each major future direction may be considered separately, they are unlikely to evolve in isolation and are likely to cross over and interact with each other such that they cannot and should not be considered independently.

14.2 Global REIT markets

The review of developed, developing and emerging REIT markets in Part II of this book highlighted a range of common themes, including the following:

- the role of the diversified REIT;
- the role of REITs where property companies predominate;
- the interplay between REIT initiatives and government policy;
- the catalysts for REIT market emergence;
- the significance of REIT tax treatment; and
- the role of leverage constraints.

Each of these points are considered in the following sub-sections.

14.2.1 The role of the diversified REIT

Case (Chapter 8) notes a commonality in North American REIT evolution with each REIT sector in the region starting its evolution with a greater proportion of diversified REITs plus a sector specific REIT, then broadening to other large sectors, then further broadening to other smaller sectors.

The US REIT sector started with diversified REITs, then broadened into industrial and office REITs before further broadening into infrastructure, data centres and timber REITs. Similarly, the Canadian REIT sector started with diversified REITs and retail REITs, then broadened to residential and office REITs before further broadening into health care, industrial and hotel REITs. The Mexican REIT sector, at a much earlier stage if its evolution, started with diversified and industrial REITs, then broadened to retail and hotel REITs but with the further broadening stage yet to come as the sector grows and evolves.

Conversely, in the UK, Moss (Chapter 10) notes that, given the size of the existing property company and unlisted fund market pre-REITs in the UK, REITs do not appear to have gone through the initial diversified portfolio phase, commonly found in other regions, but have moved swiftly into sector specific REITs such that further broadening into other sectors may be anticipated as the market evolves.

Accordingly, for emerging REIT markets, a pattern appears to be evident whereby a REIT market starts by offering diversified REIT products plus one sector-specific REIT product before broadening to offer REITs in other sectors.

14.2.2 The role of REITs where property companies predominate

There is an apparent commonality around the world that where a jurisdiction has a well-established property company or *fideicomiso* market, a REIT market may have difficulty gaining traction and becoming established.

In Hong Kong, Ooi and Wong (Chapter 12) note that the REIT market is structurally challenged through its unique combination of a long-standing and well-understood competing property company sector and the absence of tax transparency in its REIT legislation and regulation.

Similarly, in Latin America, Taltavull et al. (Chapter 9) note that recent legislative and regulatory changes in various Latin American countries have facilitated a clearer regulatory definition of investment funds across Latin America as a form of indirect vehicle based on the popular, long standing and well recognized legal entity, the *fideicomiso*, which is common across Latin America and combines attractive characteristics in risk management and investment capability. With modernization, the *fideicomiso* may coexist with advanced investment fund structures and facilitate potential conversions into REITs, though the *fideicomiso* appears likely to predominate as the preferred structure for the foreseeable future.

Indeed, in Uruguay, Taltavull et al. (Chapter 9) note that REITs, as such, do not yet exist with the Real Estate Finance Fideicomiso structure providing most REIT roles to facilitate investment in real estate. Given the long-standing and widely accepted nature of the *fideicomiso* structure, it may some considerable time before REITs become established in Uruguay.

Conversely, in the UK, Moss (Chapter 10) notes that the UK developed a strong REIT sector despite starting its REIT era with an existing and substantial property company and unlisted fund market, which was acknowledged through transition legislation that was then amended as REITs became established and transition arrangements were no longer required.

Similar supportive regulations to facilitate the transformation of Singapore property companies to a REIT structure is suggested by Liow and Yuting (Chapter 11), so that the local securitized real estate industry is able to move more cohesively towards the REIT direction and develop the popularity of the trust structure in the Singapore securitized real estate sector.

14.2.3 The interplay between REIT initiatives and government policy

There is little consistency around the world's REIT markets concerning whether government policy drives REIT development or REIT development drives government policy. Liow and Yuting (Chapter 11) note the high level and positive nature of government and regulatory intervention and support for the REIT sector in Singapore, with government regularly improving the operating and regulatory environment for SGREITs so that they can continue to grow while, at the same time, operating with good corporate governance and embracing best industry practices. Similarly, in Japan, Ooi and Wong (Chapter 12) note that growth in the J-REIT sector benefitted from the Bank of Japan including buying J-REITs as part of its quantitative easing programme.

However, Case (Chapter 8) observes that the US REIT sector illustrates how evolution in the REIT sector can drive legislative change and also how legislative change can drive evolution in the REIT sector. Significantly, given the maturity of the US REIT sector, a deep body of academic research into REITs now exists, with each legislative change subject to further academic research into both its impact on the REIT sector and the response of the REIT sector.

Conversely, Moss (Chapter 10) comments that the focus on residential property, lower levels of development and increased equity issuance anticipated by government at the establishment of REITs in the UK did not transpire. In Australia, Lee (Chapter 13) observes that the introduction and successful operation of an innovative and original futures market for the REIT sector in Australia was industry driven, not government driven, being an acknowledgement not only of the quality of the Australian REIT sector but also its forward-looking attitude and risk management focus. Further, in Hong Kong, Ooi and Wong (Chapter 12) note that the deliberate non-provision of the usual tax benefits for REITs by government has stifled the development of the sector.

A major test for the role of government in the evolution of a REIT market will, of course, be the development of REITs in China. The Chinese government could choose, like Singapore, to foster and encourage the development of REITs or, like Hong Kong, to stifle such development.

14.2.4 The catalysts for REIT market emergence

For a REIT market to grow rapidly, experience around the world suggests that a catalyst is needed. Such catalysts make take various forms and may be repeated in differing forms over the life of a REIT market. Case (Chapter 8) observes that the REIT industry in Canada emerged very quickly out of a property market downturn that suddenly generated a supply of suitable investment stock. Moss (Chapter 10) notes that a well-established equity market, access to both a supply of quality property investment assets and a pool of domestic capital provide a sound basis upon which a REIT market may emerge in Poland.

In the US, Brounen and de Koning (2012) record that growth in the US market has been driven by a range of catalysts including structural legislative change in 1986, collapsing direct property prices in the property market crash of the early 1990s, legislative change to unitholder restrictions in 1993 and the aftermath of the dotcom crash of the early 2000s.

Conversely, Moss (Chapter 10) observes that growth in the Spanish REIT market was effectively due to good luck and fortunate timing, when institutions were structurally underweight equities due to the global financial crisis and conveniently followed a debt-fuelled commercial real estate bust preceding the introduction of enabling REIT legislation.

14.2.5 The significance of REIT tax treatment

It is now readily apparent that the principal attraction of the REIT sector for investor capital is the tax transparency offered within the conventional REIT market structure. As Ooi and Wong (Chapter 12) note, the growth of the REIT sector in Hong Kong is structurally challenged by the absence of tax transparency in its REIT legislation and regulation.

Ironically, a REIT sector may only grow based on its tax transparency if a country does not offer other, more attractive tax transparent investment structures. As Taltavull et al. (Chapter 9) observe, in Brazil, where high taxation can discourage investing, the tax exemptions offered to the FIIs and their investors are so many and so significant that other real estate-based portfolio securitization solutions, such as REITs, end up being ignored, even if they present a more advanced structural design than that of the FIIs.

In China, it may be contended that the principal determinant of the success of the introduction of a REIT sector will be the willingness of government at various levels to provide a level of tax transparency considered sufficiently attractive by domestic and international investors to warrant participation in the sector.

14.2.6 The role of leverage constraints

The global financial crisis of 2007–2009 may be contended to have been a watershed for the role of debt and the extent of acceptable leverage for REITs globally. Prior to the global financial crisis, some REITs had leverage in excess of 90% which, when coupled with a freezing of global debt markets, had catastrophic outcomes for their unitholders.

It is interesting that various developing and emerging REIT markets have sought to manage debt risk in the sector through controlling the maximum level of acceptable leverage and the provision of further controls and constraints. Moss (Chapter 10) notes that Spain has introduced constraints on leverage and Liow and Yuting (Chapter 11) observes that Thailand has progressively relaxed its debt controls for REITs, moving from 10% debt under PFPO to 60% for a REIT with investment grade rating.

In Oceania, Lee (Chapter 13) comments that South Africa has a 60% gearing limit and requirement for a risk-monitoring committee together with a prohibition on using derivatives, other than in the ordinary course of business, while India has a gearing limit of 49% of the value of assets, together with a requirement to provide a credit rating and obtain unitholder approval if gearing is to exceed 25% of the value of assets.

With the adverse impact of high leverage on REITs in the event of a market downturn being well-experienced and well-recorded, the introduction of increasing layers of risk management around debt by regulators can only be beneficial for investors, particularly in emerging and developing markets.

14.2.7 Summary – global REIT markets

The review of developed, developing and emerging REIT markets in Part II of this book highlighted a range of common themes, including the following:

- the role of the diversified REIT;
- the role of REITs where property companies predominate;
- the interplay between REIT initiatives and government policy;
- the catalysts for REIT market emergence;
- the significance of REIT tax treatment; and
- the role of leverage constraints.

These common themes may be contended to give rise to a range of guiding principles for REIT evolution that may be summarized as follows:

- REIT markets may be likely to start their evolution with a greater proportion of diversified REITs plus a sector specific REIT, then broaden to other large sectors, then further broaden to other smaller sectors;
- REIT markets may have difficulty gaining traction and becoming established where a jurisdiction has a well-established property company or *fideicomiso* market;

- there is no single model for the interplay between REIT initiatives and government policy, with either potentially dominant depending on the individual jurisdictional circumstances;
- there are a range of catalysts for REIT market emergence, with direct property market collapse, structural legislative change and good luck with fortunate timing among the principal catalysts;
- that tax transparency is essential for a successful REIT market, except where the jurisdiction offers other, more attractive tax transparent investment structures; and
- while leverage is tempting as a fast route to growth in assets, capital market participants have bitter experiences of excessive leverage in the past and an increasing number of jurisdictions are seeking to constrain leverage and implement associated risk management measures.

14.3 Primary focus on people not property

With continuous change and evolution a certainty for the REIT sector around the world, the contributing chapters to this book serve to highlight five major directions for the future of international equity REITs, each of which may be expected to contribute to a change in the risk/return relativities of the REIT sector, comprising the following:

- primary focus on people not property;
- primary focus on cash flow not property;
- societal impacts on REITs;
- global flagship property REITs; and
- global regulation of REITs,

which will be considered, sequentially, in the following sections. However, while each major future direction may be considered separately, they are unlikely to evolve in isolation and are likely to cross over and interact with each other such that they cannot and should not be considered independently.

Concerning the primary focus on people not property, as REITs grow and their total assets under management rise, the implications of decision-making by REIT managers become even more significant with good decisions potentially maximizing investor returns and minimizing investor risk and bad decisions having the potentially disastrous opposite effect.

With property management, asset management and other downstream activities either outsourced or relegated to regional support offices, the size of the REIT head office management team may be expected to fall and their focus on investment strategy, portfolio management and funding become paramount. Such a trend is consistent with the senior management of REITs being drawn from investment banking, finance, accounting and law with property skills pushed down the corporate ladder. Such a trend is similarly consistent with the notion of the REIT as a business rather than the REIT as a property portfolio, with the requisite management skills being drawn from the former discipline rather than the latter (Parker, 2011).

Ooi and Wong (Chapter 12) note that, as more and more REITs are listed, the market will start to distinguish between well-managed REITs and poorly managed REITs with a poorly managed REIT penalized through a deep discount to its share price.

In the context of REIT management and decision-making, Wofford and Troilo (Chapter 6) investigate human cognition and cognitive psychology in treating the broad issue of how a limited human brain handles seemingly infinite information, noting the role of filters and time saving devices such as heuristics and biases. The authors identify five heuristics and biases of particular relevance to REIT decision-making:

- representative heuristic, with the risk of incorrect extrapolation from a sample to a general population;
- availability heuristic, with the risk of greater weight being given to information easily recalled;
- anchoring and adjustment bias, with the risk of fixating on previous information when considering a subsequent problem;
- framing, with the risk that phrasing of a choice can bias the selection of the course of action; and
- confirmation bias, with the risk of overreliance on information that reinforces existing beliefs.

Each and all of these are exacerbated by overconfidence, which may commonly manifest in the senior decision-making ranks within REIT managers.

Wofford and Troilo (Chapter 6) define cognitive risk as this:

> *variability in perceiving potential outcomes and the influence of cognition on actual outcomes created by cognitive activity.*

which may manifest in a wide range of REIT management activities including assessment and evaluation, planning, problem framing, problem-solving, decision-making and execution.

As Wofford and Troilo (Chapter 6) succinctly note:

> The cognitive sources of outcome variability include the familiar biases and heuristics of cognitive psychology, but there are others worth considering. Communication problems, organizational problems and dynamics, inattention, distractions, poor procedures, fatigue, systems issues, ergonomics, stress, information problems and time issues all find their locus in human cognition. When placed within the context of a real estate environment characterised by chaos or complexity, dynamism and ambiguity, the variability produced by these cognitive variations can be magnified in a nonlinear manner to produce significant risk exposure.

Given such scope for variability and the potentially massive positive or negative return and risk outcomes that may arise from REIT decision-making, the authors investigate how cognitive risk might be managed, noting Wofford, Troilo and Dorchester's (2010) advocacy for the use of continuous planning and strategic foresight to reduce the occurrence and magnitude of such events arising from cognitive risk and the use of evidence-based practices advocated by Wofford, Troilo and Dorchester (2011) and Wofford and Troilo (2013).

Wofford and Troilo (Chapter 6) continue by advocating a customizable framework (adjustable for the size and nature of the REIT to which it is being applied) and associated process for REIT problem-solving and decision-making that is effective in reducing cognitive risk. The authors identify the key contextual elements of the REIT problem-solving and decision-making landscape in order to provide a perspective as to how these contribute to sources of cognitive risk, proposing the following responses to the respective contextual elements:

- address significant problems ahead of urgent or easy problems and focus on the decisions that matter;

- develop and maintain ordered internal decision-making environments, especially when dealing with complex or chaotic external problem contexts;
- understand the essentials of how systems operate, the importance of identifying relevant enterprise systems and the impact of systems on enterprise operation, including problem-solving and decision-making;
- recognize the impact of stress on cognitive risk arising from time in dimensions of long-term situations, long implementation periods and dynamic environments;
- recognize that the business model shapes the organizational design decisions a firm will make on an on-going basis and how they affect the decision-making ability of the firm to create and capture value;
- recognize the distinction between enterprise growth (portfolio size, revenue, etc.) and enterprise development (improvements in the infrastructure to support enterprise growth) and the decision-making implications. As the business goes through the boundaries between stages of growth, the experience will include a sense of time pressure, inadequate numbers of competent managers, lack of clear goals and poor coordination and communications, manifest as growing pains, with many of the growing pains adversely affecting problem-solving and decision-making abilities. The number and severity of growing pains present at any point in time are signals that the gap between growth and development are affecting the enterprise and the urgency of needed adaptation in order to maintain organizational health, including cognitive risk management;
- recognize that the contextual nature of strategy affects the determination of which enterprise decisions are more important, how those decisions are made and the potential for the existence of cognitive risk and its management;
- increase order in internal decision-making environments by clearly delineating organizational structure related to problem-solving and decision-making roles;
- understand that enterprise culture can have strong impacts on how decisions are made. Assessing and managing culture is an important element in managing cognitive risk as misalignments between formal organizational design and culture produces organizational stresses, ambiguity with regard to decision-making roles and erratic and unpredictable problem-solving and decision-making behaviours, increasing cognitive risk.

The authors contend that:

> If a REIT executive were to consider a given problem or decision-making situation through the lens of the fundamental elements considered previously, managing cognitive risk may become more feasible. The likelihood of successfully managing cognitive risk would certainly be higher than if using an ad hoc approach.

which recognizes that such fundamental elements collectively map a significant portion of the problem-solving/decision-making landscape within a REIT. Accordingly, these elements are all relevant to the creation and, therefore, the management of cognitive risk facing a REIT with the overarching principle in managing cognitive risk being alignment. As the authors insightfully observe:

> *Alignment simply means ensuring that the right people in the right roles are pursuing the right problem, supported by the right systems and the right culture and pursuing the right vision, mission, goals and objectives. If all of these pieces are in place, cognitive risk will be substantially reduced.*

as:

> *Achieving alignment is the goal and enterprise cognitive risk management is the outcome.*

Accordingly, as REITs grow, their total assets under management rise and the implications of decision-making by REIT managers become even more significant, the recognition of behavioural finance and the impact of cognitive risk in the REIT investment decision-making process may be expected to increase.

14.4 Primary focus on cash flow not property

The contributing chapters to this book serve to highlight five major directions for the future of international equity REITs, each of which may be expected to contribute to a change in the risk/return relativities of the REIT sector, comprising the following:

- primary focus on people not property;
- primary focus on cash flow not property;
- societal impacts on REITs;
- global flagship property REITs; and
- global regulation of REITs.

Concerning the primary focus on cash flow not property, it is ironic, given the massive body of international research concerning whether REITs have similarity in their risk/return profile to direct property and so may be a substitute within a portfolio, that the future growth of assets under management for REITs may be more focused on cash flow than on property.

In the early decades of REIT evolution, great attention was paid by both REIT managers (who were then mainly property people, rather than investment bankers, financiers, accountants and lawyers) and equity market analysts to the quality of the physical real estate within a REIT's portfolio. Originally seeking prime property and discounting non-prime property, the REIT market evolved such that new developments became super-prime property and commanded a premium over prime property.

As large proportions of super-prime, prime and non-prime traditional property within the world's property markets became owned by either REITs or other forms of institutional ownership, other forms of real estate such as bulky goods retail parks and suburban office campuses became acceptable as investments for REITs.

Gradually, over time, it is now becoming apparent that categories such as prime and non-prime, retail and commercial and so forth are simply heuristics for income and capital risk and return potential, effectively just being the vessel that holds the key ingredient being the cash flow.

As Moss (Chapter 2) notes, within the REIT sector there is growing demand for "non-correlated" asset classes, following the increasing correlation between REITs holding assets in traditional property sectors and equities. Accordingly, there has been an increased demand for REITs holding alternative assets such as health care and student accommodation which can demonstrate lower levels of correlation and therefore increased diversification benefits. However, such increased demand is premised on those assets providing a secure and stable cash flow capable of being priced, relatively, by the capital markets.

Concerning the secure and stable cash flow, Moss (Chapter 2) further notes the requirement for increasing maintenance capex to maintain values within REIT portfolios which, while acknowledged, is an unknown amount that may impact free cash flow and therefore distributions and/or valuations.

Reddy and Cho (Chapter 3) survey the wide range of non-traditional sectors now subject to investment by REITs including timberland, health care, self-storage and so forth. The authors observe that such non-traditional sectors may be subject to the same growth opportunities and risks as traditional sectors, such as the state of the economy, availability of debt, state of the financial markets and so forth but are also subject to a wide range of idiosyncratic positive influences such as the following:

- growth in demand for cosmetic surgery for health care REITs;
- ability to reschedule harvesting to meet demand for timber REITs;
- demand for IT services for infrastructure REITs and data centre REITs;
- demographics for child care REITs;
- government permits for health care facility REITs;
- lifestyle changes/CBD living/empty nesters for self-storage REITs; and
- greater use of technology by society and business for infrastructure and data centre REITs

and idiosyncratic negative influences such as these:

- government policy for health care REITs;
- capex requirements for hotel REITs;
- fire/flood/earthquake risk for timberland REITs;
- tech stock sell off risk for infrastructure and data centre REITs; and
- accentuation as assets may be more closely grouped within economic/geographic areas for residential REITs,

being a range of strategic or longer term issues that influence the security and stability of cash flows and that potentially create a very different risk/return profile to that for traditional property investments with the REIT framework.

As the conventional determinants for property such as location quality and building quality give way to tenant quality and cash flow quality, the types of properties that are capable of securitization through REIT structures is potentially limitless if the source of the cash flow can be structured on some form of long-term contractual basis with the potential for periodic increase.

From the viewpoint of the vast corporate real estate holdings, this suggests that large amounts of non-traditional property could be made available for acquisition by REITs if such corporates are able to manage their businesses to produce secure and steady rental income streams while retaining responsibility for maintenance and capex, effectively quarantining the business risk within the corporate and so not infecting the REIT.

Accordingly, as REITs grow and their total assets under management rise to include more non-traditional property sectors, a greater focus on cash flow rather than on property may be expected.

14.5 Societal impacts on REITs

The contributing chapters to this book serve to highlight five major directions for the future of international equity REITs, each of which may be expected to contribute to a change in the risk/return relativities of the REIT sector, comprising the following:

- primary focus on people not property;
- primary focus on cash flow not property;

- societal impacts on REITs;
- global flagship property REITs; and
- global regulation of REITs.

Concerning the societal impacts on REITs, looking forward over the next 20 to 50 years, it may reasonably be expected that REITs will be impacted by changes in society across the world. As the developed world ages, becomes wealthier and becomes more demanding for services, opportunities for REITs may be expected to increase in aged care, affinity groups and infrastructure. Similarly, the scope for growth in REITs in the Islamic world may also reasonably be expected to grow strongly over the next 20 to 50 years.

14.5.1 Health and aged care

An ageing population in developed countries across both the Western world and the Eastern world will create enormous opportunities for property developers and investors, including REITs, in the provision of health care and aged care properties. Reddy and Cho (Chapter 3) note that, in 2016, there were 21 health care REITs with a market capitalization of US$95.4 billion in the FTSE EPRA/NAREIT Index in developed markets, being a weight of 7.3%.

Reddy and Cho (Chapter 3) further note that health care REITs invest in a wide range of properties including hospitals, skilled nursing facilities, medical office buildings, acute care and rehabilitation hospitals, purpose-built health care facilities and aged care housing. While the demand for such facilities may be expected to grow strongly into the future and underwrite returns, such facilities are also subject to a high level of government supervision and licencing regimes, adding a further level of risk.

14.5.2 Affinity groups

An increasingly wealthy population across the developed world provides an increasing pool of funds for discretionary investment while, as Moss (Chapter 2) notes, there is an increased focus on solutions and specific investment characteristics rather than generic sector allocations, leading to increasing interest in sub-sector specific or specialist REITs offering an investment solution rather than sector specific REITs offering a diversified sector exposure

Accordingly, the opportunity to create REITs targeted at particular affinity groups would appear substantial. Reddy and Cho (Chapter 3) note that timber REITs in the US accounted for US$28.4 billion in 2016, being 3.11% of the FTSE NAREIT All Equity REIT Index and offering an ideal investment opportunity to the environmentally friendly affinity group. Similar opportunities exist in many forms of property such as sports stadia, golf courses, country clubs and even religious buildings, where an affinity group may be prepared to invest in an affinity REIT offering a secure and stable cash flow from the buildings therein.

As Moss (Chapter 2) noted, an increasing focus on specific investment characteristics and solutions, rather than generic asset allocations, provides enormous opportunities for REIT promoters to develop and offer REITs focusing on particular sub-sectors designed to meet particular risk/return requirements. Reddy and Cho (Chapter 3) document a wide range of specialist REITs already on offer including self-storage and data centres with the availability of REITs based on property in other sub-sectors only limited by the promoter's ability to create a secure and stable underlying cash flow from such properties.

14.5.3 Infrastructure

As the developed world becomes more demanding for services, opportunities for REITs may be expected to increase in infrastructure with Reddy and Cho (Chapter 3) noting that infrastructure REITs in the US accounted for US$76.4 billion in 2016, being 8.35% of the FTSE NAREIT All Equity REIT Index and including communication towers, distributed antenna systems, pipelines, storage terminals, outdoor advertising facilities and renewable power generators.

While other forms of infrastructure such as freeways, railways, light rail, tunnels and so forth have been developed around the world as a joint venture between government and the private sector, the funding structure for such projects often results in significant ongoing user costs to the consumer. Accordingly, for that infrastructure that was previously developed in varying forms of public-private partnership, there may be opportunities for REITs to work with government and fund such infrastructure through a REIT structure, where the cost of capital and time frame for returns may be different to those of a public-private partnership, leading to lower ongoing user costs to the consumer.

14.5.4 Islamic REITs

Razali and Sing (Chapter 5) note that the Muslim population of 1.52 billion constitutes 24% of the world's population, that in 2007/08 Islamic banking assets under management were estimated at US$750 billion, that the global Islamic finance industry is growing at a rate of 15% to 20% pa and that there are more than 250 Islamic financial institutions operating in 75 countries and 100 Islamic equity funds managing assets in excess of US$5 billion, concluding that there is

> *enormous potential for Islamic financial products and fund management services.*

The authors go on to note that Islamic law encourages entrepreneurship, trade and commerce that bring societal benefits and development, but prohibit investments that are speculative in nature such as derivative instruments, interest yielding or involved in the sale of tobacco, alcohol and pork, betting, gambling or pornography production above a specified limit.

Islamic REITs operate within Islamic law and require the involvement of a Shariah Committee but are otherwise similar in structure and operation to conventional REITs, with 15 Islamic REITs in Malaysia in 2014 and Islamic REITs also trading in Kuwait, Bahrian, UAE, Qatar and Singapore with a REIT regime established in Saudi Arabia.

Interestingly, the foundation of Islamic REITs in the principles of Islamic law which advocates fairness and equity in wealth distribution (Razali and Sing, Chapter 5) may address some of the governance and ethical issues that arise in conventional REITs and could offer a broader group of investors a REIT alternative that is more consistent with their religious beliefs and core life values.

Given the enormous potential depth in the Islamic investor base around the world and the extent of investment property not yet securitized in the Gulf States, the prospects for the growth of Islamic REITs in that part of the world alone is significant. When added to the potential for growth in the rest of the world, the future for Islamic REITs looks extremely positive.

14.5.5 Summary – societal impacts on REITs

REITs, as an investment structure, offer the opportunity for growth from a range of ongoing changes in society. As the developed world ages, becomes wealthier and becomes more demanding for services, opportunities for REITs may be expected to increase in aged care, affinity groups and infrastructure. Similarly, given the size of the world's Muslim population, the opportunity for growth in Islamic REITs would appear to be very significant.

14.6 Global flagship property REITs

The contributing chapters to this book serve to highlight five major directions for the future of international equity REITs, each of which may be expected to contribute to a change in the risk/return relativities of the REIT sector, comprising the following:

- primary focus on people not property;
- primary focus on cash flow not property;
- societal impacts on REITs;
- global flagship property REITs; and
- global regulation of REITs.

Concerning global flagship property REITs, as noted in section 14.4, large proportions of super-prime, prime and non-prime traditional property within the world's property markets have, over the decades, become owned by either REITs or institutional groups. Super-prime was a name created to connote something superior to prime property, such that it may be expected that, as properties are developed that are superior to super-prime properties, a further name will be created.

Moss (Chapter 2) cautions that a structural change is occurring in the perception of shopping centre and office assets, with e-commerce, shopping centre saturation and store closures suggesting a structural shift that may impact the pricing of shopping centres which are an overweight component of many REIT portfolios. It is contended that the same may be said of office assets where the effects of technology and the impact of sustainability may be far reaching and the proportion of REIT portfolio assets is significant.

Interestingly, the response of the Westfield Group to the potential demise of the super-prime shopping centre was the creation and development of the flagship shopping centre. Through the development of such centres as Westfield London, Westfield Century City and Westfield World Trade Center, Westfield created a "great experience for retailers, consumers and brands" specifically designed to minimize the adverse impact of e-commerce through "the addition of food, leisure and entertainment, and a broader mix of uses including many new concepts, emerging technologies and online brands" (Source: www.westfieldcorp.com/news/detail/half-year-2017-results-funds-from-operations-of-343-million, accessed 24 April 2018). By definition, most major cities around the world would only have one such flagship shopping centre with the scope for a REIT to aggregate such flagship centres into a single portfolio capable of being held by a wide and diverse range of international investors.

Concerning office properties, Eichholtz et al. (Chapter 4) found that the sustainability of REITs' portfolios is strongly property driven, with office REITs being the most sustainable. Further, Eichholtz et al. (Chapter 4) found a statistically significant relationship between the cost of equity capital and REIT greenness. Overall, a 38bps average reduction in the cost of equity was found when a REIT had a 100% certified portfolio, as compared to a completely uncertified one, with the results being robust to different specifications and approaches. As Eichholtz et al. (Chapter 4) noted, their findings could be yet another incentive for real estate investors and developers to consider "going green" and integrating sustainable practices into their projects.

Accordingly, in the same way that Westfield created the flagship shopping centre with the scope for a REIT to aggregate such flagship centres into a single portfolio, scope exists to create a REIT portfolio of newly developed, massive green office buildings in major cities around the world which would also potentially benefit from a lower cost of equity.

Accordingly, as was nascent with Westfield Corporation before the Unibail-Rodamco takeover, the evolution of the global flagship property REIT may be contended to be likely to occur in the future in the retail and office sectors together with, possibly, the hotel and logistics sectors. In many respects, this parallels the development of the specialized REIT investing in student housing, data centres and timberland, except that the global flagship property REIT may be likely to be denominated in the hundreds of billions of dollars and the specialized REIT in the hundreds of millions of dollars.

14.7 Global regulation of REITs

The contributing chapters to this book serve to highlight five major directions for the future of international equity REITs, each of which may be expected to contribute to a change in the risk/return relativities of the REIT sector, comprising the following:

- primary focus on people not property;
- primary focus on cash flow not property;
- societal impacts on REITs;
- global flagship property REITs; and
- global regulation of REITs.

Concerning the global regulation of REITs, there is considerable structural commonality in REITs across the various jurisdictions of the world, as noted by Parker et al. (Chapter 1):

- corporate structure – widely held company, trust or other legal form;
- investment focus – investing in immovable property;
- time horizon – for the long-term;
- profits distribution – distributes most of the profits annually; and
- tax treatment – does not pay income tax on the income related to immovable property that is distributed to investors,

with such structural elements often accompanied by some level of limitation on debt or gearing.

One negative aspect of greater structural commonality internationally is the risk of global contagion in the event of a world-wide problem emerging, such as the global financial crisis of 2007–2009. While such bodies as the International Monetary Fund and the World Bank have a role in managing the risk of global contagion, REITs probably face the greatest risk in the statement of the value of property assets on the equity side of the balance sheet and the extent and nature of debt on the liabilities side of the balance sheet.

Concerning the statement of the value of property assets on the equity side of the balance sheet, many countries who have a REIT jurisdiction are adopters of the international valuation standards developed and promoted by the International Valuation Standards Council (IVSC). As adherents to IVSC standards, their REIT portfolios should be valued in accordance with such standards, such that the basis of the valuation and credibility of the valuation should be consistent around the world. While this does not stop the contagion of falling asset values globally due to a catalyst financial, economic or political event, it provides a high level of transparency around property values for the purposes of managing and containing the contagion.

It is probably in the extent and nature of debt on the liabilities side of the balance sheet that REITs are at the greatest risk from international contagion. The global financial crisis highlighted the extent of massive leverage held by REITs around the world, with some in Australia exceeding 90% leverage, exacerbated by the nature of debt being commercial mortgage backed securities which were, at that time, incapable of being renewed. Despite the dire impact of such high debt levels on REIT unit trading prices, few if any countries sought to tighten their regulation of REIT leverage or to reduce maximum debt limits after the global financial crisis.

It is particularly interesting that some jurisdictions, when introducing REIT regimes, have chosen to regulate the level of debt or leverage to be held by a REIT. Moss (Chapter 10) notes that Spain places constraints on leverage with Lee (Chapter 13) noting that India has a gearing limit of 49% with a requirement to provide a credit rating and obtain unitholder approval if gearing is to exceed 25% of the value of assets. Somewhat less prudent, but a limit nonetheless, is found in Thailand (Liow and Yuting, Chapter 11) and South Africa (Lee, Chapter 13) who maintain a 60% limit, with the latter requiring a risk-monitoring committee.

As REITs continue to grow as a proportion of the equity markets of various countries around the world and as the level of cross border investment continues to increase, the risks of global contagion and its impact on national financial markets also continues to grow exponentially. It may, therefore, be expected that there will be greater cooperation in the future between regulatory authorities in those countries with REIT regimes to manage the extent and nature of debt held by their respective REIT sectors through global regulation of REITs as a way of managing contagion risk to their national economies.

14.8 Implications for risk and return

With continuous change and evolution a certainty for the REIT sector around the world, the contributing chapters to this book serve to highlight five major directions for the future of international equity REITs, comprising the following:

- primary focus on people not property;
- primary focus on cash flow not property;
- societal impacts on REITs;
- global flagship property REITs; and
- global regulation of REITs,

which were considered, sequentially, previously. While unlikely to evolve in isolation and likely to exhibit interaction, each of the identified major directions may be expected to contribute to a change in the risk/return relativities of the REIT sector which may be contended to be likely to manifest in the following areas:

- impact of traditional variables changing;
- size of the REIT sector;
- polarity in the REIT sector;
- integration between global markets; and
- pricing cognitive risk in REIT management,

each of which are considered, respectively, in the following sub-sections.

14.8.1 Impact of traditional variables changing

As a bridge between the property asset class and the equities asset class, REITs have traditionally being impacted by property market variables, economic variables, financial markets variables and capital markets variables. While this may be expected to continue, as the size of the REIT sector grows and its composition varies to include greater sector specialization, the impact of such variables on the REIT sector may be expected to change.

For example, Moss (Chapter 2) notes that one of the key drivers for growth in the Post-Modern Era was identified to be a rising risk-free rate, which the author observed would raise the cost of debt and force future earnings and dividend growth to be sourced from rental growth and/or an improvement in operational efficiency rather than a reduction in interest costs.

It may be expected, therefore, in the future as in the past, that changes in property market variables, economic variables, financial markets variables and capital markets variables may impact the risk/return profile of the REIT sector and its various sub-sectors together with the relativity of the risk/return profile of the REIT sector to the direct property sector, the equities sector and the bond sector but that the nature of the impact may be expected to change.

14.8.2 Size of the REIT sector

Moss (Chapter 2) argues that, following the separation of Real Estate from Financials in the MSCI Global Industry Classification Standard, REITs have become increasingly relevant as a stand-alone sector in the equity market which will become a significantly larger sector over time, with an expectation of risk/return impacts including that:

- there will be a decline in volatility as the REIT sector "de-merges" from the historically more volatile Financials sector and becomes more related to the lower volatility of the underlying real estate; and
- fund managers will choose to utilize a separate listed real estate allocation as it contributes a true real estate exposure to the portfolio over the long-run and contributes beneficial risk-adjusted returns to a multi-asset portfolio.

Moss (Chapter 2) further notes a blurring of the distinction between companies and funds, with some REITs now exhibiting unlisted fund characteristics including external management structures, changing remuneration structures and finite life structures. Accordingly, as the size of the sector grows, the risk/return characteristics of REITs may be expected to evolve over time.

14.8.3 Polarity in the REIT sector

The emergence of flagship-REITs and the growth in specialized sector REITs may be contended to be likely to create a polarity in the risk/return profile of the REIT sector, with flagship-REITs at one end, specialized sector REITs at the other end and conventional REITs somewhere in between.

A trend towards the creation of global flagship property REITs, holding shopping centre assets created to provide a "great experience for retailers, consumers and brands" or newly developed, massive, sustainable green office buildings in major cities around the world, may be expected to exhibit different risk/return characteristics to REITs with more conventional prime shopping centre and office portfolios.

For example, Eichholtz et al. (Chapter 4) found a statistically significant relationship between the cost of equity capital and REIT greenness. Overall, a 38bps average reduction in the cost of equity was found when a REIT had a 100% certified portfolio, as compared to a completely uncertified one, with the results being robust to different specifications and approaches.

A lower cost of equity for such assets together with greater durability and resilience to property market, economic and financial markets cycles may be expected to positively contribute to returns with less volatility, so reducing the risk of such global flagship property REITs which may exhibit more bond-like characteristics.

An interesting impact on the risk/return profile of the REIT sector may be expected to arise from growth in specialized sector REITs, where the additional risks may be easily envisaged but the requisite level of compensating return less easy to envisage and the potential correlation with conventional property sectors difficult, if not impossible, to determine.

With such REITs much more closely based on particular businesses, such as self-storage, data centres and so forth, while the cash flow may be designed to be secure and stable, investors may be expected to hold residual concerns about the robustness of the underlying business which may reflect in the target risk premium.

Further, such concerns may be compounded by the nature of the physical building assets, which may often be purpose-built with few, if any, alternative uses and limited liquidity in the sales market. Possible contributors to an increased risk premium are, therefore, potentially numerous and varied, requiring an appropriate level of compensating return which may be expected to show greater volatility, so increasing the risk of such REITs which may exhibit more equity like characteristics.

As specialized sector REITs become more numerous, it may be expected that the capital markets will become more skilled at pricing their units for trading. Further, as time passes and more data becomes available, the analysis of returns, risk and correlations will lead to a much better understanding of the risk/return dynamics of specialized sector REITs.

Therefore, with global flagship property REITs potentially exhibiting more bond-like characteristics and specialized sector REITs potentially exhibiting more equity like characteristics, conventional REITs may lie somewhere in between, offering investors a broader spectrum of risk/return opportunities within the REIT sector.

14.8.4 Integration between global markets

An area for significant debate going forward concerns the appropriate level of integration between the various REIT markets at both a regional level and at a global level. Analysis by Liow and Yuting (Chapter 11) led to the conclusion that SGREITs needed to improve their current return integration level and volatility connectedness with the regional/global developed REIT markets and that the low level of return integration between MAREITs and the regional/global players was an obstacle to market growth.

While a certain level of integration may be beneficial for the purposes of market growth, too much integration could inadvertently facilitate contagion in the event of concurrent global economic or financial market events, leading to the risk of markets spiralling upwards or markets collapsing downwards close to simultaneously around the world.

Accordingly, therefore, further research would be beneficial to investigate the appropriate level of integration, both regionally and globally, whereby the benefits of integration may be realized without the potential risks of contagion.

14.8.5 Pricing cognitive risk in REIT management

Potentially one of the largest impacts on REIT risk /return in the future and therefore on future REIT pricing may come from a deeper understanding of the role of REIT management. Parker (2011) summarizes the REIT management process as comprising the enunciation and statement of:

- the vision of the REIT;
- the style of the REIT;
- the goals of the REIT;
- the strategic plan of the REIT; and
- the objectives of the REIT,

being pyramidal such that the successful achievement of the objectives will fulfil the strategic plan which, in turn, will achieve the goals which will then result in the REIT attaining its vision.

Further, Parker (2011) outlines the people involved in the various levels of management within a traditional REIT structure, including:

- CEO;
- CFO;
- fund managers;
- portfolio managers;
- strategy and research managers;
- capital transaction managers;
- investor relations managers;
- compliance and risk managers;
- asset managers;
- property managers; and
- facilities managers

and discusses the role of each in contributing to attainment of the various pyramidal levels of the REIT management process.

However, the fundamental problem in REIT management is its dependency on people as managers. Given the high order roles and significant intellectual contribution in decision-making by the relevant manager at each level of REIT management, the risk inherent in REIT management can only be marginally reduced by a greater focus on process.

Wofford and Troilo (Chapter 6) identify the key issue for REITs in their dependency on people to be cognitive risk, focusing on the management of cognitive risk within a REIT and proposing a framework for approaching problem-solving and decision-making to successfully manage cognitive risk. The authors note that recognizing the existence of cognitive risk is a necessary step, but not sufficient guidance, for managing REITs. In order to make decision-making effective, a holistic approach is essential for guarding against cognitive risk with the overarching principle in managing cognitive risk being alignment.

The authors contend that "alignment simply means ensuring that the right people in the right roles are pursuing the right problem, supported by the right systems and the right culture and pursuing the right vision, mission, goals and objectives. If each of these pieces are in place, cognitive risk will be substantially reduced".

Currently, REIT analysis focuses heavily on the cash flow and the assets with little focus on the management. As REITs become bigger and more global and the key decision-making executives become fewer with each decision becoming of greater magnitude, the role of cognitive risk may be expected to increase. Accordingly, in assessing the risk profile of a REIT and forecasting its likely returns in the future, analysts may be expected to pay greater attention to REIT management and its role in increasing or reducing the risk/return profile of the REIT in their pricing process.

14.8.5 Summary – implications for risk and return

From the five major directions for the future of international equity REITs highlighted in the contributing chapters to this book, implications for risk and return in the REIT sector may be contended to be likely to manifest in the following areas:

- impact of traditional variables changing;
- size of the REIT sector;
- polarity in the REIT sector;
- integration between global markets; and
- pricing cognitive risk in REIT management,

with the challenge being that the individual, cumulative, interactive and relative effects on risk and return are unknown.

14.9 Summary – directions for the future of international REITs

This book aims to identify key areas for research in the REIT discipline for the next five to ten years by surveying the current state of the REIT discipline around the world and identifying emerging and cutting-edge research areas through a thematic review of current contextual issues and a regional analysis based on case studies.

This book comprises two parts, the first part being a thematic review of emerging and cutting edge global research into current contextual issues in REITs internationally and the second part being a regional analysis of REITs around the world, each written by authoritative academic authors from the world's leading Universities and REIT industry experts.

Part I included six chapters each reviewing a current theme of REIT evolution through the lens of contemporary research. Chapter 1 focused on critical contextual issues in international REITs while Chapter 2, the *Post-Modern REIT Era*, examined the evolution of the global REIT market, what is deemed to be best current market practice and how the market is expected to change going forward and Chapter 3, *Emerging Sector REITs*, investigated such sectors as residential, health care, self-storage, timber, infrastructure and data centres generally and then specifically through a case study of US REITs.

Chapter 4, *Sustainable REITs*, analysed REIT environmental performance and the cost of equity and Chapter 5, *Islamic REITs*, examined the evolution of global Islamic REITs through a study of the Malaysian market. Chapter 6, *Behavioural Risk in REITs*, addressed the management of behavioural risk in global REITs and Part I concluded with Chapter 7 which reviewed recent research into *REIT Asset Allocation*.

Part II includes six chapters each analysing REITs in a region of the world, using case studies from a developed, developing and emerging REIT sector in the region. Chapter 8 analysed

REITs in the North American region, using case studies from the USA, Canada and Mexico, with Chapter 9 analysing REITs in the Latin American region, using case studies from Brazil, Argentina and Uruguay for the developed, developing and emerging REIT sectors in the region, respectively.

Chapter 10 analysed REITs in the European region, using case studies from the UK, Spain and Poland, with Chapter 11 analysing REITs in the South East Asian region, using case studies from Singapore, Malaysia and Thailand. Chapter 12 analysed REITs in the North Asian region, using case studies from Japan, Hong Kong and China and Chapter 13 analysed REITs in the Oceania region, using case studies from Australia, South Africa and India for the developed, developing and emerging REIT sectors in the region, respectively.

This chapter concluded Part II and considered *Directions for the Future of International REITs*. The review of developed, developing and emerging REIT markets in Part II of this book highlighted a range of common themes, including the following:

- the role of the diversified REIT;
- the role of REITs where property companies predominate;
- the interplay between REIT initiatives and government policy;
- the catalysts for REIT market emergence;
- the significance of REIT tax treatment; and
- the role of leverage constraints,

which may be contended to give rise to a range of guiding principles for REIT evolution that may be summarized as follows:

- REITs markets may be likely to start their evolution with a greater proportion of diversified REITs plus a sector specific REIT, then broaden to other large sectors, then further broaden to other smaller sectors;
- REIT markets may have difficulty gaining traction and becoming established where a jurisdiction has a well-established property company or *fideicomiso* market;
- there is no single model for the interplay between REIT initiatives and government policy, with either potentially dominant depending on the individual jurisdictional circumstances;
- there are a range of catalysts for REIT market emergence, with direct property market collapse, structural legislative change and good luck with fortunate timing among the principal catalysts;
- that tax transparency is essential for a successful REIT market, except where the jurisdiction offers other, more attractive tax transparent investment structures; and
- while leverage is tempting as a fast route to growth in assets, capital market participants have bitter experiences of excessive leverage in the past and an increasing number of jurisdictions are seeking to constrain leverage and implement associated risk management measures.

Further, the contributing chapters to this book highlighted five major directions for the future of international equity REITs:

- a primary focus on people not property, because as REITs grow, their total assets under management rise and the implications of decision-making by REIT managers becomes even more significant, such that the recognition of behavioural finance and the impact of cognitive risk in the REIT investment decision-making process may be expected to increase;

- a primary focus on cash flow not property as it is now becoming apparent that categories such as prime and non-prime, retail and commercial and so forth are simply heuristics for income and capital risk and return potential, effectively just being the vessel that holds the key ingredient being the cash flow;
- the societal impacts on REITs as the developed world ages, becomes wealthier and becomes more demanding for services, opportunities for REITs may be expected to increase in aged care, affinity groups and infrastructure. Similarly, the scope for growth in REITs in the Islamic world may also reasonably be expected to grow strongly over the next 20 to 50 years;
- the global flagship property REITs, as was nascent with Westfield Corporation before the Unibail-Rodamco takeover. The evolution of the global flagship property REIT may be contended to be likely to occur in the future in the retail and office sectors together with, possibly, the hotel and logistics sectors; and
- the global regulation of REITs as the level of cross border investment continues to increase, so do the risks of global contagion and its impact on national financial markets,

which, both separately and by interaction, may be expected to contribute to a change in the risk/return relativities of the REIT sector.

It may be contended that such change in risk/return relativity may be likely to manifest in the following areas:

- impact of traditional variables changing, with changes in property market variables, economic variables, financial markets variables and capital markets variables impacting the risk/return profile of the REIT sector and its various sub-sectors in different ways together with the relativity of the risk/return profile of the REIT sector to the direct property sector, the equities sector and the bond sector;
- size of the REIT sector, with a decline in volatility as the REIT sector "de-merges" from the historically more volatile financials sector and a greater use of separate allocations for real estate exposure emerges;
- polarity in the risk/return profile of the REIT sector, with global flagship property REITs at one end, specialized sector REITs at the other end and conventional REITs somewhere in between;
- integration between global markets which may be beneficial for the purposes of market growth but, conversely, too much integration could inadvertently facilitate contagion; and
- pricing cognitive risk in REIT management with the need for recognition and an assessment of the degree of alignment exhibited by the REIT in the management of cognitive risk.

However, the challenge for REIT investors, analysts and managers is determining which of the aforementioned risk/return manifestations may be significant and which may be trivial and which may operate independently and which may have interaction effects. While the response of the REIT market may become evident through the passage of time, there is considerable scope for further academic research in the interim to provide assistance.

References

Brounen, D. and de Koning, S. 2012, '50 Years of real estate investment trusts: An international examination of the rise and performance of REITs', *Journal of Real Estate Literature*, Vol. 20, No. 2, pp. 197–223.

Parker, D. 2011, *Global real estate investment trusts: People, process and management*, Wiley-Blackwell, Chichester.
Wofford, L.E. and Troilo, M.L. 2013, 'The academic-professional divide: Generating useful research and moving it to practice', *Journal of Property Investment & Finance*, Vol. 31, pp. 41–52.
Wofford, L.E., Troilo, M.L. and Dorchester, A.D. 2010, 'Managing cognitive risk in real estate', *Journal of Property Research*, Vol. 27, pp. 260–286.
Wofford, L.E., Troilo, M.L. and Dorchester, A.D. 2011, 'Real estate valuation, cognitive risk, and translational research', *Journal of Property Investment & Finance*, Vol. 29, pp. 385–408.

INDEX

Note: Page numbers in *italics* indicate figures and in **bold** indicate tables on the corresponding pages.

Printed in the USA
CPSIA information can be obtained
at www.ICGtesting.com
LVHW081538150324
774517LV00042B/1846